museo galileo
Istituto e Museo di Storia della Scienza

GALILÆANA
Journal of Galilean Studies

www.museogalileo.it

BIBLIOTECA DI
GALILÆANA
VIII

COPERNICUS BANNED

The Entangled Matter of the anti-Copernican Decree of 1616

edited by

Natacha Fabbri and Federica Favino

LEO S. OLSCHKI EDITORE
MMXVIII

Casa Editrice Leo S. Olschki
Viuzzo del Pozzetto, 8
50126 Firenze
www.olschki.it

ISBN 978 88 222 6584 5

CONTENTS

INTRODUCTION *

I. This volume deals with the anti-Copernican decree of 5 March 1616, on the occasion of the four centenary of its promulgation. The Congregation of the Index, that issued the decree, then stated the falsity of the heliocentrism and underlined its absolute inconsistency with Holy Scripture. It condemned the *Letter on the Opinion of the Pythagoreans and Copernicus* by the Carmelite friar Paolo Antonio Foscarini without appeal, and prohibited Copernicus' *De revolutionibus orbium coelestium* and Diego de Zúñiga's *Commentaria in Job* 'until corrected'. The decree also included a general ban on Copernican books, that were all prohibited, condemned and suspended.[1]

No matter how one evaluates it, the anti-Copernican decree of 1616 marked a pivotal moment in the history of European culture. The prohibitive general ban placed at the end of the document appeared in all editions of the Index of forbidden books from 1619 to 1758, when it was omitted, probably on the initiative of the 'enlightened' Pope Benedict XIV. Nevertheless, the decree was left out but not annulled in 1758, what accounts for the fact that the works by Copernicus, de Zuñiga, Foscarini, Galileo and Kepler (the works by the last two authors had been prohibited later, for the same reason) remained suspended and condemned until the 1819 edition of the Index.[2] Well a century after the discovery of the starlight aberration by James Bradley (1728) – the first indirect proof of the earth's motion – the Master of the Sacred Palace Filippo Anfossi was still able to brandish the 1616 decree in justification of his refusal to grant the *imprimatur* to an innocuous University textbook of astronomy, the *Elementi di ottica e astronomia* (*Element of optics and astronomy*), by the Roman canon Giuseppe Settele. After Anfossi was defeated by the new generation of the

---

* This text is the result of a close collaboration between the editors. In particular, Federica Favino is responsible for writing section I of this Introduction, Natacha Fabbri of section II.

1 *Decree of the Index (5 March 1616)*, in Maurice Finocchiaro (ed.), *The Galileo Affair*, Berkeley and Los Angeles, University of California Press, 1989, pp. 148-150.

2 Id., *Retrying Galileo, 1633-1992*, Berkeley and Los Angeles, University of California Press, 2005, pp. 126-153; Annibale Fantoli, *Galileo e la Chiesa. Una controversia ancora aperta*, Roma, Carocci, 2010, pp. 198-201.

Dominican theologians who were now leading the Congregation of the Holy Office,[3] granting the *imprimatur* to Settele's book marked the implicit recognition of Copernicus' theory by the Church and therefore even the books by Copernicus, Zuñiga, Foscarini, Galileo and Kepler were finally removed from the Index (1835), although with the requirement ordered by the Pope Gregory XVI that no judgment should be added to the matter in order to safeguard the prestige of the Roman Curia.

In the year marking the third centenary of the decree – 1916 – as war was raging on the battlefields of Europe, echoes of the ideological conflicts fuelled in the nineteenth century by the passions of the Risorgimento against the intolerance of the Church towards free thought, had not yet faded.[4] The ritual of reviewing scientific books 'hot-of-the-press' provided then the occasion for confrontations – in resolute, when not harsh, tones – between supporters of opposing interpretations of the political and cultural significance of the decree, on both the Catholic and the secular sides of the argument. That same year, Antonio Favaro published his biographies of Matthias Bernegger, the printers Elzevier, Mario Guiducci, as part of his collection of essays on Galileo's correspondents,[5] whilst he launched the series *Adversaria Galilæana* following his collection of essays *Scampoli*.[6] Also in 1916, he wrote an article for the journal *Scientia* (founded by the mathematician Federigo Enriques and the engineer Eugenio Rignano) vigorously refuting the thesis proposed by the German Jesuit Adolf

---

[3] On the '*Affaire* Settele', see WALTER BRANDMÜLLER – EGON JOHANNES GREIPL (eds.), *Copernico, Galilei e la Chiesa: fine della controversia (1820): gli atti del Sant'uffizio*, Firenze, Olschki, 1992; PIERRE-NOËL MAYAUD, *La condamnation des livres coperniciens et sa révocation à la lumière de documents inédits des Congrégations de l'Index et de l'Inquisition*, Roma, Pontificia Università Gregoriana, 1997; FRANCESCO BERETTA, *L'affaire Settele, 1820-1835: fine della controversia?*, in MASSIMO BUCCIANTINI – MICHELE CAMEROTA – FRANCO GIUDICE (eds.), *Il caso Galileo. Una rilettura storica, filosofica, teologica*, Firenze, Olschki, 2011, pp. 387-402.

[4] BUCCIANTINI, *Galileo e le passioni del Risorgimento*, *ibid.*, pp. 347-366; MICHELE CILIBERTO, *Galileo e Bruno martiri del libero pensiero*, *ibid.*, pp. 367-386; BUCCIANTINI, *Campo dei fiori: storia di un monumento maledetto*, Torino, Einaudi, 2015. By the way, in 1916 Pietro Pagnini did not fail to draw the attention of the readers of the liberal-catholic Florentine journal *Rassegna nazionale* on some of Galileo's scientific contributes that proved useful in the ongoing war (*Galileo in guerra*, «La Rassegna Nazionale», Issue 16 Apr. and 1 May, 1916, pp. 1-28).

[5] Respectively in «Atti dell'I. R. Istituto Veneto di scienze, lettere ed arti», s. 8, 75, 18/2, 1915-1916, pp. 29-53, pp. 481-514, pp. 1357-1418. This collection has been newly published: ANTONIO FAVARO, *Amici e corrispondenti di Galileo*, ed. by P. Galluzzi, Firenze, Libr. editrice Salimbeni, 1983.

[6] ID., *Adversaria Galilaeiana*, s. I. III, *Galileo e le leggi del Keplero*, «Atti e memorie della R. Accademia di scienze, lettere ed arti di Padova», n.s., 32, 2, 1916, pp. 129-132. The collection has been later republished as an anastatic reprint thanks to the Centro per lo studio dell'Università di Padova along with the Accademia Patavina di scienze lettere e arti: ID., *Adversaria Galilaeiana*, ser. I-VII, ed. by Lucia Rossetti and Maria Laura Soppelsa, Padova, LINT, 1992.

Müller in his books on Galileo, which had just been translated into Italian.[7] Based on his research, which relied heavily on Favaro's own National Edition of Galileo's work, Müller denied that the 1616 decree and Galileo's condemnation in 1633 prevented and delayed progresses in European science and philosophy. To prove him wrong, Favaro pointed out Descartes' self-censorship when he was preparing his treatise *Le Monde* for publication soon after he had been made aware of the harsh fate that had befallen Galileo and his *Dialogue*.[8]

Müller's voice was not isolated in the Catholic world. For instance, the Jesuit Bellino Carraba used the occasion of the third centenary of the decree to declare in the *Rivista di apologia cristiana* that: «[it was] not thanks to the rebellion against the decree – as is wrongly believed, said, and published – that the progress of the same Copernican astronomical system was not blocked but, with all due homage to that decree, it was thanks to the Catholics themselves that the study of that system progressed until it arrived at its complete triumph».[9] At the same time the Jesuit of the journal *La Civiltà Cattolica* seized the opportunity, when reviewing some recently published editions of documents pertaining to Galileo – vol. 2 of *Dal carteggio e dai documenti. Pagine di vita di Galileo*, edited by Isidoro Del Lungo and Antonio Favaro,[10] and the last volumes of the National Edition of Galileo's works edited by Antonio Favaro, which contained Galileo's correspondence[11] – to denigrate the biased selection criteria adopted by the editors, as well as Galileo's moral and human stature (his reputation as a scientist being unassailable), that was the necessary precondition to the apologetic

---

7 ADOLF MÜLLER, *Galileo Galilei: studio storico scientifico*, Roma, M. Bretschneider, 1911, translated from (chap. 1-18): *Galileo Galilei und das kopernikanische Weltsystem*, Freiburg i. Br., Herder, 1909; and from (chap. 19-38): *Der Galilei-Prozess (1632-1633) nach Ursprung, Verlauf und Folgen*, Freiburg i. Br., Herder, 1909. For a wide analysis of Müller's thesis see FINOCCHIARO, *Retrying Galileo* (cit. note 2), pp. 270-272.

8 FAVARO, *La condamnation de Galilée et ses conséquences pour le progrès des sciences*, «Scientia», 20, 51, 7, 1916, pp. 1-11. On the journal *Scientia. Rivista internazionale di sintesi scientifica*, see SANDRA LINGUERRI, *La grande festa della scienza. Eugenio Rignano e Federigo Enriques. Lettere*, Milan, F. Angeli, 2005.

9 BELLINO CARRABA, *Il 5 marzo 1616: ossia il decreto della Sacra congregazione dell'Indice intorno al sistema copernicano*, «Rivista di apologia cristiana», 86, 1916, pp. 81-105 (translation mine).

10 ISIDORO DEL LUNGO, FAVARO (eds.), *Dal carteggio e dai documenti. Pagine di vita di Galileo*, 2 vols., Firenze, Sansoni, 1915.

11 *Le opere di Galileo Galilei: Edizione Nazionale sotto gli auspici di Sua Maestà il Re d'Italia*, direttore A. Favaro, 20 vols., Firenze, Barbera, 1890-1909. The volumes 10-19 concerned Galileo's correspondence. On Favaro's accomplishment see BUCCIANTINI, *Scienza e filologia: l'Edizione Nazionale delle Opere di Galileo*, «Giornale critico della filosofia italiana», 76, 1997, pp. 424-445.

thesis which attributed (and still attributes) to the scientist himself no small part in the guilt for his conviction.[12]

In the following hundred years the Church undertook several attempts to overcome this apologetic position: from the 'semi-official rehabilitation' of Galileo by authoritative persons and institutions in the Catholic word (such as Agostino Gemelli, president of the Pontifical Academy of Sciences, and Pio Paschini, president of the Lateran University), to the deploring of any misunderstanding of the legitimate autonomy of science (with the constitution *Gaudium et Spes* issued by Pope Paul VI at the close of the Second Vatican Council), and culminating in the decision announced by John Paul II in 1979 to open an official investigation by a special study commission to reconsider the 'Galileo case' and eventually lay the foundations for a future desirable alliance between science and faith.[13]

In 2016, despite the genuine efforts made by the two sides, Catholic and secular, to agree on a position that represented a balanced equilibrium between their respective views, it would appear that the fundamental disagreements that led to the Galileo *affaire* in the first place have not been completely resolved.[14] In the meantime, much historical research remains to be done in order to deepen our understanding of the evolution of the Church's attitude towards Copernicanism and Galileo, and to shed light on the nature, role and motivations of the institutes and the individuals who played a central role in the events leading to the banning of Copernicus' work and the trial of Galileo. And it is in this perspective that we conceived this volume; that is not to launch just one more debate on the *vexata quaestio* of the relationships between science and faith, nor to support or attack well-entrenched historical positions. Our aim is rather to open up new ho-

---

12 On the attitude of the Jesuits of the *Civiltà Cattolica* toward Galileo's case on a later context, see: Luciano Malusa, *Un confronto ricorrente nella cultura cattolica: il caso Galilei ed il caso Rosmini*, «Trans/Form/Ação», 2014, 37, pp. 117-134, https://dx.doi.org/10.1590/S0101-3173201400ne00008

13 Finocchiaro, *Retrying Galileo*, *passim* (to whom I owe the quote, p. 275); Fantoli, *Galileo per il copernicanesimo e per la Chiesa*, Città del Vaticano, Specola vaticana, 2010^3, ch. 7; Alberto Melloni, *Galileo al Vaticano II: storia d'una citazione e della sua ombra*, in Bucciantini – Camerota – Giudice (eds.), *Il caso Galileo* (cit. note 3), pp. 461-490.

14 Mariano Artigas – Melchor Sánchez de Toca, *Galileo e il Vaticano: storia della Pontificia Commissione di studio sul caso Galileo, 1981-1992*, Venezia, Marcianum press, 2009; Paolo Galluzzi, *Il 'caso Galileo'*, in Gian Mario Bravo – Vincenzo Ferrone (eds.), *Il processo a Galileo Galilei e la questione galileiana. A vent'anni dalla morte di Luigi Firpo. Convegno internazionale (26-27 marzo 2009)*, Roma, Edizioni di storia e letteratura, 2010, pp. 3-16; Fantoli, *Il caso Galileo. Una questione chiusa?*, *ibid.*, pp. 229-257; Massimo Firpo, *History and Historiography in the Condemnation of Galileo. The Judgment of the Church*, «Galilæana», 14, 2017, pp. 115-130, article first delivered as a lecture at the Accademia dei Lincei on 15 December 2016.

rizons for future studies based on hitherto unknown or insufficiently unexplored documentary sources, using broader methodological approaches and adopting fresh points of view.

Our initiative stems from some general historiographical observations. Even the title chosen by the Cultural Commission of the Italian Chamber of Deputies for what was perhaps the most solemn public event dedicated to the anti-Copernican decree – *Il quattrocentenario del primo processo a Galileo* [15] – held in 2016, seemed to us to clearly epitomize what appears to be a basic notion shared by many historians but also, perhaps, by the public opinion, that the decree of 1616 concerned almost exclusively Galileo Galilei and mostly overlapped with his 'first trial'. Moreover, while taking into account pertinent exceptions, most scholars seem to regard this trial as merely the first step towards what they consider the real clash between science and faith – the 1633 condemnation of Galileo. In fact, the legal instruments generated by the activity of the Congregations appointed in 1616 to examine the Copernican issue represent the main links between the two phases of the same trial, which is why their authenticity remains the bone of lively contentions among historians (the admonition issued by the cardinal-inquisitor Robert Bellarmine to Galileo on 26 February 1616, the injunction against Galileo emitted by the Inquisition commissary Michelangelo Seghizzi, and even the anti-Copernican decree itself) that now take up almost the whole literature on the matter.[16]

In our view, this mainstream approach runs the risk of distorting the perspective from which we look at the events surrounding the decree of 1616 and measure the significance that the condemnation of Copernicanism had for the Church, as well as the actual role played by Galileo in that

---

[15] The workshop took place in Rome, at the Library of the Chamber of Deputies, on 4 March 2016. In 2016 other celebrative events were organized, by the Center of Medieval and Renaissance Studies at UCLA, *The Roman Inquisition in the time of Galileo*, 26 February 2016 (to celebrate Thomas F. Mayer's three-volume study, *The Roman Inquisition*, Philadelphia, University of Pennsylvania Press, 2013-2015), and by the National Academy of the Lincei, *Cesi e Galileo*, Rome, 15 December 2016.

[16] A review of the literature on Bellarmine's admonition, in Mario D'Addio, *Considerazioni sui processi a Galilei*, Roma, Herder, 1985, pp. 49-51 and, more recently, Sergio Pagano, *Il precetto del cardinale Bellarmino a Galileo un falso?: con una parentesi sul radio, madame Curie e i documenti galileiani*, «Galilæana», 7, 2010, pp. 143-203; on the decree: Vittorio Frajese, *Il processo a Galileo Galilei: il falso e la sua prova*, Brescia, Morcelliana, 2010; Id., *Il decreto anticopernicano del 5 marzo 1616*, in Bucciantini – Camerota – Giudice (eds.), *Il caso Galileo* (cit. note 2), pp. 75-89; on Seghizzi's injunction, Finocchiaro, *Galileo's First Confrontation with the Inquisition (1616): Four Orders and Three Issues*, «Galilæana», 13, 2016, pp. 29-60; Herry A. Kelly, *Galileo's Non-trial (1616), Pre-trial (1632-33), and Trial (May 10, 1633): A Review of Procedure, Featuring Routine Violations of the Forum of Conscience*, «Church History», 85, 2016, pp. 724-761.

affaire. As Massimo Bucciantini wrote in 1995 in a work whose premises are entirely in line with ours: «We are dealing here with an intellectual and political case [...] made up of many courses of action and lines of thinking, that all need to be followed and kept together step-by-step, if we really want to try to understand one of the crucial stages in the *affaire* and in the history of Copernicanism [...]».[17]

In the light of all these considerations, we chose for this volume a title that is intended to draw the attention of the reader not immediately to Galileo but rather to the ban on Copernicus' cosmology as laid out in his *De Revolutionibus orbium coelestium*, as the event that epitomizes a much more complex affair than has hitherto been recognized. By both relying on the conclusions and dialoguing with the results of important other studies, we tried to consider the *affaire* of 1616 as an independent event from the process of 1633, interpreting it as an 'entanglement' of cultural, philosophical, doctrinal, theological, political and personal factors. As is literally the case for the physical phenomenon alluded to in the subtitle – i.e., that the quantum states of two or more objects must be described with reference to one another even though the individual objects may be spatially separated – it was the reciprocal interactions of all the elements mentioned above that brought the situation to a breaking point in the late winter of 1616, forcing the Church to adopt an official position concerning Copernicanism, Galileo's philosophy and – as we shall see – much more besides.

This volume consists of nine essays that can be divided into three sections. The papers in the first section, by Roberto Bondì, Natacha Fabbri and Franco Motta, draw attention to certain aspects of the late Renaissance philosophical and theological-political debate that have (in our view) been mistakenly considered to have little to do with the promulgation of the decree. The studies in the second section, by Luigi Guerrini, Federica Favino and Giovanni Pizzorusso, shed light on some less well known or completely unknown figures who were in fact involved in the network of personal and intellectual interests, as well as in the entanglement of political and cultural strategies relating to the issuing of the decree. The contributions to the third section by Édouard Mehl, Rienk Vermij and Steven Vanden Broecke focus on some key chapters in the intellectual debate that was conducted across Europe *post*-1616, in order to grasp the impact of the decree beyond the Alps.

---

[17] Bucciantini, *Contro Galileo. Alle origini dell'*affaire, Firenze, Olschki, 1995, pp. 19-23: 21-22 (translation mine).

II. Rome, 5 March 1616. The decree issued by the Congregation of the Index declared the Pythagorean doctrine of the mobility of the earth and the stability of the sun to be «false and totally contrary to the Holy Scripture». In that session Copernicus' *De Revolutionibus* and Zuñiga's *In Job commentaria* were suspended *donec corrigatur*, whereas Foscarini's *Letter* was prohibited and condemned. The greater threat posed by the latter also emerges from the cardinals' choice to draw from its title – *Letter on the Opinion of the Pythagoreans and Copernicus on the Motion of the Earth and the Stability of the Sun* – the two principles of the Pythagorean doctrine that were defined as false and contrary to the Bible. Eager to prevent the spread of such a dangerous opinion, they also decided that any other book aiming to teach that doctrine would have to be «prohibited, condemned and suspended». The *Edictum Librorum Qui post Indicem Felicis Recordationis Clementis VIII Prohibiti Sunt*, published in 1619,[18] reaffirmed the same precept: it included the books by Copernicus, Zuñiga and Foscarini – to which Kepler's *Epitome Astronomiae Copernicanae* was added –, whilst generically mentioning «all other books» that taught the motion of the earth and the immobility of the sun.

The emendation of *De Revolutionibus* – which was initially entrusted to Cardinal Bonifacio Caetani and, after his death, was continued by Francesco Ingoli – culminated four years later with the drafting of the *Monitum ad Nicolai Copernici Lectorem* (1 March 1620). The 1616 decree was formally enacted as the wisest way to safeguard Catholic truth, the very same objective that led Ingoli's corrections to *De revolutionibus*. In his report,[19] dated 2 April 1618, Ingoli stressed the benefit of *De revolutionibus* for the Christian Republic, as that work provided the observations and calculations on which the reform of the Gregorian calendar was grounded. The crucial role that *De Revolutionibus* played in Christian life persuaded Ingoli to preserve the original text as far as possible, to the point of maintaining the hypothesis of the mobility of the earth: if the latter were to be expurgated, all the other vastly useful computations on celestial movements would have also been

---

[18] *Edictum Librorum Qui post Indicem Felicis Recordationis Clementis VIII Prohibiti Sunt*, Romae, Ex typographia Camerae Apostolicae, 1619.

[19] Francesco Ingoli, *De emendatione sex librorum Nicolai Copernici De revolutionibus*, already published in Bucciatini, *Contro Galileo* (cit. note 17), pp. 207-209 and in Engl. trans. in Owen Gingerich, *The Censorship of Copernicus' "De revolutionibus"*, «Annali dell'Istituto e Museo di Storia della Scienza di Firenze», 6, 1981, pp. 45-61: 51-56. See now the recent edition, with parallel French translation, provided in the unvaluable Nicholas Copernicus, *De revolutionibus orbium coelestium / Des révolutions des orbes célestes*, edited and translated into French by Michel-Pierre Lerner, Alain-Philippe Segonds, Jean-Pierre Verdet, *et alii*, 3 vols., Paris, Les Belles Lettres, 2015, vol. III, Appendix, Text 5, pp. 636-643.

destroyed. Therefore, Ingoli was inclined to follow a middle path: he limited his corrections to the sentences that claimed the realistic movement of the earth, given that this statement would have prejudiced the truth and the Holy Scripture. Moreover, Ingoli's choice reflected the epistemological status of astronomy, whose aim consisted merely in providing a mathematical explanation for astronomical phenomena rather than in revealing the true structure of the universe.

The 1620 *Monitum*[20] reiterated the usefulness of *De revolutionibus* for the *Respublica Christiana*, by pointing out that the same aim had been shared by the Fathers of the Holy Congregation of the Index when they unanimously decided to emend Copernicus' text, although they actually would have preferred to prohibit it altogether.[21]

Ingoli's education and activities are retraced in this volume by Giovanni Pizzorusso, who also examines his first works devoted to astronomical matters – *De stella Anni 1604* and *De cometa anni 1607* – which had been written under the influence of Tychonic astronomy. The figure of Ingoli that emerges from this chapter contributes to a better understanding not only of the redaction of the *Monitum* and its emendations, but also of the interplay between scientific culture and the networks in the Curia at the beginning of the seventeenth century in general and in the Copernican affair in particular.

Despite the fact that the *Monitum* had established that any future edition of *De Revolutionibus* was required to respect the corrections introduced and to add the *Monitum* itself as a forward, the only seventeenth-century edition of Copernicus – namely Nicholas Mulerius' *Astronomia Instaurata* (*Restored Astronomy*) – did not bear any trace of Ingoli's work[22] since it came out in

---

[20] *Monitum ad Nicolai Copernici Lectorem, eiusque emendatio*, OG, XIX, pp. 400-401. See also Copernicus, *De revolutionibus orbium coelestium / Des révolutions des orbes célestes* (cit. note 19), vol. III, Appendix, Text 6, pp. 644-649. On the *Monitum* see Lerner, *Copernic suspendu et corrigé: sur deux décrets de la Congrégation romaine de l'Index 1616-1620*, in *I primi Lincei e il Sant'Uffizio: questioni di scienza e di fede*, Roma, Bardi, 2005, pp. 321-403: 336-345. A comparison between Ingoli's emendations and those accepted by the censors of the Congregation of the Index was undertaken in Mayaud, *La condamnation des livres coperniciens* (cit. note 3), pp. 77-79.

[21] The Fathers of the Congregation of the Index did not express themselves in the way Ingoli reported, as they had only declared the suspension of *De revolutionibus* "donec corrigatur". On this contradiction, see Favaro, *Per il terzo centenario del Decreto contro la dottrina del moto della terra (5 marzo 1616-5 marzo 1916)*, now in Id., *Adversaria Galilaeiana: serie I-VII* (cit. note 5), p. 41.

[22] See Rienk Vermij, *The Calvinist Copernicans. The Reception of the New Astronomy in the Dutch Republic, 1575-1750*, Amsterdam, Koninklijke Nederlandse Akademie van Wetenschappen, 2002, pp. 45-52; Djoeke van Netten, *Astronomia instaurata? The Third Edition of Copernicus's "De revolutionibus" (Amsterdam, 1617)*, «Journal for the History of Astronomy», 43, 2012, pp. 75-91.

1617, that is, three years before the publication of the *Monitum*.[23] As Rienk Vermij shows in his chapter, the echo of the 1616 decree in the Protestant world was indeed very limited, if not entirely absent. Mulerius published a new edition of Copernicus, which became the most widely disseminated and frequently quoted version, despite the fact that he did not agree with the heliocentric theory mainly because of his religious convictions, even if he did not raise any theological arguments.

The Dutch context sketched out by Vermij emphasizes the fact that the 1616 decree was subjected to two different interpretations. In the countries beyond the Alps, when reading the text of the decree or the transcriptions of it provided by Adam Tanner and Marin Mersenne,[24] no one had an inkling of the charges that had been formulated against Galileo in those years. In 1621, for instance, Michael Maestlin[25] claimed that the members of the Congregation of the Index did not have any knowledge of Copernicus' writings and that they had acquired a general idea about *De revolutionibus* and its content, as well as about Zuñiga's work, by reading Foscarini's *Letter*. Maestlin did not notice that the only work condemned by the decree, namely Foscarini's *Letter*, was dedicated to Galileo, Kepler, and to the members of the Accademia dei Lincei,[26] and that in those years the attempt to reconcile Copernicanism and the Scriptures had involved other protagonists.

Beyond the Italian mainland, the debates that arose around the decree focused especially on the definition of Copernicanism as a temerarious – and not heretical – doctrine, and on the "healthy" nature of the edict.

---

23 Mayaud thought that the specific reference to future editions of *De revolutionibus* was due to the publication of Mulerius' edition. Mayaud, *La condamnation des livres coperniciens* (cit. note 3), p. 77. The fourth edition of *De Revolutionibus* would come out more than two centuries later in 1854.

24 Adam Tanner, *Dissertatio Peripatetico-Theologica de coelis*, Ingolstadii, ex typographeo Gregorii Haenlin, 1621, p. 212. Marin Mersenne, *Quaestiones in genesim*, Paris, Cramoisy, 1623, col. 904. See Lerner, *Copernic suspendu et corrigé* (cit. note 20), pp. 359-363.

25 Maestlin's reaction to the condemnation of Copernicus has been published in Johannes Kepler, *Opera omnia*, ed. by Ch. Frisch, 8 vols., Frankfurt and Erlangen, Heyder et Zimmer, 1858-1878, vol. I, pp. 56-58. See Copernicus, *De revolutionibus orbium coelestium / Des révolutions des orbes célestes* (cit. note 19), vol. III, Appendix, Text 8, pp. 654-657: 657. According to Frisch and Mayaud (*Le conflit entre l'Astronomie Nouvelle et l'Ecriture Sainte aux XVIe et XVIIe siècles*, 6 vols., Paris, Honoré Champion, vol. III, pp. 189-191), Maestlin's intention was to prepare this text as the preface to a new edition of *De Revolutionibus*. Michel-Pierre Lerner (*Copernic suspendu et corrigé*, cit. note 20, p. 364), instead, claimed that Maestlin wished to add this short writing to the re-edition of Rheticus' *Narratio Prima* which was included in the second edition of Kepler's *Mysterium Cosmographicum* (1621).

26 Paolo Antonio Foscarini, *Lettera sopra l'opinione de' pittagorici e del Copernico*, In Napoli, per Lazzaro Scoriggio, 1615, p. 13.

For instance, although the friar Minim Marin Mersenne only deemed the Copernican hypothesis temerarious,[27] he however admitted that Copernicanism could still have prejudiced the authority of the Catholic Church, as it might have been used by the enemies of the Church to damage its power. In his heartfelt defence of the Church's interests, Mersenne went even further: should Euclid's *Elements* and other mathematical writings be used by heretics to overthrow the Catholic creed, the Church might also suspend them.[28] If the Church could be justified in its attempt to prevent the spread of works that contained true propositions and that did not touch upon religious or theological matters, she could not be blamed for promulgating the anti-Copernican decree. Indeed, the latter concerned works that hinged on hypotheses and had displayed rashness when they diverged from the universal agreement of the Church Fathers over the interpretation of Scriptures.

Within the Catholic world, such a reading came into greater prominence following Galileo's trial. In this regard, the words that the Jesuit Pierre Le Cazré wrote to Gassendi are enlightening:

> Then Genesis will be placed in doubt, when it says that the earth was made before the other heavenly bodies and that these latter were created on the fourth day to light the earth [...] The whole economy of the incarnate Word and the truth of the Gospels will thereby be rendered suspect. Even more, the whole Christian Faith will be suspect [...]. It is not undeservedly that since the time of Copernicus, the Church has always opposed this error and that again recently, it was not a few Cardinals (as you say) but the supreme leader of the Church who, by pontifical decree, has condemned it in the person of Galileo.[29]

In the text that Michael Maestlin had been working on as the foreword for a new edition of Rheticus' *Narratio Prima*, he wondered why Copernicus' work had been banned seventy years after its publication and not before. His answer identified the publication of Foscarini's *Letter* as the

---

[27] Marin Mersenne, *Quaestiones in Genesim* (cit. note 24), cols. 901 sq. This *Articulus* is titled «An quis saltem sine haeresis nota, aut erroris, temeritatisque periculo possit opinari, atque tueri terram mobilem esse, coelos vero immobiles».

[28] Id., *La Vérité des Sciences*, Paris, Du Bray, 1625, book I, chap. IX, p. 112: «[...] les Prelats, et les Pasteurs de l'Eglise peuvent deffendre la lecture des livres, desquels les Heretiques et les autres ennemis de l'Eglise de Dieu se servent pour renverser notre creance, bien que tels livres ne continssent rien qui ne fût vray».

[29] Pierre Le Cazré to Gassendi, 3 November 1642, in Pierre Gassendi, *Opera omnia*, 6 vols., Lugduni, sumptibus Laurentii Anisson & Ioan. Bapt. Devenet, 1658, vol. VI, p. 451 (the translation is mine). The reference to the «time of Copernicus» seems more a generic hint at the 1616 decree, than an allusion to the hostility that had already spread within the Church in the sixteenth century.

event that prompted the Church to censure Copernicus. Although Foscarini played a crucial role in the events leading to Copernicus' ban, upon examining the Florentine and Roman context, as well as the readings of the 1616 decree provided by Galileo's supporters and opponents, it is possible to give Maestlin's question a more multifaceted response.

Between the fourth decade of the sixteenth century and the first decade of the seventeenth century, the temerity of *De Revolutionibus* did not go unnoticed: Bartolomeo Spina, Giovanni Maria Tolosani, Jean Lorin, and Lodovico Delle Colombe all denounced the rashness of this hypothesis, and some of them its heresy. In the meantime the second edition of *De revolutionibus*, which was published in Basel in 1566 by Heinrich Petri, lent new impulse to the spread of Copernicus' ideas, while several 'heretical authors' such as Rothman, Rheticus, Maestlin, and Kepler placed an even stronger emphasis on the truthfulness of heliocentrism. Indeed the interest and intellectual activity surrounding *De revolutionibus* was quite lively outside of Italy. For instance, by 1595 the volumes of the first and second editions were out of print[30] and in 1614 Maestlin wrote to Kepler about his intention to prepare a new edition for the printer Sebastian Heinricpetri[31] in Basel; in 1612 an edition of *De revolutionibus* prepared for teaching purposes and entitled *Sphaera Nicolai Copernici*[32] was circulating around Paris; and lastly, the new edition by Mulerius came out in 1617.

As is well known, Osiander's *Preface* had attempted to conceal the overturning of the traditional hierarchy among the disciplines of knowledge that the text of *De Revolutionibus* had traced by bringing Copernicus' statements back to the core of the long-established instrumentalist interpretation of astronomy.[33] Conversely, Galileo's observations and works drew attention to Copernicus' sentences claiming the truthfulness of heliocentrism. Before Galileo, some 'heretical' philosophers – such as Bruno and Kepler – had underscored the contradiction between Osiander's preface, on one hand, and Copernicus' dedicatory letter to Pope Paul III and the text of *De revolutionibus*, on the other. Their statements, however, had only been discussed in the context of philosophical debates. The barriers

---

30 See Gingerich, *An Annotated Census of Copernicus' "De revolutionibus" (Nuremberg, 1543 and Basel, 1566)*, Leiden, Brill, 2002, pp. xiv, xxvi.

31 Maestlin to Kepler, 17 June 1614, KGW XVII, p. 114.

32 *Sphaera Nicolai Copernici. Seu systema mundi secundum Copernicum*, Paris, Jean Liber, 1612. See Lerner, *Copernicus in Paris in 1612: A Teaching Text Edition of De Revolutionibus*, «Journal for the History of Astronomy», 31, 2000, pp. 55-67.

33 See for instance Lerner – Alain-Philippe Segonds, *Sur un avertissement célèbre: l'*Ad lectorem *du* De revolutionibus *de Nicolas Copernic*, «Galilæana», 5, 2008, pp. 113-148.

erected around *De Revolutionibus* and lasting almost seventy years suffered a final blow once the telescope pierced the dome of the sky: the striking consequences of Galileo's observations have been permeating every field of knowledge (as well as the public imagination) ever since, demanding that astronomical discoveries be accorded a place within the spheres of the Christian cosmos.[34]

Following the publication of *Sidereus Nuncius*, it became even more evident that Copernicus' supporters were no longer confined to the heretics of the Reformed countries. They also included various prominent members of the Medici court, which was a very Catholic household, having furnished three popes just in the previous century: Leo X, born Giovanni di Lorenzo de' Medici, who served as pope from 1513 to 1521; Clement VII, born Giulio de' Medici, who was pope from 1523 to 1534; and Leo XI, born Alessandro de' Medici, who served as pope for just one year, in 1605.

The Florentine context is addressed in the chapters by Federica Favino and Luigi Guerrini, who retrace the complex relationships among several of the leading personalities in the cultural and political life of the city. Luigi Guerrini's chapter provides an in-depth examination of the anti-Galileo faction that sprang up around the bishop of Florence Alessandro Marzimedici and which included not only theologians, but also men of letters such as Michelangelo Buonarroti the Younger and Carlo Marucelli. Federica Favino focuses on the opposite front by examining documents that have hitherto been largely ignored and shedding new light on two figures within Galileo's entourage: the parish priest of Castel Fiorentino, Giannozzo Attavanti, and the nobleman Cosimo Ridolfi, who were both key witnesses to the events that led the Dominicans friars of Santa Maria Novella submitting a denunciation of Galileo to the Holy Office. Their correspondence with Cardinal Alessandro Orsini reveals close links between Attavanti and Ridolfi, the cardinal, and the supposed 'party' of Galileo's enemies at the Medici court, as well as their simultaneous (and apparently not contradictory) involvement in the practice of metallurgical alchemy and in the program of the Accademia dei Lincei, of which Ridolfi had been a member since 1613.

The anti-Copernican factions looked upon Galileo as the head of a group that could almost be regarded as a sect – the *galileisti*[35] –, one «who

---

[34] See for instance Christoph Clavius, *In sphaeram*, in *Opera mathematica*, 10 vols., Moguntinae, sumptibus Antonij Hierat, excudebat Reinhard Eltz, 1611-1612, vol. V, p. 75.

[35] Niccolò Lorini to Paolo Sfondrati, February 1615, in *Processo di Galileo*, OG, XIX, pp. 297-298; Tommaso Caccini, 20 March 1615, in *Processo di Galileo*, OG, XIX, p. 311.

tramples the whole philosophy of Aristotle, on which the Scholastic theology relies»[36] and who spread his ideas through writings and manuscripts, an example of which is his letter to Benedetto Castelli. The importance of Galileo in the tangled web of events that led to the anti-Copernican decree is further attested by what happened within the Academy of the Lincei in the aftermath of the decree.

The supporters of Copernicus referred to in Foscarini's *Letter* included, alongside Galileo and Kepler, the members of the Academy. Indeed, Foscarini dedicated his work «to the Illustrious, and very virtuous academy of lords of the Lynx, who universally (if I am not mistaken) follow this opinion».[37] As Prince Federico Cesi wrote to Galileo in March 1615, Foscarini's statement was incorrect and the Lincei did not necessarily share Galileo's Copernicanism, but instead the «freedom to philosophize *in naturalibus*».[38] Indeed, while Cesi was inclined to adhere to the Tychonic system,[39] in 1605 the Academy published *De nova stella disputatio* by Jan van Heck, which instead presented a critique of Brahe's thesis of the fluidity of the sky.[40] Besides the attention devoted to Brahe and Galileo – whose *Sunspot Letters* (1613) were published under the auspices of the Academy – Cesi was so interested in the work of Telesio and Antonio Persio that he drew Galileo's attention to Persio's treatises[41] and planned the publication of his *De natura ignis* – a project that, however, was never realized.

For our purposes, it is worth recalling how the Accademia dei Lincei reacted to the 1616 decree, as the minutes of the 14 March 1616 report. In that session – in which Cesi, Galileo, Francesco Stelluti, Angelo de Filiis and Johannes Faber participated – the Academy did not accept the resignation of Luca Valerio, who had been eager to dissociate himself from Galileo and who had accused him of making mistakes and of serious misconduct («erroris et magni delicti»). They did however decide to strip Valerio of

---

36 Niccolò Lorini, February 1615 (cit. note 35), p. 298.

37 Foscarini, *Lettera* (cit. note 26), pp. 13-14.

38 Cesi to Galileo, 7 March 1615, OG, XII, p. 150.

39 See Galluzzi, *"Libertà di filosofare* in naturalibus*": i mondi paralleli di Cesi e Galileo*, Roma, Scienze e lettere, 2014, pp. 79-103.

40 Before publishing Heck's work, Cesi emended the text: see Saverio Ricci, *Federico Cesi e la Nova del 1604: la teoria della fluidità del cielo e un opuscolo dimenticato di Johannes van Heeck*, «Atti della Accademia nazionale dei Lincei. Classe di scienze morali, storiche e filologiche», ser. 8, 43, 1988, pp. 111-133; Galluzzi, *"Libertà di filosofare* in naturalibus*"* (cit. note 39), pp. 79-83.

41 See Cesi to Galileo, 22 March 1612, OG, XI, p. 285; 19 May 1612, *ibid.*, p. 298; 26 May 1612, *ibid.*, p. 301.

several academic prerogatives, such as the right to vote and to participate in meetings:

[...] firstly, because without any reason he [Valerio] desired to absent himself and to withdraw as a member of the Academy; secondly, because with this act of not wishing to be a Lincean, he rendered the Academy itself blameworthy, as if it had committed an offence or professed a manifest error in opining the movement of the earth: an opinion that he affirmed was held by Lord Galileo who is a member of our Academy; in the third place, because he accused the same Lord Galileo of error and serious culpability, notwithstanding the fact that Lord Galileo had professed this same opinion only as such, and despite the fact that he [Valerio] had never acted as anything other than a friend to Lord Galileo.[42]

Greater emphasis must be laid on the Accademia's attempt to dissimulate the core of Galileo's research in the aftermath of the anti-Copernican decree. In that session, the Lincei – and Galileo himself, since he was present at the meeting – claimed that Galileo had defended Copernicanism simply as a hypothesis, thus pretending to be unaware of the fact that the pages of the so-called Copernican Letters revealed a completely different reading and that the *Sidereus Nuncius*, and even more the *Letters on Sunspots*, had already upheld the truthfulness of heliocentrism.[43]

The crucial role Galileo found himself playing in the trial of 1616 has already been broadly shown by several seminal works: [44] Galileo's discoveries and his philosophy, his lack of prudence, as well as the entangled rela-

---

[42] Ms. Archivio Linceo 30 bis, Roma, Biblioteca dell'Accademia Nazionale dei Lincei e Corsiniana, cc. 17r-v (the translation is mine). On Luca Valerio and this episode, see Giuseppe Gabrieli, *Contributi alla storia della Accademia dei Lincei*, 2 vols., Roma, Accademia Nazionale dei Lincei, 1989, vol. I, pp. 835-855. See also Ugo Baldini – Pier Daniele Napolitani, *Per una biografia di Luca Valerio: fonti edite e inedite per una ricostruzione della sua carriera scientifica*, «Bollettino di storia delle scienze matematiche», 11, 1991, pp. 3-157.

[43] Galileo, *Sidereus Nuncius*, OG, III, pp. 24, 65, 73. See also, for instance, Id., *Istoria e dimostrazioni intorno alle macchie solari*, OG, V, pp. 99, 195.

[44] See especially Bucciantini, *Contro Galileo* (cit. note 17); Francesco Beretta, *Les dominicains et le procès de Galilée ou de l'Inquisition comme instrument de promotion sociale et d'hégémonie intellectuelle*, in Carlo Longo (ed.), *I domenicani e l'Inquisizione romana. Atti del III Seminario internazionale di studi su "I domenicani e l'Inquisizione"*, Roma, Istituto storico domenicano, 2008, pp. 483-498; Luigi Guerrini, *Galileo e la polemica anticopernicana a Firenze*, Firenze, Polistampa, 2009; Id., *Cosmologie in lotta: le origini del processo di Galileo*, Firenze, Edizioni Polistampa, 2010; Id., *Galileo e gli aristotelici: storia di una disputa*, Roma, Carrocci, 2010; Id., *Echoes from the pulpit: a preacher against Galileo's astronomy, 1610-1615*, «Journal for the History of Astronomy», 43, 2012, pp. 377-390; Id., *Features and trends of the polemic against Copernicus in Tuscany before Galileo*, in Allì Caracciolo (ed.), *Filippo Salviati, filosofo libero: atti del Convegno nel IV Centenario della morte*, Macerata, Eum, 2016, pp. 341-363. See also Mario Biagioli, *Galileo, Courtier: the Practice of Science in the Culture of Absolutism*, Chicago, The University of Chicago Press, 1993.

tionship between members of the court and the Dominicans beneath the Florentine skies were all key elements that led to the 1616 decree. With this volume, we intend to broaden the range of topics in order to determine whether other factors – extending beyond the figure of Galieo – also paved the way for the ban on Copernicus' work.

As Natacha Fabbri's chapter elucidates, it became increasingly clear that the new image of the world introduced by Copernicanism not only involved cosmological matters, but also the doctrinal, moral and religious system on which the Church was established, such as the existence and localization of hell, the opposition between the Empyrean and hell, and the doctrinal question of the ascent of Christ into heaven.

The other factor deserving of consideration is that the threats to Christian values and messages were emerging from a discipline that had in general been regarded as temerarious: studying the movements and positions of the planets, stars, and comets was considered by many as synonymous with taking one step closer to the divine. In this context, the charges of temerity brought against Copernicus and Galileo were threefold: the first was doctrinal and referred to Melchior Cano's definition of rashness; the second concerned the epistemological statute of astronomy, which needed to not outreach the bonds of hypotheses; the third involved the evaluation of the most appropriate behavior to be followed in a time of strained relations between the Church and *novatores*.

Indeed, at the end of the sixteenth century several new philosophical doctrines and cosmological models came into play, provoking a strong reaction from the Church. Among them, it is worth mentioning not only the execution of Giordano Bruno – whose burning at the stake shocked the whole of Europe –, but also the death sentence passed on Francesco Pucci (who was executed in 1597), the trials against Girolamo Cardano, Francesco Patrizi, and Bernardino Telesio, as well as the twenty-seven year detention of Tommaso Campanella. Cardano's trial ended with his abjuration, the prohibition of *De varietate* and the long process of emending his works that only ended in 1607.[45] In 1592 the inquisitorial proceedings against Patrizi's *Nova de Universis Philosophia* began, and the work was in the end condemned by the Congregation of the Index of Prohibited Books 'until corrected'.

---

[45] See, among others, Baldini, *Cardano negli archivi dell'Inquisizione e dell'Indice. Note su una ricerca*, «Rivista di storia della filosofia», 4, 1998, pp. 761-766; Id., *L'edizione dei documenti relativi a Cardano negli archivi del Sant'Ufficio e dell'Indice. Risultati e problemi*, in Marialuisa Baldi – Guido Canziani (eds.), *Cardano e la tradizione dei saperi*, Milano, Franco Angeli, 2003, pp. 457-515; Ricci, *Inquisitori, censori e filosofi sullo scenario della Controriforma*, Roma, Salerno Editrice, 2008, pp. 304-350, 377-389.

In one of his many enlightening works, Luigi Firpo called attention to the fact that Galileo's condemnation was only the epilogue to a much longer history marked by the *libertas philosophandi*'s attempt to stand up to its opponents, as well as by the strong repression it met along the way.

[...] it will appear clearly how in the last decade of the century [i.e., the Cinquecento] free thought in Italy fought its most decisive battle, in the end succumbing – yet in its martyrs unconquered – only because it was stifled materially. Thirty years later, in the silence of Arcetri, the last, posthumous act of this not inglorious tragedy would be staged.[46]

As Roberto Bondì clearly shows in his chapter, in order to fully understand the 1616 decree, another anti-Aristotelian cosmological model has to be examined: Bernardino Telesio, whose works were included in the Index of Prohibited Books, was often mentioned by the foes of Copernicanism, by whom he was not only regarded as heretical, but even as a heresiarch – exactly as would happen to Galileo several years later.

The intellectual affinity between Telesio and Galileo was pointed out by Tommaso Campanella, who wrote the *Apologia pro Galileo* during his detention in 1616, after having already prepared two other apologies, both of which were in support of Telesio. In his 1609 letter[47] to Pope Paul V, Emperor Rudolph II and King Philip III of Spain, Campanella claimed to have penned *De unica animae substantia contra Galenistas apologia pro Telesio*[48] – a medical treatise against the Galenic doctrine – and the *Defensio Telesianorum ad Sanctum Officium libro uno*. According to Luigi Firpo,[49] the *Defensio* was written in 1595, just before Campanella had to face charges «for very serious suspicion of heresy» and his consequent decision to abjure *de vehementi*. Campanella's determination to present himself as a champion of the anti-Aristotelian cosmology exacerbated the strains between theology and natural philosophy, while at the same time strengthening the connection between Telesio's case and the Galileo affair.

---

46 Luigi Firpo, *Filosofia italiana e controriforma*, extract with additions from «Rivista di Filosofia», 41, 1950, pp. 150-173; *ibid.*, 42, 1951, pp. 30-47, Milano, without year, p. 5 (the translation is mine).

47 Campanella to Pope Paolo V, the Emperor Rudolph II and the King Philip III of Spain, March 1609, in Tommaso Campanella, *Lettere*, ed. by G. Ernst, on unpublished papers by L. Firpo, with L. Salvetti Firpo and M. Salvetti, Firenze, Olschki, 2010, n. 33, p. 170.

48 See Firpo, *Appunti campanelliani. III: La perduta "Apologia pro Telesio"*, «Giornale critico della filosofia italiana», 21, 1940, pp. 435-445.

49 Id., *Bibliografia degli scritti di Tommaso Campanella*, Torino, Reale Accademia delle scienze, 1940, pp. 179-181. According to Firpo, this defense corresponds to the *Apologia pro philosophis Magnae Greciae ad Sanctum Officium*.

After touching on the hapless fate of the *novatores*, let us turn to the other side of the barricade, where Roberto Bellarmino stood. In addition to being one of the protagonists of the 1616 trial, he played a crucial role in the events that led to the banning of Telesio's writings[50] and to Bruno's condemnation, as well as to Campanella's trial. Franco Motta's chapter focuses on Bellarmino and the 1616 censure of Copernicanism, framing it within a broader historical context and a wider set of processes that influenced the political and theological stance of the Catholic Church. Among them, he examines the questions of the monopoly over the interpretation of the Scriptures, which was fiercely disputed with James I; the controversy over grace and free will that opposed the Jesuits and the Dominicans; and lastly, the wave of anti-Jesuitism that swept across France after the murder of Henry IV in 1610, giving greater prominence to the thesis of the Holy See's indirect power in political matters.

Bellarmino's intransigence even dealt with the «infected and pernicious books»[51] coming from Reformed countries. Among them we find works by Kepler who, following the 1616 decree, undertook a second Copernican campaign in 1618, which was marked by his *Responsio* (1618) to Ingoli's *Disputatio*, the publication of the *Epitome* and the *Harmonices Mundi Libri V*, as well as the condemnation of the *Epitome* (which was listed in the Index of Prohibited Books in 1619). In those years Kepler claimed – in a tone that expressed both boldness and something closely resembling the commercial promotion of his works – that the *Harmonice Mundi* would have shown the truth of the Copernican system both to learned philosophers and to strict theologians (primarily in Italy).[52] Kepler seemed confident that he would succeed where Galileo had failed: that is, in providing evidence in favor of heliocentrism and redeeming Copernicus. As Édouard Mehl argues in his chapter, the search for the annual stellar parallax played a crucial role in the effort to prove the earth's movement: both Kepler (1618) and Galileo (1624) attempted to refute the erroneous conclusions reached by Francesco Ingoli in *De situ et quiete Terrae* (1616) concerning the absence of a measurable

[50] See Peter Godman, *The Saint as Censor. Robert Bellarmine between Inquisition and Index*, Leiden, Brill, 2000, p. 274. *De rerum natura* of 1586 and the booklets *Quod animal universum* and *De somno* were included in the Index of Prohibited Book in 1596.

[51] Bellarmino to the Inquisitor of Modena, 26 July 1614, quoted in Antonio Rotondò, *Nuovi documenti per la storia dell'Indice dei libri proibiti: 1572-1638*, «Rinascimento», 3, 1963, pp. 145-211: 196. See Eugenio Garin, *Scienza e vita civile nel Rinascimento italiano*, Bari, Laterza, 1965, pp. 115-119; Delio Cantimori, *Storici e storia. Metodo, caratteristiche e significato del lavoro storiografico*, Torino, Einaudi, 1971, pp. 664-665.

[52] Kepler, *Admonitio ad Bibliopolas exteros*, KGW VI, pp. 543-544.

parallax, just as Michael Maestlin did in the second re-edition of Rheticus' *Narratio Prima* (1621). Although they shared the Copernican claim that the annual parallax may appear, they absolutely disagreed with regard to how it might be seen, given their diverging views on the magnitude of the sphere of the fixed stars. Moreover, this topic also had religious implication: since the survey around the stellar parallax involved the hotly debated issue of the infinite space of the world, it also ran the risk of dragging Copernican theory into the minefield of theology.

Besides the issue of the theological concerns that ensued from the ever more pervasive interest in Copernicanism, another vehement reaction deserves to be discussed at length: namely the «harsh controversy» that arose among the astrologers who saw «astrology ruined and a great part of medicine dismantled».[53] Galileo was aware of the difficulty of bringing the heliocentric system and Jupiter's satellites in line with traditional astrological doctrine, and only one year after the publication of the *Sidereus* he dealt with this topic in his letter to Piero Dini.[54] The significant tension between astrology and Copernicanism is confirmed as such by Girolamo da Sommaia, *provveditore* of the University of Pisa: «The opinions of Copernicus, and of Galileo dismantle the whole of Philosophy, most of Theology, leave out Astrology and the influences, and create extravagant consequences, equating the earth with the stars, and the sky, and other things».[55]

Indeed, astrology and Copernicanism seemed to be theoretically incompatible and the acknowledgement of the latter might have played a part in the process of marginalization of the former.[56] At the same time, one cannot ignore the fact that astrology benefitted from the computations of the ephemerides provided by *De Revolutionibus*. For instance, Ismaël Boulliau incorporated Copernicanism into his astrological works.[57] Steven Vanden Broecke's chapter deals with the issue of the relationship between astrology and Copernicanism after the 1616 decree. The astrological defence of geocentrism that Jean Baptiste Morin undertook is a case in point and the chapter tackles it by examining the astrological underpinnings of Morin's strong opposition – on both religious and scientific

---

53 Giovanni Battista Manso to Paolo Beni, March 1610, OG, X, p. 295.

54 Galileo to Dini, 21 May 1611, OG, XI, pp. 107-108.

55 Quoted in Bucciantini, *Contro Galileo* (cit. note 17), p. 27.

56 See Vermij and Hiro Hirai, *The Marginalization of Astrology: Introduction*, «Early Science and Medicine», 22, 2017, pp. 405-409.

57 See Robert Alan Hatch, *Between Astrology and Copernicanism: Morin – Gassendi – Boulliau*, «Early Science and Medicine», 22, 2017, pp. 487-516.

grounds – to the heliocentric models proposed by Philip and Jacob Lansberg, Kepler and Galileo.

This volume does not seek to cover the whole range of topics and personalities involved in the 1616 decree; rather the case studies presented are intended to shed further light on several aspects of the discussions that arose around Copernicus' 'temerity' and, in the end, saw the banning of his work. The result is a mosaic whose pieces we hope will contribute not only to the completion of the picture outlined by other studies – both by building on their conclusions and engaging in dialogue with them – but that will also give new meaning to the entire scenario.

This work could not have been undertaken without the continual support of the Museo Galileo – Institute and Museum for the History of Science of Florence. We therefore owe particular thanks to the director of the Museo, Paolo Galluzzi, for his intellectual guidance and fruitful insights. We would also like to express our sincere gratitude to the editors of the journal *Galilæana* – Massimo Bucciantini, Michele Camerota and Franco Giudice – for raising important questions and making it possible for us to realize this project. We are furthermore greatly indebted to Germana Ernst for her endless supply of enlightening advice.

Our gratitude goes to Patrizia Ruffo and Maria Conforti for their valuable comments and to Marta Stefani for supporting us in the final phase of seeing this volume through the press. We must express our thanks as well to the staff at the library of the Museo Galileo, whose help was – as always – unstintingly and competently provided. We would like to thank Lisa Chien for her careful copy editing of the final drafts, as well as Sabrina Guzzoletti and the entire staff at Leo Olschki Editore for their meticulous attention to countless production details.

The enormous on-line archive of Galilean resources – Galileo//thek@ – created by the Museo Galileo (https://galileoteca.museogalileo.it) proved to be invaluable to the authors in the pursuit of their research for *Copernicus Banned*.

Roberto Bondì

DANGEROUS IDEAS: TELESIO, CAMPANELLA AND GALILEO

At the beginning of *Quaestiones celeberrimae in Genesim*, dealing with the first six chapters of the Bible and published in Paris in 1623, the «secrétaire de l'Europe savante»,[1] Marin Mersenne, juxtaposed, among others, Telesio, Campanella and Galileo. He emphasized that their disciples shared a «false» belief, namely «that Catholic doctors and theologians only follow Aristotle and swear by his word, even if experience and phenomena reveal the opposite».[2] Referring to Campanella in particular, Mersenne spoke of an groundless complaint (*iniusta querimonia*): had it been necessary to admit the existence of a sea within the Sun or even to attribute a vegetative soul to the Earth, the theologian, claimed Mersenne, would not have flinched, so much so that when a better understanding of the «mysteries of the faith» is at stake, he «willingly» uses other forms of knowledge.[3] In *Paralipomena*, Mersenne launched another attack on Campanella, who, the Minim friar reiterated, was completely wrong in believing that the Church or the Catholic doctors (*Catholice agentes*) give credit to Aristotle, Plato or other philosophers to the point of embracing their precepts even when these are false or implausible.[4]

These are well known and relatively sanguine passages[5] taken from a complex work which is cautious, but at times also daring. Indeed, it is dif-

---

I wish to thank Clare Tame for her translation of the Italian text.

[1] Barthélemy Hauréau, *Histoire littéraire du Maine*, I, Paris, Julien, Lanier et Cie, 1852, p. 370. See, in particular, Robert Lenoble, *Mersenne ou la naissance du mécanisme*, Paris, Vrin, 1971[3] (1st ed.: 1943).

[2] Marin Mersenne, *Quaestiones celeberrimae in Genesim, cum accurata textus explicatione*, Lutetiae Parisiorum, sumptibus Sebastiani Cramoisy, 1623, *Praefatio et prolegomena ad lectorem*, p. 1 unnumbered. Besides Telesio, Campanella and Galileo, Mersenne also cites Bruno, Kepler and Gilbert.

[3] *Ibid.*, p. 2 unnumbered.

[4] *Ibid.*, col. 1837.

[5] Michel-Pierre Lerner, *Tommaso Campanella en France au XVIIe siècle*, Naples, Bibliopolis,

ficult to exaggerate the influence of widespread anti-Aristotelianism in the second half of the sixteenth century on the progressive exacerbation of religious life in Italy. I hardly need mention that when Mersenne published his impressive work Telesio's treatises had been on the Index of Prohibited Works for thirty years, Campanella had been in prison for almost twenty-four years, and Galileo had been facing the consequences of the anti-Copernican edict (5 March) for seven years. The edict did not mention heresy directly, but it established that the «Pythagorean doctrine» of the Earth's motion and the motionlessness of the Sun was «false» and went «against the divine Scriptures».[6] At the same time, exactly twenty-three years had passed since Giordano Bruno, whom Mersenne also placed alongside Telesio, Campanella and Galileo at the start of *Quaestiones*, had been burnt at the stake in Campo de' Fiori «with his tongue imprisoned on account of his wicked words».[7] Yet for all their differences, Telesio, Campanella and Galileo did indeed share some dangerous ideas. Not only were they anti-Aristotelian, but they thought that even the Bible was anti-Aristotelian and were determined to demonstrate it. They were on slippery ground and sooner or later would come to realize it.

Regarding Galileo, the letter of 30 November 1612 to him from the prince and natural philosopher Federico Cesi, then busy trying to obtain a *nihil obstat* for Galileo's *Istoria e dimostrazioni intorno alle macchie solari e loro accidenti* (1613), is extremely clear. Cesi reminds Galileo that «one cannot remove the grip of the Peripatetics, if not very slowly» before adding:

> I wrote *in iure* (so to speak), adducing from ten passages in the Scriptures, and as many Holy Fathers, in confirmation of Your Worship's words, that celestial aptitude to corruption was in accord with the Scriptures and indicated by them. It was not enough; and they replied that the passages had been very well interpreted by others in a peripatetic sense. I had to be patient: in conclusion, they do not want you to say anything about the Scriptures in this passage.[8]

1995, pp. 19-26. See also, Natacha Fabbri, *Cosmologia e armonia in Kepler e Mersenne. Contrappunto a due voci sul tema dell'*harmonice mundi, Florence, Olschki, 2003 and Id., *Galileo between Christian Virtuosi and Catholic Apologetics: John Wilkins and Marin Mersenne*, in Massimo Bucciantini – Michele Camerota – Franco Giudice (eds.), *Il caso Galileo. Una rilettura storica, filosofica, teologica, convegno internazionale di studi* (Firenze, 26-30 maggio 2009), Florence, Olschki, 2011, pp. 289-310.

[6] OG, XIX, p. 323. The text of the edict is reproduced by Mersenne in *Quaestiones celeberrimae in Genesim* (cit. note 2), col. 904.

[7] Luigi Firpo, *Il processo di Giordano Bruno*, ed. by D. Quaglioni, Rome, Salerno Editrice, 1993, pp. 355-356.

[8] Cesi to Galileo, 30 November 1612, OG, XI, p. 439.

As for Telesio, his «patience» was only apparent. He threw himself into trying to demolish Aristotle's natural philosophy and did not refrain from using the sort of biblical exegesis that probably formed the grounds for his condemnation in 1593 and that was to be so heavily censured in Padua in November 1600. Burdened by doubt and uncertainty, Telesio waited a long time before having his first work printed, with the challenging title of *De natura iuxta propria principia*. Yet once he had decided to do so in 1565, it was a full frontal attack on Aristotle and the principle of authority and left little room for mediation. It was, in fact, a point of no return. Campanella was well aware of this and he began to revere Telesio, «both for [his] freedom to philosophize, and because he derived his doctrines from the natural world, and not from the words of men».[9] The young friar from Stilo understood that for Telesio «Our discourses must relate to the sensible world and not to one on paper», to use a well-known Galilean expression.[10]

Campanella cherished the lesson of the philosopher who had managed to pierce the «tyrant of wits»,[11] and who had had no «patience». He was not yet twenty-one years of age when he wrote his immense tome in which he mocked the Neapolitan lawyer Giacomo Antonio Marta[12] who had dared criticize Telesio in his defence of Aristotle, and made a broad critique of the Aristotelian tradition. The text was published in Naples in 1591 under the title of *Philosophia sensibus demonstrata* by Orazio Salviani, the publisher of the third and last edition of Telesio's *De rerum natura* (1586). Campanella's problems began with this, his first printed work, and took concrete and dramatic form with his 'first trial' the following year.

It was precisely the experiences of those like Telesio and Campanella that played a decisive role in creating a climate that led to the events of 1616. These events have often and incorrectly been «studied and reconstructed in relation to the sentence of 1633».[13] Indeed, the proximity of

---

[9] Tommaso Campanella, *Sintagma dei miei libri e sul corretto metodo di apprendere – De libris propriis et recta ratione studendi syntagma*, ed. by G. Ernst, with an iconographic note by E. Canone, Pisa-Rome, Fabrizio Serra Editore, 2007, p. 33.

[10] Galileo, *Dialogo sopra i due massimi sistemi del mondo*, OG, VII, p. 139 (Stillman Drake's translation: G. Galilei, *Dialogue Concerning the Two Chief World Systems – Ptolemaic & Copernican*, Eng. trans. by S. Drake, foreword by A. Einstein, second edition, Berkeley and Los Angeles, University of California Press, 1967, p. 113).

[11] Campanella, *Le poesie*, critical text, introduction and commentary by F. Giancotti, Milan, Bompiani, 2013, p. 182: «Telesio, il telo della tua faretra / uccide de' sofisti in mezzo al campo / degli ingegni il tiranno senza scampo; / libertà dolce alla verità impetra».

[12] On Marta, see Paul F. Grendler, *The University of Mantua, the Gonzaga and the Jesuits, 1484-1630*, Baltimore, The Johns Hopkins University Press, 2009, pp. 83-126.

[13] Bucciantini, *Contro Galileo. Alle origini dell'"affaire"*, Florence, Olschki, 1995, p. 21. Buc-

Galileo and Cesi[14] to Telesio and his followers was a fact and had been noted by both the 'anti-Aristotelian party' and by the 'Aristotelian party' (and not just for polemical reasons). There were, indeed, objective points in common, and more than a few convergences emerged that may strike us as strange today. As Cesi confirmed to Galileo on 22 March 1612 about the Telesian Antonio Persio, who had passed on «to a better life, to the distress of all», the abbot «was entirely [in agreement with] Your Worship except in the opinion of Copernicus».[15] A few weeks later Cesi promised to let Galileo have «some of S.r Persio's things» which would have been «very irritating for the Peripatetics».[16]

1. Italian Frustrations

The transalpine optimism flaunted by Mersenne at the beginning and end of *Quaestiones celeberrimae in Genesim* was not shared by the disciples of Telesio and Campanella, and even less by those of Galileo. What had happened to their masters in the relatively recent past merely confirmed this. From the outset Telesio, Campanella and Galileo all found themselves hedged in by circumstances that caused them intense frustration, and which are worth recalling briefly.

1.1. *Telesio between the Dominicans, Franciscans and Jesuits*

Since the publication in 1993 of a previously unedited letter by Telesio, we now know for certain that his experience with censorship began well before the publication of the last edition of *De rerum natura*. On 28 April 1570 Telesio was in Naples for the publication of the second edition of

ciantini is right to devote ample space to figures such as Tommaso Caccini and Francesco Ingoli, and cites the relevant judgements of Girolamo da Sommaia, *provveditore* of the University of Pisa: «L'opinioni del Copernico e del Galileo distruggono tutta la Filosofia, molta Teologia, levan l'Astrologia et gli influssi, e fanno stravaganti conseguenze, pareggiando la terra alle stelle, et a' cieli, et altro»; «Il Tilesio fu gran nemico d'Aristotele, grand'ingegno, l'opere sue sono prohibite [...] Il Campanella ancora è grandissimo nimico d'Aristotele» (p. 27). On the 'caso Galileo' in general, see the contributions in Bucciantini – Camerota – Giudice (eds.), *Il caso Galileo* (cit. note 5).

14 See Paolo Galluzzi's important work, *Libertà di filosofare* in naturalibus. *I mondi paralleli di Cesi e Galileo*, Rome, Scienze e Lettere, 2014. See also, Eugenio Garin, *Fra '500 e '600: scienze nuove, metodi nuovi, nuove accademie*, «Nuncius», 1, 1986, pp. 3-23.

15 Cesi to Galileo, 22 March 1612, OG, XI, p. 285.

16 Cesi to Galileo, 19 May 1612, OG, XI, p. 298.

his *magnum opus* and sent a letter to Cardinal Flavio Orsini, the head of the Archdiocese of Cosenza. The letter relates how Giambattista di Benedetti, the archbishop's deputy, informed Telesio that someone in Cosenza had noticed the presence of «other propositions counter to religion»[17] in the first edition of what was his most important work. Understandably alarmed, Telesio was swift to respond by listing the authors who had examined the 1565 edition.

Twice in the space of three lines Telesio affirms that the text had been sanctioned by the Dominican friar Eustachio Locatelli. This reiteration was not accidental; to Telesio the name of Locatelli was a guarantee for he was the famously severe inquisitor of Bologna – an erudite theologian, procurator general of his order from 1560, and renowned for the rigor of his work. In 1559 he ruthlessly presided over a trial that ended with the sentencing to death of four women accused of witchcraft, despite it being their first sentence and despite their avowal of penitence.[18] In the meantime he had become, perhaps not exactly by chance, confessor to Pope Pius V, who appointed him Bishop of Reggio Emilia in 1569.[19] Locatelli's approval of *De rerum natura*, however, was based on the evaluation of an unidentified «Theologian» according to whom Telesio's text «does not contain anything against religion». Telesio added emphatically that his work had been vetted by «endless Theologians, and Philosophers», including the Dominican friar, Tommaso de Vio, «heir not only to the name, but [also] the doctrine of the cardinal of [Santa Maria sopra] Minerva», to wit the famous Gaetano, who «not only approved it, but [who] said that, were he so authorized it would be read in all convents of his religion». Moreover,

---

[17] Girolamo de Miranda, *Una lettera inedita di Telesio al cardinale Flavio Orsini*, «Giornale critico della filosofia italiana», 72, 3, 1993, p. 373.

[18] Adriano Prosperi, *L'inquisizione romana. Letture e ricerche*, Rome, Edizioni di Storia e Letteratura, 2003, p. 231: «Non sappiamo bene quale cavillo fu usato: un cenno su quanto era emerso dal processo a proposito del rapporto tra streghe e demonio forse ne offrì uno sufficiente alla volontà per "dare un esempio". Le streghe sarebbero state ribattezzate dal demonio ("a daemonio rebaptizari patientes") e questo poteva in qualche modo configurarle come uno di quei casi di eresia estrema (antitrinitarismo, negazione della divinità di Cristo) per i quali una bolla di Paolo IV aveva stabilito che non valesse il principio del perdono in prima istanza ai pentiti». Prosperi alleges that the timing and ways used were of an extreme harshness, arbitrary and cruel.

[19] On Locatelli see Guido Dall'Olio, «Locatelli, Eustachio», *Dizionario storico dell'Inquisizione*, directed by A. Prosperi, with the collaboration of V. Lavenia and J. Tedeschi, 4 vols., II, Pisa, Edizioni della Normale, 2010, p. 929. On Locatelli's «harshness», see also Adriano Prosperi, *Tribunali della coscienza. Inquisitori, confessori, missionari*, Turin, Einaudi, 2009, p. 131. The Dominican continued to deal with book censorship, and was an advisor to the cardinal's council for the revision of the Tridentine Index.

it had also been seen by Eustachio Locatelli's brother, Giovanni Antonio Locatelli, who was the bishop of Venosa. To conclude, «No-one ever saw anything against religion» in it.

We do not know who in Cosenza possessed the «very acute eyes» that spotted «errors» which had previously gone undetected «in Rome, and the rest of Italy», but from the letter we learn what «errors» had been made. From these «propositions», writes Telesio, «one can infer that I believe the soul to be mortal, and that I deny that the Sky is moved by the intelligences».[20] It is no accident that, unlike the first edition of *De rerum natura*, the new edition contains a passage which distinguishes between a soul of divine origin and one of material origin, and the reader is advised that only and exclusively the second is referred to throughout the book.[21] In any case, Telesio's priority now was to offer reassurances regarding the new edition of *De rerum natura* which was already «with the printer». We learn that it had been seen by the Franciscan Baldassarre Crispo, renowned professor at the Neapolitan *Studio* of San Lorenzo, but above all it had been seen by Gaspar Hernández, Rector of the Jesuit College of Naples, and by the mathematician Giovanni Paolo Vernalione: «They all approved it without difficulty». Furthermore, as regards the thorny question of the soul, the two books of the new edition «say very little» and «what they do say only regards the matter of principles, and the sensitive and motive soul».[22] Telesio had reckoned on publishing new parts to his book dealing specifically with this argument, and considered ready to print, in Rome by the end of the year. In fact, it took sixteen years before this material finally appeared, as part of the third Neapolitan edition of *De rerum natura* (1586). The letter tells us that these texts as well, which Telesio hoped would be considered to «conform strictly» to religion, had been endorsed by the Jesuits, in the person of Hernández and, above all, the influential Alfonso Salmerón, provincial superior of the Jesuit College in Naples, one of the first companions of Ignatius of Loyola, and papal theologian to the Council of Trent.

According to Telesio, both Hernández and Salmerón had also bestowed their approval on «other things»: in particular, his theories regarding the number of primary bodies in the world and the «substance, and nature

---

20 de Miranda, *Una lettera inedita di Telesio al cardinale Cardinale Flavio Orsini* (cit. n. 17), p. 373.

21 Bernardino Telesio, *De rerum natura iuxta propria principia, liber primus, et secundus, denuo editi – Opuscula*, Preface by N. Ordine, Introduction and analytical index by R. Bondì, Rome, Carocci, 2013, p. 47. See Roberto Bondì, *Introduzione a Telesio*, Rome-Bari, Laterza, 1997, pp. 43-44.

22 de Miranda, *Una lettera inedita di Telesio al cardinale Flavio Orsini* (cit. n. 17), p. 374.

of the Sky». Significantly, it was during this period that Salmerón took the young Giacomo Antonio Marta under his protection, the same Marta who was to become one of Telesio's detractors, and who would consequently come under attack from Campanella. Here Telesio focuses on the question of motion and relates that he has been assured by the two Jesuits that the thesis according to which – «as per Aristotle» – the skies are moved by the intelligences «is not to be found in the Scriptures». Moreover, Hernández went so far as to judge the Aristotelian position as «absurd, and ridiculous».[23]

Telesio marshaled a small army of impressive names to vindicate his work and suggested to Cardinal Orsini that the attack on his writings was more a question of «jealousy, and malice» than of genuine «religious zeal». He had no intention of asking that the Church suspend its work of checking and controlling works before their publication: «If I have erred, I shall be supremely thankful to see, and to correct my errors», given that «my mind is, and shall always be, by the grace of Our Lord God, entirely subject to, and in obeisance to the true, Catholic religion». He then added: «I would be quite ready to burn all my works, were it demonstrated that they are lacking in Christian piety». He had one sole request to make to Cardinal Orsini: that any revision be entrusted to «a discreet person, and one not heavily devoted to Aristotle».[24]

1.2. *A twenty-three-year-old on trial*

The young Tommaso Campanella was not in the least «devoted» to the philosophy of Aristotle. Indeed, he was as much disappointed by the doctrine of the Stagirite and his commentators as he was enthusiastic about Telesian thought. His enthusiasm must not have pleased his superiors, who packed him off to a convent in Altomonte. Here, while he cultivated his study of Telesio, friends who shared his passion for the «truth of things» introduced him to a book published in Rome in 1587 entitled *Pugnaculum Aristotelis adversus principia Berardini [sic] Telesii* by the same Giacomo Antonio Marta who had grown up under the protection of Salmerón and now laid claim to pompous titles, but whom Campanella described as the «peripatetic know-it-all» who had dared to attack the «greatest of philosophers» despite not knowing «any truth».[25]

---

23 *Ibid.*

24 *Ibid.*, p. 375.

25 Campanella, *Philosophia sensibus demonstrata*, ed. by L. De Franco, Naples, Vivarium, 1992, p. 9.

The young Dominican friar only needed a few months (this is what he reports and we have good reason to believe him) to write five hundred pages in which, using vivid epithets for the Neapolitan lawyer (he calls him *sciolus* forty-two times, *homuncio* fifteen times, *philosophaster* thirteen times and *homunculo* 'only' twice),[26] he literally «tore to shreds» the defence of Aristotle in *Pugnaculum*.[27]

Marta had accused Telesio of speaking about nature without having seen it and without giving definitions and demonstrative syllogisms, hence without being able to recognize it. But does nature really (*O infelix homuncio*) reveal itself through logic? Or does it instead reveal itself through the senses? Is it perhaps not true that Aristotle himself considered «sensible experience» as *magistra rerum*? The truth is that Marta was struggling to defend a philosopher who was both indefensible and overrated. It is difficult to understand why «Marta the philosophaster» and many others considered Aristotle to be a god when, if read carefully, he appears for what he is – that is, far from being a purveyor of the truth, he was contradictory, «impious» and «very ignorant» as regards «divine things». As for the «pure Peripatetics», these followers of Aristotle are simply written off as «pure heretics».[28]

It was in these terms that Campanella presented readers with the eight *disputationes* of *Philosophia sensibus demonstrata* (1591), a work devoted to Telesian topics integrated with pre-Socratic, neo-Platonic and hermetic themes, and that included a critique of Aristotle. The adherence to Telesianism revealed by this weighty tome was the real reason for the first of Campanella's many trials. He was imprisoned later that same year and his case was closed on 28 August 1592 with the admonition to adhere to Thomist doctrine and to renege Telesian notions (*...ut disputando, praedicando et legendo teneat doctrinas sancti Thomae et reprobet doctrinas Telesii*).[29] Some weeks later Campanella was to feel the direct consequences of his writings. In October he left Florence, disappointed not to have been accepted into the service of the Grand Duke.[30] The information requested by Ferdinando I before making his decision described the friar as a «very great

---

[26] *Ibid.*, pp. 12, 15, 16, 17, 18, 21, 93, 98, 99, 110, 118, 138, 139, 140, 143, 195, 196, 265, 279, 292, 339, 341, 344, 388, 438, 459, 461, 464, 503, 548, 561, 566, 584, 588, 589, 608, 619, 625, 691, 732). Marta is then, among many other things, «idiotis rudior» (*ibid.*, p. 313).

[27] *Ibid.*, p. 21.

[28] *Ibid.*, pp. 10, 15, 16.

[29] Luigi Firpo, *I processi di Tommaso Campanella*, ed. by E. Canone, Rome, Salerno Editrice, 1998, p. 52.

[30] Campanella, *Lettere*, ed. by G. Ernst, on unpublished papers of L. Firpo, with L. Salvetti Firpo and M. Salvetti, Florence, Olschki, 2010, p. 7.

man of letters, a very excellent philosopher, albeit rather Telesian»,[31] and «a young man of mature judgment, and who held numerous and recondite ideas, as one sees from his learned reasoning, no less than from his work printed with the title *de philosophia sensibus demonstrata*».[32] However, it was noted that there was a shadow hanging over Campanella: «The truth is that today some in Rome are now trying to ban Telesio's philosophy, given that it seems to harm the scholastic Theology based on Aristotle, who is much criticized by Campanella; he too runs the same risk owing to his allegiance to the same school».[33]

1.3. *Galileo, Telesio and the Bible*

At the turn of the seventeenth century being compared to Telesio was enough to create problems for anyone. Even more so after *De rerum natura* and the two pamphlets *De somno* and *Quod animal universum ab unica animae substantia gubernatur* were placed on the Index of Forbidden Books in 1593 and, even more significantly, in 1596. Galileo had direct experience of this and had seen himself placed alongside Telesio during the first attacks on his work in Florence in 1610-1611, and hence before the serious allegations made against him by the Jesuit Orazio Grassi during the famous controversy over comets. In *Libra astronomica ac philosophica* (1619), Grassi provocatively likened Galileo's positions to those of other authors such as Telesio and Girolamo Cardano who were themselves – and not coincidentally – condemned by the Church. As followers of a «sterile and unhappy» philosophy, they are described as leaving to posterity not «sons» (*liberos*), but only «books» (*libros*).[34] Yet beyond the polemics, Grassi seized on a real similarity between Galileo and Telesio (and Cardano) in their explanation of the phenomenon of comets, and it is difficult to imagine that the Pisan scientist would have marveled at this juxtaposition. In fact, he replied that

---

31 Germana Ernst, *Tommaso Campanella. Il libro e il corpo della natura*, Rome-Bari, Laterza, 2002, p. 22. This was a letter sent on 2 October 1592 by the Secretary of State, Lorenzo Usimbardi, to the librarian of the Laurenziana, Baccio Valori.

32 Campanella, *Opere*, 2 vols., I, ed. by A. D'Ancona, Turin, Pomba, 1854, p. lxxv. The letter is from Baccio Valori and dated 15 October 1592.

33 *Ibid.*

34 Orazio Grassi, *Libra astronomica ac philosophica*, OG, VI, p. 118. See also the references to Galileo and Telesio in the Aristotelian Fortunio Liceti, *De novis astris et cometis libb. sex*, Venetiis, Apud Io. Guerilium, 1623, pp. 47-48. On the relations between Galileo and Liceti, see Luigi Guerrini, *Galileo e gli aristotelici. Storia di una disputa*, Rome, Carocci, 2010. See also, Id., *Galileo e la polemica anticopernicana a Firenze*, Florence, Edizioni Polistampa, 2009 and Id., *Cosmologie in lotta. Le origini del processo di Galileo*, Florence, Edizioni Polistampa, 2010.

he had never read the two condemned authors, but did not flinch from defending them explicitly in *The Assayer*, describing Telesio and Cardano as two «venerable Fathers of natural philosophy».[35]

Some years earlier the man of letters, Aristotelian philosopher, and one of Galileo's sharpest critics, Ludovico delle Colombe, juxtaposed Galileo and Telesio in a treatise entitled *Contro il moto della Terra* (1610-1611). This is an important piece of work insofar as it raises the question of the relationship between Copernicanism and the Sacred Scriptures. It also emphasizes that «all theologians, with no exceptions» share the belief that «when Scripture can be interpreted literally, it ought never to be interpreted differently».[36] The treatise began thus:

> The ambitious human spirit, urged on beyond every advisable limit by the desire for immortality, quite sick with navigating the ocean of truth, plunged into a sea of lies, scorning the Columns of Hercules, mocking Aristotle and jeering at Plato, it cries *plus ultra*, until he comes to unknown shallows, and smashes to pieces against unseen rocks. It happens that some, despairing of understanding Aristotle, or being able to say something that gives them notoriety in his philosophy, are resolved [...] to rediscover or dream a new philosophy and a new way to philosophize. Yet such were some Ancients, and in our age the Telesians. Others, having no grounding in any philosophy, turn to mathematical studies and declare them to be sovereign over all other things.[37]

For delle Colombe, the difference was clear: on the one hand there was Aristotle, whose work contains «the sum of philosophical truths»,[38] and on the other there were those, such as the Telesians, who misled themselves into thinking they had discovered a «new philosophy» and a «new way to philosophize», and those who, like Galileo, deluded themselves that mathematics was the key to reality. Both the Galileans and the Telesians were deemed to be not only dangerous, but also far from the truth. In the face of the danger of Copernicanism in particular, one could and indeed should draw up an exhaustive list of «sound foundations of the Scriptures» that confirm the old cosmology and confute the new one. This

---

35 GALILEO, *Il Saggiatore*, OG, VI, p. 236. See BONDÌ, *The First of the Moderns: Telesio between Bacon and Galileo*, in PIETRO DANIEL OMODEO (ed.), *Bernardino Telesio and the Natural Sciences in the Renaissance*, Leiden, Brill (forthcoming). On comets, see the Introduction by MIGUEL A. GRANADA in BERNARDINO TELESIO, *Sobre los cometas y la Vía Láctea*, edición del texto latino, introducción, traducción y notas de M.A. Granada, Madrid, Editorial Tecnos, 2012, pp. IX-CXXV.

36 LODOVICO DELLE COLOMBE, *Contro il moto della Terra*, OG, III/1, p. 290.

37 *Ibid.*, p. 253.

38 DELLE COLOMBE, *Discorso apologetico*, Firenze, Appresso il Pignoni, 1612, OG, IV, p. 317.

is what delle Colombe thought, and indeed he ended his pamphlet with just such a list.[39]

The problem of comparing scientific work with the Holy Scriptures, which Copernicus himself had faced, became virtually unavoidable but extremely difficult to tackle. Galileo was well aware of this when he published *Istoria e dimostrazioni intorno alle macchie solari e loro accidenti* (1613). In the autumn of 1612 he was already testing reactions to a work dealing with a «novelty», that would be in his words «the funeral or rather the extreme and last judgment of pseudo-philosophy», as he announced to Cesi on 12 May 1612 before the appearance of this work.[40]

In its first version, Galileo wrote:

> Who will, having observed and considered these things, still persist in an opinion [which is] not only false, but erroneous and repugnant to the indubitable truths of the Holy Scriptures? which tell us [that] the heavens and the entire world not only are generable and corruptible, but [are also] generated, dissoluble and transitory.

The passage was rejected by the censors and Galileo revised it as follows:

> Who [after] seeing, observing and considering these things, will not embrace [...] the opinion so true to the unquestionable truths of the Sacred Scriptures, which in many places very openly and clearly lead us to the unstable and short-lived nature of celestial matter? Yet meanwhile not defrauding of deserved praises those sublime wits that with subtle speculation were able to fashion, to the sacred dogmas, the apparent discord of the physical discourses. Which is now good reason for them to give way, once likewise the supreme theological authority is removed, to the natural reason of other very serious authors and even more to sensible experiences, to which I have no doubt that Aristotle himself would have conceded, since we are aware of having not only acknowledged experience as one

39 Delle Colombe, *Contro il moto della Terra* (cit. note 36), pp. 289-290: «Perché, *Fundasti Terram super stabilitatem suam*, dice il Salmo 103: *Deus fundat orbem immobilem*, intendendosi della Terra, dice l'Abulense, Paralipom. 16. Ma che la Terra sia nel centro, dicalo Iobbe, 26: *Qui appendit Terram super nihilum, idest super centrum*. Che ella sia grave: *Necdum montes gravi mole constiterant*, Proverb. 8; and Esaia, cap. 40: *Quis libravit in pondere montes? quis appendit tribus digitis molem Terrae?* e ne' Proverbi, cap. 27: *Grave est saxum et onerosa arena*. Ecco di più, ne' Proverbi, cap. 30, la Terra essere nel centro del Mondo: *Caelum sursum, et Terra deorsum*. Ora, se la Terra fusse, come vuole il Copernico, nella grossezza di un Cielo, ella non sarebbe *deorsum*, poiché il Cielo si dice esser *sursum*; e conseguentemente il Sole non è nel centro, poiché è nel Cielo che si dice esser *sursum*. Ma che esso Sole non sia immobile, ecco l'Ecclesiastico, c. 1: *Oritur Sol, et occidit et ad locum suum revertitur, ibique renascens girat per meridiem, et flectitur ad aquilonem*. Che più? non si fermò, perché Iosuè ottenesse la vittoria? non retrogradò al tempo del re Ezecchia? Che poi la Luna non sia un'altra Terra, non dice la Scrittura: *Fecit Deus duo luminaria, idest luminare maius, et luminare minus et stellas ut lucerent super Terram?...*».

40 Galileo to Cesi, 12 May 1612, OG, XI, p. 296.

of the powerful means to settle natural questions, but once more granted first place: where if he argues the immutability of the skies on the basis of not having seen any sensible change in them over time, it is credible that when the senses let him see what for us is manifest, he would have followed the contrary opinion, to which with such wonderful discoveries we are called to.[41]

Yet even this passage was rejected and in the end Galileo was forced to remove all references to the Scriptures and to thus abandon his attempt to find corroboration and support in the Bible for his theories.

2. «Sending Aristotle packing»

Telesio managed to do just this and did so on the theme of the nature of the heavens. One of the cornerstones of the cosmology of *De rerum natura* was the rejection of the qualitative distinction between the celestial world and the sublunary world, and Telesio made no bones about resorting to biblical texts to legitimate his anti-Aristotelian orientation. The Copernicans really did «send Aristotle packing»[42] and destroyed the dogma of the immutability of the heavens, so that Galileo commented, «I do not know where it can be saved and concealed».[43] Telesio, although no Copernican, had also made a significant contribution to subverting Aristotle's cosmological model. This was well understood by, among others, the Aristotelian Giulio Cesare Lagalla, Galileo's adversary, who had openly and not coincidentally rejected the Telesian notion of the fiery nature of the heavens.[44]

---

41 Galileo, *Istoria e dimostrazioni intorno alle macchie solari*, OG, V, pp. 138-139. See William R. Shea, *La controriforma e l'esegesi biblica di Galileo Galilei*, in Albino Babolin (ed.), *Problemi religiosi e filosofia*, Padua, La Garangola, 1975, pp. 47-48 (see Shea, *Copernico, Galileo, Cartesio. Aspetti della rivoluzione scientifica*, It. trans., Rome, Armando, 1989, pp. 113-29); Paolo Rossi, *Galileo Galilei e il Libro dei Salmi*, «Rivista di filosofia», 69, 1978, pp. 47-49 (see Paolo Rossi, *La scienza e la filosofia dei moderni. Aspetti della rivoluzione scientifica*, Turin, Bollati Boringhieri, 1989, pp. 67-89). See also Mauro Pesce, *L'ermeneutica biblica di Galileo e le due strade della teologia cristiana*, Rome, Edizioni di Storia e Letteratura, 2005, pp. 24-26.

42 Delle Colombe, *Contro il moto della Terra* (cit. note 36), p. 254.

43 Galileo to Federico Cesi, 12 May 1612 (cit. note 40), p. 296.

44 Giulio Cesare Lagalla, *De phaenomenis in orbe lunae*, OG, III/1, pp. 363-366. Lagalla wrote: «Hanc sententiam, quod Coeli et astra sint ex igne conflata, quamvis prisca omnis philosophia ratam habere visa fuerit (omnes enim ante Aristotelem philosophi, aut Coelum ex igne, aut ex igne et terra, vel etiam ex quattuor elementis, quod Plato, Aristotelis praeceptor, constare existimarunt), hanc, inquam, sententiam (quam etiam paulo ante nostram aetatem Bernardinus Telesius, vir priscae philosophiae studiosus, nulli sectae addictus, et novus philosophiae post multa saecula exarchus, renovavit) licebit nobis libere et sine offensione confutare...».

Delle Colombe was of the opinion that the question remained squarely in the terms posed by the Aristotelian tradition:

> Is it not very clear to the senses, God has with marvelous providence in all things created, ordered by degrees, distinguishing the lowest and the most supreme in every place and every located thing? Is water not purer and simpler than the Earth? Air, than water? Fire, than air? And consequently is not each of these elements located in the highest place, corresponding to its noble nature? The sky above them is thus located, since it is far more excellent than all the elements in its substance, being a fifth nature or substance not communicating in matter with the other natural bodies, so it is proper that it have a supreme place over all others, on account of its dignity and nobility.[45]

Faced with the insistence of the Copernicans, delle Colombe went in search of support to confirm the traditional vision of the world. This led to his enthusiasm for a passage in the text that the mathematicians of the Collegio Romano had sent to Cardinal Roberto Bellarmino on 24 April 1611 in reply to his request for information. The «attestazione de' PP.ⁱ Giesuiti al Rever.ᵐᵒ Card. Bellarmino», as Galileo called him, signed by four «unworthy servants in Christ» (Clavio, Grienberger, Maelcote, Lembo), stated that while «one cannot deny the great inequality of the Moon» it is also true that «for Father Clavio it appears more likely that its surface is not unequal, but rather that the lunar body is not uniformly dense and that parts of it are more dense and [other] parts more rare, as are the ordinary spots that one can see with the naked eye».[46] It was enough to encourage delle Colombe to send Clavio a letter of congratulations in which he stresses that «in celestial bodies, where there is no mixing, there is no reason for inequality of the figure, particularly the spherical figure being the most perfect, it is appropriate that the celestial bodies and globes have it; and even more, as they are more supreme».[47]

The qualitative distinction between what lies below the sphere of the Moon and what is found above it, obstinately defended by delle Colombe, had been swiftly rejected by Telesio in the first edition of *De rerum natura* (1565). Delle Colombe must have had this in mind when, at the beginning of his pamphlet *Contro il moto della Terra*, he refers to the «lies» of those who, like the «Telesists», flatter themselves on having found a "new phi-

---

45 Delle Colombe, *Contro il moto della Terra* (cit. note 36), p. 276.

46 The Mathematicians of the Collegio Romano to Roberto Bellarmino, 24 April 1622, OG, XI, p. 93.

47 Lodovico delle Colombe to Christoph Clavius, 27 May 1611, OG, XI, p. 118.

losophy" and a "new way to philosophize". As delle Colombe was certainly aware, Telesio was not one of the «very famous authors» who said that *natura locuta est ex ore illius* when referring to Aristotle.[48] He was instead perturbed, as was Galileo, that one could speak of someone in such terms, even if the person in question was Aristotle, and deplored the fact that Aristotelian precepts were accepted *fide ac religione*, as pouring from nature's mouth (*ut ex ipsius naturae ore prolata*).[49] Telesio was obsessively insistent on this point. He believed that the world was still open to discovery only if men were prepared to reject bookish culture and return to observing real objects. Men needed the courage to dissent (*dissentire audeamus*) from the pre-eminent interpreter of nature. Aristotle was no god and did not merit, just as no man merits, to be revered more than the truth. He needed to be debunked and the «vow» that bound so many seekers after truth to his authority, convinced that all one had to do was to verify Aristotle's opinion on any particular thing (*non igitur amplius rei ullius natura inspicienda indagandaque videtur cuipiam, at tantum quid de quaque Aristoteles senserit speculandum*) had to be broken once and for all.[50]

A change of mentality was needed and this called for a superhuman effort. In the words of Galileo: «The idea of whoever believes in introducing a new philosophy by passing judgment on this or that author is utterly futile: we must first learn to reconstruct men's brains, and make them able to distinguish true from false, something that only God can do». This would certainly take a very long time. It was sufficient to enter, as Sagredo did, «into the heart of Sig. Simplicio» to realize that there was, on the one hand, the «strength» of the «even overly convincing arguments» of the new philosophy, but on the other hand there were the «great universal authority acquired by Aristotle», the «number of famous interpreters who had toiled to expound his meaning», and the «other sciences» which build «a great part of their esteem and reputation on the credit of Aristotle». There was good reason to be «puzzled» and «scared». The ensuing disorientation led to a spasmodic search for who would take his place, «once Aristotle had been dethroned».[51] In response to the question «Who?» Galileo replies, «No-one», because nature must be studied directly. This was also Telesio's response.

---

48 Delle Colombe, *Discorso apologetico* (cit. note 38), pp. 317-318.

49 Telesio, *De natura iuxta propria principia – Ad Felicem Moimonam iris*, Preface by N. Ordine, Introduction and analytical index by R. Bondì, Rome, Carocci, 2011, p. 201.

50 *Ibid.* See also Telesio, *De rerum natura iuxta propria principia* (cit. n. 21), p. 87; Telesio, *De rerum natura iuxta propria principia, libri IX*, Preface by N. Ordine, Introduction and analytical index by G. Giglioni, Rome, Carocci, 2013, p. 87.

51 Galileo, *Dialogo sopra i due massimi sistemi del mondo*, OG, VII, pp. 81-82.

3. The Sky and the Bible

De natura iuxta propria principia was published in Rome in 1565 and opened with the polemical and challenging words: «Bernardinus Consentinus haec cogitabat».[52] The no longer young Calabrian philosopher presented his thinking immediately, in an abrupt and peremptory way: «*Summus Caelo inesse apparet calor*».[53] In other words, the world is not as Aristotle described it. Whoever studies nature directly, seeking in it the principles of things, whoever affirms the primacy of the senses, whoever refuses to give credit to reason without sensory data, and whoever does not impose merely logical principles (*quae omnino logica sunt*)[54] on reality will discover a new world which is unitary and homogeneous.

For Telesio, and later for Galileo, the book of nature was not to be interpreted, but to be read. Telesio thought, and Galileo would later think, that the Aristotelians simply did not do this. Indeed, they took care not to. In his defence of Telesio, the young Campanella, sharing in his maestro's impatience, stresses that he has never seen one of Aristotle's followers «examine things, go to the countryside, the sea, the mountains»; Aristotelians only cared about Aristotle's books and spent their days studying them.[55] Telesio believes that it is sufficient to observe the world and its parts, since they spontaneously manifest their properties.[56]

3.1. *Between the Heavens and the Earth*

The world is plainly the theatre of a perpetual war between heat and cold. All things originate from the unrelenting opposition of these two principles, which reside respectively in the Sky and on the Earth, that are not organs of some other substance, and that seek to conquer matter.

Aristotle denied the igneous nature of the Heavens and the Sun, together with the substantiality of the principles. This is Telesio's starting point. It is one of those preliminary questions (*quibus ignotis nihil fortassis in natura notum esse queat*), which have to be broached when one debates what Galileo's critic, Giorgio Coresio, calls the «best interpreter

---

[52] Telesio, *De natura iuxta propria principia – Ad Felicem Moimonam iris* (cit. n. 49), p. 45.

[53] *Ibid.*, p. 50.

[54] Telesio, *Delle cose naturali libri due – Opuscoli – Polemiche telesiane*, Preface by N. Ordine, Introduction and analytical index by A.L. Puliafito, Rome, Carocci, 2013, p. 491.

[55] Campanella, *Philosophia sensibus demonstrata* (cit. n. 25), p. 5.

[56] Telesio, *De rerum natura iuxta propria principia, libri IX* (cit. n. 50), p. 14.

of nature».[57] Telesio would have commented ironically on this definition, judging it to be far too hesitant. If Aristotelian arguments were «solid, true and necessary», then Aristotle should indeed be revered like a god, since thanks to his superhuman talents he would have understood that the Sun is not fiery, as common mortals fondly imagined, clinging to their senses, and he would have discovered the surprising way in which something that is not hot manages to produce heat continually. Yet it turns out that these arguments are anything but «solid, true and necessary». Hence, Aristotle must be taken for what he is: a writer who does not speak about the nature of things, but only about his vague conjectures about such things.[58]

Since the first edition of *De rerum natura*, the question had been put very clearly. Aristotle holds that the sky differs from the elements because their motion is different: circular in the case of the sky, and straight for the elements. His conclusion is based on the conviction that one can deduce the substance of a thing from the way it operates: indeed, the operation of a substance is unique. Once generated, fire operates immediately, in other words, it moves. If for some reason fire is prevented from moving, it extinguishes itself as if it had been stripped of its nature. It is correct to deduce the nature of a thing from its operation; were we to find a difference in operation between the sky and the sublunary world we would have to conclude the existence of a diversity of substance. Yet that diversity of operation is not given and Aristotle errs when he claims the contrary. Straight and circular motions are not different or contradictory operations, but one and the same operation (*eadem omnes existentes operatio*). Movements towards opposite places differ for the same reason that the road from Megara to Athens differs from the one that goes from Athens to Megara. Indeed, perhaps the places are not opposites either, if not for the bodies they contain. The real opposition is thus not between types of movement, but between movement on the one hand, and an absence of motion on the other. The real opposition is between the highly mobile sky and the stationary earth. Circular motion belongs, by its very nature, to fire. Consequently, there is no real difference between fire and the sky.[59]

Conversely Aristotle thought that the stars and the sky were not made of fire but of ether. Given that the Sun is not fiery it is clear that heat is

---

57 Giorgio Coresio, *Operetta intorno al galleggiare de' corpi solidi*, OG, IV, p. 238.

58 Telesio, *De rerum natura iuxta propria principia, libri IX* (cit. n. 50), p. 59.

59 Id., *De natura iuxta propria principia – Ad Felicem Moimonam iris* (cit. n. 49), pp. 143-144. See also Telesio, *De rerum natura iuxta propria principia, liber primus, et secundus, denuo editi – Opuscula* (cit. n. 21), pp. 58-60.

not derived from it, but from the friction between the fifth element and the higher reaches of the atmosphere. Yet how can the Sun generate friction on sublunary air from such a distance? Surely it is impossible that the Sun, fixed in a sphere and transported by it, can in some way act on the spheres beneath it. In fact the Sun only touches them at one point. Yet if the spheres that divide the Sun from the air are not rubbed, how can we explain the friction? Should we perhaps suppose that the rubbing can be transmitted by the Sun to the sublunary air without involving the intermediate spheres? This is certainly not feasible since the rubbing is a *passio penitus corporea*. Then, if we consider the position and the consistency of the Moon in relation to the position and consistency of the Sun, surely it would be reasonable to expect it to produce more heat[60] In short, for Telesio it was all one great mass of absurdity. The entire question, as posed by Aristotle, was inconceivable (*quae nulla concipi queat imaginatione*).[61] For the ardent young Campanella it was sheer «madness» (*insaniam!*) that had understandably sown panic among Aristotle's interpreters. Only a «stupid man» could believe such a thesis.[62] Yet the great commentators did not distance themselves from their master, and in adding absurdity to absurdity reveal themselves as, in Telesio's words, «lovers and students not of truth, but of Aristotle's principles». The act of distancing themselves from Aristotle's assertions was tantamount to committing a horrible «crime» (*scelus*).[63]

By way of example, in his comment on *Meteorologica* Alexander of Aphrodisias uses the instance of the torpedo fish that numbs the hands of the fisherman when caught in his net but without numbing the net to explain how the Sun can rarefy the air without touching it. Yet Alexander himself is dissatisfied with this explanation, and eventually admits defeat when faced with the «difficulty of things», conceding that «perhaps» the divine body is not «impassible» after all: Aristotle has shown not that it is «immutable»,

---

60 Id., *De natura iuxta propria principia – Ad Felicem Moimonam iris* (cit. n. 49), pp. 333-334, 338. See also Id., *De rerum natura iuxta propria principia, liber primus, et secundus, denuo editi – Opuscula* (cit. n. 21), pp. 156-157.

61 Id., *De rerum natura iuxta propria principia, liber primus, et secundus, denuo editi – Opuscula* (cit. n. 21), p. 156. See also Id., *De rerum natura iuxta propria principia, libri IX* (cit. n. 50), p. 75.

62 Campanella, *Philosophia sensibus demonstrata* (cit. n. 25), pp. 10-11, 42-43, 292 (see, in general, the *disputatio tertia*, entitled *De coelo et mundo*, pp. 283-432, and *disputatio septima*, entitled *De calore et luce coeli*, pp. 651-691). See also Id., *Epilogo magno*, ed. by C. Ottaviani, Rome, Reale Accademia d'Italia, 1939, pp. 194-205; Id., *Del senso delle cose e della magia*, ed. by G. Ernst, Rome-Bari, Laterza, 2007, p. 5.

63 Telesio, *De rerum natura iuxta propria principia, liber primus, et secundus, denuo editi – Opuscula* (cit. n. 21), pp. 157, 159. See also Id., *De rerum natura iuxta propria principia, libri IX* (cit. n. 50), p. 153.

but merely that it is «incorruptible». Therefore it is not «absurd» that the sky nearest the Sun «suffers» due to the movement of the latter, albeit not to the point of changing its substance, and thus inflames so that the alteration is transmitted to the body beneath it. Alexander thus recognized that heat comes from the sky, but in order to remain faithful to Aristotle he denies that the sky has its own heat.[64]

At the start of the third edition of *De rerum natura*, Telesio presents his subversion of Aristotelian cosmology in unambiguous terms: it is not just the Sun which is hot, but the entire sky; that is, the vast «body» placed «between the earth and the sea and the stars and between the stars themselves», that has «no internal distinction» and that manifests itself as «unique and completely identical».[65]

In the same period a large collection of treatises by Jacopo Zabarella entitled *De rebus naturalibus libri XXX* was published in Venice. The *Liber de calore coelesti* deals with the question without referring to the moderns. It dwells on explanations provided by Alexander and Averroes, since the terms posed by Aristotle needed to be clarified and revised. Yet on the point in question he says little. Zabarella argues that antiquity had no shortage of philosophers who, in conceding too much to sensible knowledge (*nimium sensui tribuentes*), claimed that the stars transmit heat to elements because they are fiery. There is no need to bother about confuting this thesis since Aristotle has already done so when he demonstrates that the sky does not have an elementary nature, but is a fifth essence.[66]

3.2. *The Bible is not Aristotelian*

This apparent demonstration clashes with the text of the Sacred Scriptures. This is what Telesio thinks when, from the first edition of *De rerum natura* onwards, he tries to do what Galileo would attempt to do later: that is, to invoke the Scriptures in support of his own theories and to present the Aristotelian argument as being contrary to the Bible. In Book II he describes the sky, with everything that it contains, as hot by nature, and as all

---

[64] Id., *De natura iuxta propria principia – Ad Felicem Moimonam iris* (cit. n. 49), pp. 333, 41. See also Id., *De rerum natura iuxta propria principia, liber primus, et secundus, denuo editi – Opuscula* (cit. n. 21), p. 77; Id., *De rerum natura iuxta propria principia, libri IX* (cit. n. 50), pp. 153-154. See Alexander of Aphrodisias, *In quatuor libros Meteorologicorum Aristotelis commentatio lucidissima, quam Latinitate donavit Alexander Piccolomineus*, Venetijs, Apud Hieronymum Scotum, 1540, p. 5r.

[65] Telesio, *De rerum natura iuxta propria principia, libri IX* (cit. n. 50), p. 17.

[66] Jacopo Zabarella, *De rebus naturalibus libri XXX*, Venetiis, apud Paulum Meietum bibliopolam patavinum, 1590, p. 389.

consisting of the same substance, only differing internally in density and rarity. Telesio emphasizes that all this is confirmed by the Sacred Scriptures, which describe as «fiery» the highest heavens of «Empyrean» and «glorious God's ministers». Moreover, the Scriptures also say that God spoke to Moses from a pillar of fire (Deut 4: 36; Psalms 103: 4; Hebrews 1: 7).[67]

Despite the pressure and warnings received just before the publication of the new edition of *De rerum natura*, and borne out in the letter to Cardinal Orsini, Telesio nevertheless reaffirms his theses and the biblical references in the second edition: not only sense and reason, but also the Holy Writ attests that the Sun is hot and fiery.[68] Yet even at the beginning of the text, Telesio draws on the Bible to authenticate his thesis according to which the Heavens and the Earth are the first and main components of the world from which all the rest originates.[69] Beyond the assurances that he could have received from the authoritative figures listed in the letter of 1570, the argument in question would not allow compromise. Renouncing the fiery nature of the sky would mean renouncing his new cosmology.

The negation of unmoved movers was an integral and essential part of this new cosmology. Once again Telesio was summoned to respond to Cardinal Orsini on this matter, and once again Telesio was unwilling to compromise. The negation of movers necessarily derived from the claim that the sky had a fiery nature. In the first edition of his opus magnum Telesio argues that the sky moves by virtue of its nature, with no need or opportunity to search for unmoved movers. It is a difficult thesis to grasp or even to imagine.[70] In the second edition he expresses himself in the same terms,[71] and in the third he deals at length with Aristotle's «impiety» insofar as the Greek philosopher relegates God to being the mover of the heavens, giving him an inglorious task that should instead be attributed directly to the

---

67 Telesio, *De natura iuxta propria principia – Ad Felicem Moimonam iris* (cit. n. 49), p. 374. On the theme of fire, see Thomas Litt, *Les corps célestes dans l'univers de Saint Thomas d'Aquin*, Louvain-Paris, Publications Universitaires – Béatrice Nauwelaerts, 1963, pp. 255-261; Bruno Nardi, *Saggi di filosofia dantesca*, Florence, La Nuova Italia, 1967, pp. 167-214.

68 Telesio, *De rerum natura iuxta propria principia, liber primus, et secundus, denuo editi – Opuscula* (cit. n. 21), p. 173.

69 *Ibid.*, p. 11: «Sacrae etiam divinae Litterae Coelum solum et Terram solam prima esse corpora testantur, a Deo Opt. Max. in principio constituta esse effectaque enuntiantes».

70 Telesio, *De natura iuxta propria principia – Ad Felicem Moimonam iris* (cit. n. 49), pp. 382-387. For an overview on the theme of the movers of the spheres, see Michel-Pierre Lerner, *Il mondo delle sfere. Genesi e trionfo di una rappresentazione del cosmo*, It. trans., Florence, La Nuova Italia, 2000, pp. 245-287.

71 Telesio, *De rerum natura iuxta propria principia, liber primus, et secundus, denuo editi – Opuscula* (cit. n. 21), p. 176.

nature of the sky itself. It would have been better to conceive God as doing nothing at all rather than as being a simple mover of the sky. Certainly he should not have presented a range of substances «coeval with God», with His «nature» and «almost» with His «power». It is here that Aristotle is «odious» and can be accused of «impiety».[72] Sertorio Quattromani summed this up in a book that he published soon afterwards:

> Since we have demonstrated that the sky is informed with heat and that the reasons used by the Peripatetics to counter it are of no importance, or value, who will forbid to conclude (as heat is mobile by nature) that the sky moves due to its own nature without needing Aristotle's movers? which will either be necessary for fire and air and other moving things, no less than for the sky, or if not useful for them, then they are of no use for the sky.[73]

3.3. *Neither philosopher nor theologian*

The positions assumed by Telesio were extremely shocking for the time. His attempt to mitigate this by using the solemn preventive abjuration at the start of the third edition of *De rerum natura* proved futile.[74] It was decided that *De rerum natura* and the two pamphlets published in 1590 by his pupil Persio should be placed on the Index, albeit *donec expurgentur* (1593, 1596). This was encouraged by Roberto Bellarmino,[75] who had been appointed "consultor" to the Congregation of the Index in 1587 and who had already left his mark on the censorship of printed works.

In 1951 Luigi Firpo published documents concerning two attempts to obtain free circulation for *De rerum natura* after its expurgation, but which failed «miserably».[76] The documents relate in particular to a highly critical

---

72 Id., *De rerum natura iuxta propria principia, libri IX* (cit. n. 50), pp. 182-183: «Nec tamen, si quid eorum quae nobis posita sunt sacris literis Catholicaeve Ecclesiae decretis non cohaereat, tenendum id, quin penitus reijciendum asseveramus contendimusque. Neque enim humana modo ratio quaevis, sed ipse etiam sensus illis posthabendus, et, si illis non congruat, abnegandus omnino et ipse etiam est sensus».

73 Sertorio Quattromani, *La philosophia di Berardino Telesio ristretta in brevità et scritta in lingua toscana*, in Napoli, appresso Gioseppe Cacchi, 1589, pp. 126-127.

74 Telesio, *De rerum natura iuxta propria principia, libri IX* (cit. n. 50), p. 14.

75 Ugo Baldini – Leen Spruit (eds.), *Catholic Church and Modern Science. Documents from the Archives of the Roman Congregations of the Holy Office and the Index*, 4 vols., III, Rome, Libreria Editrice Vaticana, 2009, p. 2417: «In litera B addendi similiter qui adduntur in indice Sixti V et praeterea addi possent Bernardini Tilesij libri de natura, item Liber de somno, item liber, quod animal universum ab una anima gubernetur». See also Peter Godman, *The Saint as Censor. Roberto Bellarmine between Inquisition and Index*, Leiden-Boston-Cologne, Brill, 2000, p. 274.

76 Luigi Firpo, *Filosofia italiana e Controriforma. IV. La proibizione di Telesio*, «Rivista di filosofia», 42, 1951, pp. 30-47. See also Baldini – Spruit (eds.), *Catholic Church and Modern Science* (cit.

text of 25 November 1600 drawn up by Girolamo Pallantieri, who taught theology in Padua; it was signed by diocesan consultors of the Congregation of the Index and by professors of theology and philosophy at the University of Padua, and approved by the Inquisitor, Felice Pranzini, and his Episcopal deputy Camillo Peltrari. Pallantieri's text focuses on the first twelve chapters of Book I of *De rerum natura* and criticizes the cosmology of the «arrogant» Telesio who had dared to attack the great Aristotle, starting with the «flattering» and «daring» dedication to Ferdinando Carafa, in which Aristotelian precepts are firmly presented as contradictory, at odds with the senses, and incompatible with the Scriptures (*et sensui et sibi ipsi et sacris divinis letteris passim dissentientem*).[77]

The censors in Padua, including the prominent Cesare Cremonini, concluded that when Telesio expresses his judgments on Aristotle – the «Magnus Magister» – he «lies», «errs» and above all «raves» (*hallucinatur*), particularly when he stresses the incompatibility between Aristotelian precepts and biblical claims. Passages from Saint Paul's *Letter to the Romans* (Romans 1: 19-20) testify to this (*invisibilia Dei per ea, quae facta sunt etc. quod notum est Dei manifestum est in illis etc.*). Indeed, Aristotle and the Greeks were raised in the knowledge of God. Saint Paul, argue the censors, did not judge the doctrine of Aristotle and the other Greeks as being contrary to the sacred texts, even if it is true that they did not follow up their knowledge of God with proper worship.[78]

In his study Firpo speaks of a «sour and superficial» censure, one «devoid of speculative observations» which is only interesting «as evidence of the tenacious persistence of a deaf hostility in the circles of Padua where Telesio himself was trained and that had reacted so violently to Persio's attempts at diffusion and discussion».[79] Firpo's judgment must be reconsidered, however, in the light of what subsequently has emerged from the letter to Cardinal Orsini. Today we know that this «hostility» was in fact widespread. As recalled, the letter explicitly testifies that questions on the nature and

---

n. 75), III, pp. 2417-2425. See Bondì, *Introduzione a Telesio* (cit. n. 21), pp. 131-136; Saverio Ricci, *Inquisitori, censori, filosofi sullo scenario della Controriforma*, Rome, Salerno Editrice, 2008, pp. 377-389.

77 Telesio, *De rerum natura iuxta propria principia, libri IX* (cit. n. 50), pp. 3-4.

78 Baldini – Spruit (eds.), *Catholic Church and Modern Science* (cit. n. 75), III, pp. 2418-2419. The censors also referred to Clemente Alessandrino's Stromata, VI, 5: «ideo Clemens Alex<andri>nus Stumatum libro 6° Philosophiam ait a Deo Graecis veluti proprium testamentum datam quemad<modu>m autem nec testam<entu>m Iudaeorum aut testam<entu>m Christianorum ita testam<entu>m Graecorum sine interprete minime dereliquit, et is fuit Ar<istotele>s».

79 Firpo, *Filosofia italiana e Controriforma, IV. La proibizione di Telesio* (cit. n. 76), p. 36.

movement of the sky were at the center of harsh critical observations from the outset. The documents published by Firpo target positions frequently found, even in the same form, in the two earlier editions of *De rerum natura*.

The censors insist on one point: Telesio presents ideas that have perilous consequences and tries to authenticate them with clumsy references to the Sacred Scriptures. It is improper, they argue, to involve the sacred writers to justify broadening of the use of the term «sky» to denote the entire body between the Earth and the Sun.[80] It is equally unacceptable to drag in the Sacred Scriptures to justify the fiery nature of the sky. Telesio, who proves to be neither philosopher nor theologian, does not understand that not all of the sky is empyrean, that the latter is referred to as such on account of its splendor and not its heat, that the angels do not have a fiery nature, and that the column from which God spoke to Moses has nothing to do with the sky.[81] When Telesio asks for permission to refer to the lowest reaches of the heavens as «sky», he is mixing the eternal and the temporal.[82] When he speaks of the way in which the sky is shaped from a sole and homogeneous matter and from a sole and homogeneous heat, adding that both the Sacred Scriptures and reason teach us that the sky is the work of heat directed by God, Telesio strikes himself a mortal blow with his own sword (*suo se gladio Telesius iugulat*) by attributing divine activity to heat and matter and mocking the Sacred Scriptures.[83] When he explains that God made the Sky and the Earth as opposing forces that corrupt one another, but without ever destroying themselves, he cites the example of the craftsman who, however perverse and wicked, desires that his works last as long as possible. This, he argues, applies even more when it comes to God. Moreover, these claims do not contradict the Sacred Scriptures, which affirm that the Sky and Earth are perishable. Telesio denies saying that God is impotent, and concludes by declaring that He, in creating the Sky and the Earth, took joy from doing this in such a way that, even though they opposed each other, they would preserve themselves. It is here that Telesio reveals his ignorance.[84]

---

80 See Baldini – Spruit (eds.), *Catholic Church and Modern Science* (cit. n. 75), III, p. 2420. See Telesio, *De rerum natura iuxta propria principia, libri IX* (cit. n. 50), p. 17: «...quoniam nullo id a se ipso distinctum discrimine, sed unum idemque apparet universum et, ut sequens manifestabit sermo, unum idemque est universum, uno universum et, antiquiorum sacrorumque scriptorum more, coeli nomine appellandum videtur...».

81 *Ibid.*, pp. 2420-2421. See Id., *De rerum natura iuxta propria principia, libri IX* (cit. n. 50), p. 18.

82 *Ibid.*, p. 2421. See Id., *De rerum natura iuxta propria principia, libri IX* (cit. n. 50), p. 24.

83 *Ibid.* See Id., *De rerum natura iuxta propria principia, libri IX* (cit. n. 50), pp. 26-27.

84 *Ibid.*, p. 2422. See Id., *De rerum natura iuxta propria principia, libri IX* (cit. n. 50), pp. 27-28.

The beginning of *De rerum natura* was enough to convince the censors that Telesio understood neither philosophy nor theology, and that his text contained fundamental errors that would not be easy to remove. The problems with expurgating Telesio's work became clear the following year. The philosopher's nephew, Orazio Telesio, «begged» Cardinal Agostino Valier, in the name of the city of Cosenza, to exert his influence with the Congregation of the Index to instruct «some theologians and philosophers» to review and expurgate Telesio's works «of all those errors, however minimal, that could in some way offend the pious ears of who had read them unexpurgated», especially since it had «always been the will and desire of the said author to obey the Holy Mother Church and to place his every dictum at the very holy feet of Our Lord and of this Holy Apostolic Seat, as we see from many solemn declarations made throughout all his books, always printed with revisions and ordinary licenses of superiors». On 17 November 1601 Orazio Telesio appeared (*nomine Universitatis Cosentinae et familiae Thelesiorum*) before the Congregation. From the minutes of the session it appears that, although concession was granted to proceed with a revision of his work, their expurgation was considered «impossible» (*huiusmodi librorum expurgatio censeatur impossibilis*), as was the removal of the author's name from the Index (*ex Indice auctor nequaquam deleri poterit*).[85] In the words of Firpo, «speaking of "*expurgatio impossibilis*", at the time of the sentence the men of the Roman Curia had recognized the error committed at the time of the sentence and that the conditional and partial sentence was now absolute and irrevocable».[86]

4. Campanella between Telesio and Galileo

In the years between the sentence handed down against Telesio and the condemnation of heliocentrism, Campanella put a great deal of effort into demonstrating the orthodoxy of Telesio and Galileo and the convergence between Telesianism and the Galileism. In fact, in *Philosophia sensibus demonstrata*, Campanella addressed the problem of the compatibility of Telesian doctrines with the Sacred Scriptures, spurred on by Marta's observations. In *Pugnaculum*, the Neapolitan lawyer had for example questioned Telesio's decision to class water as one of the many bodies produced by the action of the Sun on the Earth, when it was instead a *magnum corpus*

---

85 See Firpo, *Filosofia italiana e Controriforma. IV. La proibizione di Telesio* (cit. n. 76), p. 46.

86 *Ibid.*, p. 39.

elementare, as was generally asserted by theologians and also by the Sacred Tridentine Synod, which speaks of it – and this is a key point – as the matter of baptism. Marta insisted on the question, because the "perfection of the world" was at stake and above all the biblical text with which Telesio seemed decidedly in disagreement. Chapter 1 of the Book of Genesis, argued Marta, clearly affirms that "the spirit of God was hovering over the waters". Hence during the creation of the world the element of water is also present, and not just the earth that would produce it only subsequently. The Scriptures contain «infinite» instances that mention water as an element. Why, asks Marta, having the truth and being obliged to believe as a question of faith, would we prefer to «rave like the Calabrian»? [87]

Campanella replied that Telesio's doctrine did not disagree in the least with Moses, who never meant to say that water or air were primary bodies. He begins a lengthy interpretation of the Book of Genesis by mixing references to Telesio and Ficino and opening up syncretic perspectives distant from the teachings of his master. Among other things, he re-examines the biblical references that Telesio had made to support his thesis of the fiery nature of the heavens. He argues that when theologians teach that we are baptized with water and refer to it as an element, they are simply complying with a common way of expressing things, but they are not claiming that this point is a matter of faith. Finally, he emphatically affirms that examining Telesio's doctrine from a theological perspective would allow him to validate it on the basis of the Bible. One could thus present Plato and Telesio alongside the Sacred Scriptures and confute Aristotle's «fictions», starting with the sacred text itself.[88]

The ban on Telesio's work did not upset Campanella's attempts, and in his *De gentilismo non retinendo* (1609), he defends the idea that the ecclesiastical sentence was not based on the anti-Aristotelianism of *De rerum natura*. Relying on the attenuating clause *donec expurgentur* used for Telesio's works, he argued that the problem was simply tied to the existence of *quaedam dubia* requiring clarification, and was convinced that after expurgation the great philosopher's doctrine would be welcomed by all.[89]

---

[87] Giacomo Antonio Marta, *Pugnaculum Aristotelis adversus principia Berardini [sic] Telesii*, Romae, typis Bartholomaei Bonfadini, 1587, pp. 91-92. Here too Telesio wanted to protect himself against possible accusations of heterodoxy: see Telesio, *De rerum natura iuxta propria principia, libri IX* (cit. n. 50), p. 30.

[88] Campanella, *Philosophia sensibus demonstrata* (cit. n. 25), pp. 522-528. In *Monarchia di Spagna*, Campanella affirms that «la filosofia telesiana è ottima, sendo conforme a santi Padri» (Id., *Monarchie d'Espagne – Monarchie de France*, ed. by G. Ernst, Paris, Presses Universitaires de France, p. 96).

[89] Id., *De gentilismo non retinendo*, Parisiis, apud Tussanum Du Bray, 1636, p. 48.

Campanella's defence of Telesio's orthodoxy went hand-in-hand with his personal anti-Aristotelian critique, and his defence of Galileo's orthodoxy. The latter culminated in the *Apologia* (1616), which was well received by Mersenne. Among many other things, Campanella recalls the Augustinian theme of the *derisio* of the Sacred Scriptures. Galilean discoveries, he argues, are in keeping with the Bible, and that first and foremost «they redeem it from the torture of theologians and the mockery of philosophers» (*a tortura theologorum et irrisionibus philosophorum vindicare*).[90]

Here it is interesting to recall Campanella's attempt to juxtapose Galileo to Telesianism, despite the now definitive condemnation of the latter. Obviously, the Pisan scientist had everything to lose and nothing to gain from such a comparison. The fact remains that in some respects this was not an arbitrary operation. Campanella does not conceal the fact that Galilean «sensible observations» posed serious problems for Telesio's cosmology. As he writes at the end of the *Apologia*:

> In as much as for many years I too held that the sky was made of fire which gave origin to all types of fire, and that also the stars were made of fire, according to the opinion of Augustine, Basil and other Fathers, and recently of my countryman Telesio, in *Questioni* and *Metafisica* I have striven to confute the reasons put forward by Copernicus and the Pythagoreans. But since the observations of Tycho Brahe and Galileo leave no room for doubt as to the formation of a new star in the starry sky, and of comets not only above the Moon, but they too in the starry sky, and of clouds around the Sun, I am inclined to suppose that not all stars are made of fire.[91]

In Campanella's eyes neither this nor other more problematic aspects, such as earthly motion, marred the proximity of Galileanism to Telesianism. Moreover, as was recalled at the beginning, the belief in geocentrism which the Telesian Antonio Persio never repudiated did not in any way prevent him from being, with the exception of the opinion of Copernicus, in full agreement with Galileo.[92]

Campanella reasoned in the same way as Persio. What he referred to as a «way of philosophizing», on which he wrote what was to become one of his most famous sonnets, was the basis for this reasoning. We are in the

---

90 Id., *Apologia pro Galileo*, ed. by M.-P. Lerner, Pisa, Edizioni della Normale, 2006, p. 181, and pp. 97-101, 171. On the relation between the *Apologia* and the *Lettera a Cristina*, see Alfredo Damanti, *Libertas philosophandi. Teologia e filosofia nella "Lettera alla Granduchessa Cristina di Lorena" di Galileo Galilei*, Rome, Edizioni di Storia e Letteratura, 2010, pp. 225-244.

91 *Ibid.*, p. 185.

92 See n. 15.

early seventeenth century and Campanella sets the «book and temple of God», i.e. the world, against the «books and dead temples of men» which «impiously» take precedence «over the divine».[93] The return to the «original», which must be read via the senses and not interpreted, was a typical Telesian theme, and one that Campanella observed in Galileo's work. It was a guarantee: Galileo does not incur «falsehoods» since he «proceeds on the basis of sensible observations (*observationibus sensatis*) made in the book of nature, not opinions».[94]

In some cases both Telesio and Galileo reach the same conclusions using the same method. In *Apologia*, Campanella presents a key example: «As regards the "swift course of the Sun", spoken of in Hesdra, it is sufficient to understand it as referring to the motion of the Sun's rotation on itself, already demonstrated by Telesio through sensible experiences (*sensatis experimentis*), and by Galileo with the movement of solar clouds».[95] Yet in a lengthy letter dated 13 January 1611, Campanella had already clarified this sense of the proximity of Telesianism to Galileo. He relates having heard «with great pleasure» that *Sidereus Nuncius* was uncovering «God's secrets» and how he regretted not having read the «booklet» before completing his *Metaphysica*, where he proposes the existence of a multitude of «star systems» that remain hidden in the heavens. This text, continues Campanella, not only reveals the verisimilitude of the «Copernican hypothesis», but also points out the many errors made by Copernicus who had mistakenly accepted «manifestly discordant theses», based partly on the Pythagoreans and partly on Ptolemy. In particular, he should have attributed «the greatest motion» to the Sun, as demonstrated by heat, «which, provided [it is] concentrated, rarefies and rapidly moves everything that it touches». Given that the Sun does not revolve, it is inevitable that it has a motion of rotation

---

93 Campanella, *Le poesie* (cit. n. 11), p. 26.

94 Id., *Apologia pro Galileo* (cit. n. 90), p. 101.

95 *Ibid.*, p. 149. See Telesio, *De natura iuxta propria principia – Ad Felicem Moimonam iris* (cit. n. 49), p. 381; Id., *De rerum natura iuxta propria principia, liber primus, et secundus, denuo editi – Opuscula* (cit. n. 21), p. 175; Id., *De rerum natura iuxta propria principia, libri IX* (cit. n. 50), p. 75. See Lerner, *"Sicut nodus in tabula": de la rotation propre du soleil au seizième siècle*, «Journal for the History of Astronomy», 11, 1980, pp. 114-129; Id., *La physique céleste de Telesio: problèmes d'interprétation*, in *Atti del Convegno Internazionale di Studi su Bernardino Telesio* (Cosenza, 12-13 maggio 1989), Cosenza, Accademia Cosentina, 1990, pp. 83-114. On the themes of *experientia* and *experimentum sensatum*, see Charles B. Schmitt, *Filosofia e scienza nel Rinascimento*, ed. by A. Clericuzio, postfaction by C. Lohr, Florence, La Nuova Italia, 2001; Gabriele Baroncini, *Forme di esperienza e rivoluzione scientifica*, Florence, Olschki, 1992; Paolo Ponzio, *Tommaso Campanella. Filosofia della natura e teoria della scienza*, Bari, Levante editori, 2001, pp. 145-158; Marco Veneziani (ed.), *Experientia. X Colloquio Internazionale* (Roma, 4-6 gennaio 2001), Florence, Olschki, 2002.

on its own axis, as was clearly visible from the mirrors placed underwater or from its own disc when observed at dawn or sunset. Thus, continued Campanella in Telesio's style, the Sun «enjoys its own heat and with which it fertilizes the earth so that the latter then generates secondary bodies», concluding triumphantly: «In this way the principles of my learned countryman Telesio remain valid, if the Earth's motion is the product of that soul to which Origen refers».[96] Without forgetting the important passage from *Metaphysica* (1638) in which Campanella likens Telesianism to Pythagorism: according to Telesio the Earth has sense (*sensus*) and can thus be considered *ad motum apta*.[97]

It should come as no surprise that it was Antonio Persio, the Telesian, acclaimed *post mortem* as a member of the Accademia dei Lincei with Galileo's approval, who let Campanella have a copy of Galileo's *Sidereus Nuncius*.

96 CAMPANELLA, *Lettere* (cit. n. 30), pp. 593-594.

97 ID., *Universalis philosophiae, seu metaphysicarum rerum, iuxta propria dogmata, partes tres, libri 18*, Parisiis, Denis Langlois, 1638, pars III, book XI, chap. VII, art. III, p. 38.

Natacha Fabbri

THREATS TO THE CHRISTIAN COSMOS. THE RECKLESS ASSAULT ON THE HEAVENS AND THE DEBATE OVER HELL

The divergence between the Copernican world and the biblical cosmos gave rise to widely discussed theological issues, among which the debates over the location of hell, the lowly status of the earth, together with their relation to the sublime sphere of the Empyrean have thus far not received the attention they deserve.[1]

The theological issue of the site of the infernal fire, alongside that of Christ's motion upward from the earth to the heavens and downward from the earthly surface to hell, provides an insight into how sixteenth- and seventeenth-century philosophers and theologians intertwined topics concerning the salvation of mankind and cosmological matters. The Copernican system indeed seemed to shatter several cornerstones of the Catholic moral system and theology as it brought into question the largely shared vision of hell and the ontological hierarchy among God's creatures (epitomized by the opposition between the angels and the blessed on the one hand, and the damned and mortals on the other). Fostering the astronomical model of heliocentrism would have implied reframing the mechanism of social and moral order that religion had implemented and in which it was rooted.

Moreover, such a 'temerarious' upward movement of the earth was seen as the outcome of a comparably temerarious behavior on the part of astronomers: they seemingly aspired to conquer the kingdom of God and

All translations are my own unless otherwise indicated.

[1] Recently Mauro Pesce drew more attention to this topic: see Mauro Pesce, *Il copernicanesimo e la teologia: perché il caso Galileo non è chiuso*, in Massimo Bucciantini – Michele Camerota – Franco Giudice (eds.), *Il caso Galileo. Una rilettura storica, filosofica, teologica*, Firenze, Olschki, 2011, pp. 33-46: 34-37.

to grasp the true structure of the universe, without taking into consideration – thus showing an excess of *hubris* – the deep-rooted interpretations of the biblical cosmos.

This chapter shall examine these controversial implications of the 'ascent' of the earth in the sky, as well as the stance of those among Galileo's contemporaries who believed that «the way heaven goes» (*come vadia il cielo*) and «the way of going to heaven» (*come si vadia al cielo*)[2] could not be regarded as two different and independent aspects. It seemed clear to them that the Copernican model entailed rethinking some theological issues that had already been a matter of lively discussions due to their heterodox implications.

Furthermore, I shall deal with the rise of Galileo's awareness of the tight interlacing between biblical cosmology and the geocentric system – and, consequently, of the difficulty of separating the cosmos of astronomers from the cosmos of theologians – which dated back well before the public accusations that prompted him to write the well-known letters to Castelli, Dini and Cristina of Lorena. I shall also cast more light on the rhetorical and theoretical tools Galileo employed to face the charges of temerity that were raised against both the movement of the earth and his apologia for Copernicanism.

1. The Reckless Assault on Heaven

One year had just passed since the publication of *Sidereus Nuncius* when John Donne drew a quirky portrait of Copernicus and his advocates in the *Conclave Ignati*: Copernicus was imagined knocking at hell's doors in an attempt to be admitted to Lucifer's kingdom. He introduced himself as the one who has dragged hell into heaven, giving it the honor of bearing the same rank as the stars: «I am he, which pitying thee who wert thrust into the Center of the world, raysed both thee, and thy prison, the earth, up into the heavens; so as by my meanes *God* doth not enjoy his revenge upon thee».[3]

---

[2] Galileo to Cristina of Lorena, OG, V, p. 319.

[3] John Donne, *Ignatius: his conclave or his inthronisation in a late election in hell...*, London, printed by N.O. for Richard More, 1611, pp. 13-14. John Donne strongly scorned Copernicanism: see the seminal works by Marjorie Hope Nicolson, *The "New Astronomy" and English Literary Imagination*, «Studies in Philology», 32, 1935, pp. 449-460; Charles M. Coffin, *John Donne and the New Philosophy*, New York, Columbia University Press, 1937. See also William Empson, *Essays on Renaissance Literature: Donne and the new philosophy*, ed. by J. Haffenden, Cambridge,

At the end of Copernicus' introduction, Ignatius of Loyola came on the scene and pointed out that the overthrow of the physical structure of the universe was the premise of a completely new moral order. On the one hand, it was the result of a state of reckless – and deplorable – trust in human knowledge; on the other, it implied reframing one of the keystones of the Catholic doctrine, namely that of hell, along with any religious matter more or less connected to it: «What cares hee whether the earth travell, or stand still? Hath your raising up of the earth into heaven, brought men to that confidence, that they build new towers or threaten God againe? Or do they out of this motion of the earth conclude, that there is no hell, or deny the punishment of sin? Do not men beleeve? do they not live just, as they did before?».[4]

Although John Donne's depiction of Copernicus was a caricature, it nevertheless emphasized how the change in the astronomical paradigm entailed significant repercussions on the moral order of believers and on the precepts that the Holy Church had been setting for more than a thousand years.

It is significant that among the theological arguments put forward against the heliocentric reading of the universe, the most prominent upholders of catholic orthodoxy – such as Roberto Bellarmino, Francesco Ingoli, Jean-Baptiste Morin, Libert Froidmont, Melchior Inchofer – stressed the need to preserve the opposition between the place of angels and the blessed (the Empyrean) and the site of the damned (hell). Meanwhile, the leading supporters of Copernicanism and, more in general, those who aimed to undermine the alliance between Aristotelian philosophy and the biblical cosmos – such as Foscarini, Campanella, Galileo, Wilkins – did not conceal their awareness of the prominent role such a moral and cosmological order would have played in the rebuttal of the Copernican system.

The charge of rashness often raised against Galileo and the Copernicans forms part of a well-established portrait of the astronomer; throughout the sixteenth century, the effort to grasp the planetary motions had been regarded as an attempt to invade the divine kingdom, or rather a downright act of usurpation – as John Donne's *Ignatius* also attested.

This was even the framework within which many of Galileo's contemporaries acted. Mark Welser's first letter published in the *Sunspot Let-*

Cambridge University Press, 1993, pp. 207-215, 255-258; Bucciantini – Camerota – Giudice, *Galileo's telescope: a European story*, Eng. trans. by C. Bolton, Cambridge (Mass.), Harvard University Press, 2015, pp. 147-153.

[4] Donne, *Ignatius* (cit. note 3), pp. 16-17.

ters is a case in point. In the first version of that letter, Welser had quoted the Gospel of Matthew 11:12: «[...] the kingdom of heaven has suffered violence, and the violent take it by force». In the version that made it into print, this quotation was instead substituted by a paraphrase of Horace, followed by these enlightening words: «Human minds are already forcing heaven, and the hardiest are conquering it».[5] The meaning involving the idea of a violent conquest of the heavens remained unchanged, whereas the reference to the divine kingdom – which was the exclusive domain of theology – disappeared.

That stereotyped image of astronomers arose from the interplay of epistemological, moral and theological underpinnings. For many authors, the hypothetical nature of astronomy made every attempt at the cognitive conquest of the sky utterly futile. As early as the second century AD, Lucian of Samosata had reprehended the reckless endeavor of those astronomers who sought to overcome the limits of sensory knowledge by claiming to be capable of unveiling every aspect of nature.[6] In his *Praise of Folly*, Erasmus described nature as it mocked the overconfidence of the astronomers who aspired to participate in the meeting of gods and to share their secrets, when instead they were getting lost in innumerable universes that were only the fruit of their suppositions.

> There is certainly a pleasant form of madness, which sets them building countless universes and measuring the sun, the moon, stars and planets by rule of a thumb or a bit of string, and producing reasons for thunderbolts, winds, eclipses and other inexplicable phenomena. They never pause for a moment, as if they'd access to the secrets of Nature, architect of the universe, or had come to us straight from the council of gods. Meanwhile Nature has a fine laugh at them and their conjectures, for their total lack of certainty is obvious enough from the endless contention amongst themselves on every single point.[7]

This stance may be regarded as the background to Andrea Alciato's *Emblemata*, a book that was published in many editions throughout the sixteenth century and in the end was composed of roughly two hundred emblems.[8] More precisely, the emblems *In Astrologos* and *In Temerarios* em-

---

[5] GALILEO, *Istoria e dimostrazioni intorno alle macchie solari e ai loro accidenti*, OG, V, p. 93.

[6] LUCIAN OF SAMOSATA, *Icaromenippus, or the Sky-Man*, Eng. trans. by A.M. Harmon, in *Lucian*, 8 vols., Cambridge, Mass., Harvard University Press, 1913, vol. II, pp. 277-279.

[7] DESIDERIUS ERASMUS, *The Praise of Folly*, Eng. trans. by B. Radice, Harmondsworth, Penguin Classics, 1971, p. 151.

[8] On Alciato's emblems and, more in general, on the condemnation of Renaissance as-

phasized the consequences that astrologers and astronomers would suffer upon attempting to gain access to the truths held by heaven;[9] the dreadful outcome was illustrated by Icarus' ruinous fall,[10] and their rashness was symbolized by Phaeton's disastrous flight.[11]

The examples of Icarus and Phaeton bring into the foreground the role that *techne* played in their audacious assaults on the sky: it is the determining element in overcoming man's natural limits, enabling him to get closer to the celestial things. Therefore, it was not surprising that Angelo Grillo, a Benedictine abbot who was a contemporary of Galileo, defined the telescope as a «temerarious glass» that «dares to penetrate the bowels of the sky and the stars, and spy whether the Moon has the stone disease».[12]

In the same years that Alciato's *Emblemata* spread across Europe, Tycho Brahe sang Copernicus' praises in such an insightful manner that Pierre Gassendi chose to quote these passages more than sixty years later in his *Lives* of Brahe and Copernicus. Brahe – who compared his own work to Phaeton's – paid homage to Copernicus' boldness: since Copernicus had tried to reach the heavens by means of his mind, he was rewarded with success. He was to be applauded for having 'rolled the earth in Olympus' and changed the shape of the world: «He introduced himself, up where lives the pre-eminent / Which had never been allowed to mortals since the beginning of the World»;[13] «The great Copernicus alone, although less strong / Than any of them, was happier? He can not be wrong / He who moves the whole earth, with every mountain on it, and the Stars, / he whom no thunderbolts hit. Waging temerarious wars by the power / Of

---

piration to know the *arcana Dei*, see Carlo Ginzburg, *Clues, Myths, and the Historical Method*, Baltimore, Johns Hopkins University Press, 1989, pp. 60-76. See also Peter Harrison, *Curiosity, Forbidden Knowledge, and the Reformation of Natural Philosophy in Early Modern England*, «Isis», 92, 2001, pp. 265-290; Lorraine Daston, *Curiosity in Early Modern Science*, «Word and Image», 11, 1995, pp. 391-404.

9 Andrea Alciato, *Emblemata*, Leyden, Officina Plantiniana, 1591, *Qui alta contemplantur, cadere* (emblem 104).

10 *Ibid.*, *In Astrologos* (emblem 103).

11 *Ibid.*, *In Temerarios* (emblem 56).

12 Angelo Grillo, *Delle lettere volume secondo*, Venezia, Evangelista Deuchino, 1616, vol. II, p. 305, also quoted in Bucciantini – Camerota – Giudice, *Galileo's telescope* (cit. note 3), p. 240.

13 Pierre Gassendi, *Tychonis Brahei equitis Dani, astronomorum coriphaei vita, itemque Nicolai Copernici, Georgii Peurbachii & Ioannis Regiomontani celebrium astronomorum*, in *Opera omnia*, 6 vols., Lugduni, sumptibus Laurentii Anisson & Ioan. Bapt. Devenet, 1658, vol. V, book II, p. 410; Eng. trans. *The Life of Copernicus (1473-1543). The Man Who Did Not Change the World*, ed. by Oliver Thill, Fairfax, VA, Xulon Press, 2002, p. 246.

the muscles is displeasing to Jupiter. / He placidly penetrates the heavens by perspicacious mind».[14]

As Brahe mentioned, Greek mythology had already exemplified the attempt to conquer the divine kingdom: two giants, the Aloads, sought to reach the summit of Mount Olympus by piling Mount Pelion on top of Mount Ossa – where the sanctuary of the god Pheme ('Fame') was. The tragic end of the two brothers – who were killed by Apollo according to one version, or who killed each other according to another version – meant the inevitable failure of any attempt to storm the heavens.

While Brahe traced a distinction between Copernicus' theoretical survey – which was welcome to Jupiter – and the giants' brute force, the Dominican theologian Giovanni Maria Tolosani stressed their perfect overlap as they shared the same desecrating attitude. In the first known anti-Copernican work, titled *De coelo supremo immobili* (Florence 1546-1547), Tolosani indeed claimed: «Copernicus, with his imagination and his sense, changes the order of God's creatures when he acts just like a giant who wants to stack Mount Ossa on top of Mount Pelion and attempts to lift the earth – which is the heaviest element – from the lowest place where it is, all the way up to the sphere where everybody commonly puts the Sun».[15]

Moreover, the opening pages of *De coelo* mentioned the theological underpinnings of Tolosani's geocentrism: he clarified that the huge distance between heaven and earth (especially the Empyrean) stemmed from the need to oppose the seat of God, the angels and the blessed on one side, and the place where mortal beings lived, on the other.[16]

The condemnation of reckless behavior, which was inherent to Christian ethics, took on a relevant role within the framework of the 'healthy

---

[14] Gassendi, *Tychonis Brahei* (cit. note 13), *Nicolai Copernici Vita*, p. 511; Eng. trans. *The Life of Copernicus* (cit. note 13), pp. 246-247 (with some changes).

[15] Giovanni Maria Tolosani, *De coelo supremo immobili et terra infima stabili ceterisque coelis et elementis intermediis mobilibus*, chap. II. On Tolosani's work, published for the first time by Eugenio Garin (*Rinascite e Rivoluzioni*, Bari, Laterza, 1975, pp. 283-295), see Salvatore I. Camporeale, *Giovanmaria dei Tolosani O.P.: 1530-1546. Umanesimo, Riforma e teologia controversista*, «Memorie Domenicane», 17, 1986, pp. 145-252; Miguel Angel Granada, *Giovanni Maria Tolosani e la prima reazione romana al "De Revolutionibus": la critica di Copernico nell'opuscolo "De Coelo et elementis"*, in Bucciantini – Maurizio Torrini (eds.), *La diffusione del copernicanesimo in Italia. 1543-1610*, Firenze, Olschki, 1997, pp. 11-35; Michel-Pierre Lerner, *Aux origines de la polémique anticopernicienne I. L'*Opusculum quartum *(1547-1548) de Giovanmaria Tolosani*, «Revue des Sciences Philosophiques et Théologiques», 86, 2002, pp. 681-721. For a reading of Tolosani's *De Coelo* in view of the theological issues he tackled in *De purissima veritate divinae Scripturae adversus humanos errores*, see Luigi Guerrini, *Cosmologie in lotta. Le origini del processo di Galileo*, Firenze, Polistampa, 2010, pp. 19-88.

[16] Tolosani, *De coelo* (cit. note 15), chap. I.

edict' of 1616. For instance, in referring to Aristotle (*Ethics* 2,7; 3, 7), Thomas Aquinas had listed temerity among the vices, in opposition to the cardinal virtue of fortitude, as the latter was the happy medium that moderated fear and boldness. Temerity was said to derive from presumption, which in turn was connected to pride: temerarious individuals acted with haste or under the impulse of their will and passions – or even in disregard of the law – and the excess of *hubris* that characterized their rashness was also reflected in the opposition between temerity and prudence.[17]

These epistemological and moral readings of temerity were intertwined with the theological reading: the temerarious opposition of Copernicanism to Catholic doctrine was noted by many at the beginning of the seventeenth century.[18] As Lodovico delle Colombe spelled out in *Pleasant and Curious Answers* (*Risposte piacevoli e curiose*, 1608)[19] and again in *Against the motion of the earth* (*Contro il moto della Terra*, 1611), the definition of the falseness and rashness of Copernicanism – already remarked on by Juan De Pineda – relied on the classification provided by Melchior Cano in *De locis theologicis* (book 12, chapter 10),[20] which was aligned with the modern commentators of Saint Thomas: «[...] something against the universal sentence of the Fathers, it can be said that such a proposition is temerarious. Moreover, theologians say it is a universal rule that a big mistake in philosophy is suspect to theology, and especially if it is something the Scripture deals with, such as this is».[21] A few years earlier, the French Jesuit Jean Lorin

[17] See THOMAS AQUINAS, *Summa Theologiae*, I-II, qu. 127, art. 2; qu. 53.

[18] See LERNER, *La réception de l'héliocentrisme copernicien*, in NICHOLAS COPERNICUS, *Des révolutions des orbes célestes*, Fr. trans. with note, ed. by Lerner, A.Ph. Segonds, J.-P. Verdet *et al.*, 3 vols., Paris, Les Belles Lettres, 2015, vol. I, pp. 485-491.

[19] LODOVICO DELLE COLOMBE, *Risposte piacevoli e curiose alle considerazioni di certa maschera saccente nominata Alimberto Mauri*, Firenze, Caneo & Grossi, 1608, p. 17r: «Certamente non può esser se non sospetto di erronea dottrina, dicono autor gravissimi, quegli che contro la comune peripatetica, e Teologica scuola osa temerariamente stampare».

[20] On Melchior Cano, see BERNHARD KÖRNER, *Melchior Cano "De locis theologis". Ein Beitrag zur Theologischen Erkenntnislehre*, Graz, Styria Medienservice, 1994; RICHARD J. BLACKWELL, *Galileo, Bellarmino and the Bible*, Notre Dame, University of Notre Dame Press, pp. 15-20, 255-258.

[21] DELLE COLOMBE, *Contro il moto della Terra*, 1611, OG, III, p. 290. *Ibid.*: «Della qual dice il Pineda, sopra Iobbe, che questa fu sentenza de' Pittagorici, e che molti la illustraron di quei bei titoli. Diremo le sue proprie parole, per non dar sospetto ad ampliazione: "Alii certe scientiam hanc deliram dicunt, nugatoriam, temerariam et in fide periculosam dicunt, atque ex ore antiquorum illorum philosophorum a Copernico et Caelio Calcagnino revocatam, potius ad ingenii specimen, quam ad philosophiae atque astrologiae bonum et utilitatem aliquam". Conchiudesi adunque, la Terra esser nel centro del Mondo, immobile per ragion della sua gravità, e il Sole nel quarto Cielo girare intorno alla Terra, e la Luna esser sparsa di parti rare e dense, e non montuosa o ineguale, ma liscia e rotonda, come si è creduto sin qua».

had pursued the point further: he foresaw a charge against Pythagoras' followers that was echoed eleven years later in the 1616 decree. Specifically, the Pythagoreans and the sectators Copernicus, Caelio Calcagnini and Andrea Cesalpino, who adhered to Didacus à Stunica's opinion and set the earth in motion, were accused of holding a false and temerarious opinion (*Eam opinionem falsitatis, ac temeritatis...*), which was also dangerous and repugnant to faith (*periculosa et Fidei repugnans*).[22]

Charges of rashness similar to those raised by Tolosani, Lorin and Delle Colombe were the target Antonio Foscarini thought of while writing the *Defense*[23] of his *Letter concerning the Opinion of the Pythagoreans and Copernicus*, which he addressed to Bellarmino in March 1615. To show that the movement of the earth could not in any way be considered temerarious, Foscarini refuted point by point Melchior Cano's threefold definition of rashness. Firstly, the statements concerning the earth's motion were not pronounced by chance «but [were] founded on principles of the natural and mathematical sciences». Secondly, its movement was not to be listed among things that «are asserted with insolence and excessive imprudence» as it was grounded in «proper testimony and probable reason», in addition to agreeing «most fittingly with the Scriptures according to the methods and arguments used most commonly by the Holy Fathers». Thirdly, it was in no way «opposed on a doctrine of faith to any common decree of the universal Church».[24]

2. Is Hell in the Heavens?

Tolosani, in the second book of the *Compendio di sphera et machina del mondo nuouamente composto* (1514), addressed the issue of the location of hell in the center of the earth in terms akin to the description set forth in *La*

---

22 Jean Lorin, *In Acta Apostolorum Commentaria*, Lugduni, Apud Horatium Cardon, 1605, p. 215: «Ubi Didacus Stunica illi opinioni adhaerescit, quae auctores habet Pythagoricos, et sectatores Copernicum, et Caelium Calcagninum, terram naturali perenni motu moveri. Eam opinionem falsitatis, ac temeritatis convincit inter alios Clavius Noster, et Pineda in Job, capite 9. [...] Quin etiam periculosa et Fidei repugnans videri potest dicta sententia, quoniam sapiens dicit quod terra in aeternum stat [...]».

23 Paolo Antonio Foscarini, *Defensio Epistolae [...] super mobilitate terrae*, published in Emanuele Boaga, *Annotazioni e documenti sulla vita e sulle opere di Paolo Antonio Foscarini teologo "copernicano" (ca. 1562-1616)*, «Carmelus», 37, 1990, pp. 173-216: 205-214. Richard Blackwell dated back the *Defensio* to late March or very early April: Blackwell, *Galileo, Bellarmino and the Bible* (cit. note 20), p. 98.

24 Foscarini, *Defensio* (cit. note 23), Eng. trans. *Defence*, in Blackwell, *Galileo, Bellarmino and the Bible* (cit. note 20), pp. 255-256.

Sfera, a poem attributed to the Dominican friar Leonardo Dati (1360-1425): «[...] in its center the earth is empty / And therein is the infernal abyss / Such house remote to us / keeps the damned within its eternal doors / And of this region unknown to many, / Who over here hold a beastly life, / once over there they will have clear certitude of it / With torment, pain, mourning and serious damage».[25]

The issues of the existence of hell as a physical place, of its shape and of its site were not taken for granted and were a source of puzzlement in Christian thought. The notion of hell sketched out in Dante's *Divine Comedy* had strengthened the idea of a place that lay in the depths of the earth, a view that was amplified by numerous pictorial representations and literary works.[26] Yet throughout the sixteenth century discussions about that issue were still alive, both within the Catholic Church and in its challenge of Protestantism. As will be shown later on, such theological debates fitted in the broader questions over the new vision of the world put forward by Copernicanism.

Juan de Pineda, one of the most influent Jesuits of that time, reaffirmed the centrality of hell and earth, defining the latter as 'gloomy' and 'miserable' since it was the prison of the damned.[27] At the same time, Pineda railed against the attempt made by several contemporaries – among whom he named Gilbert – to bring the Pythagoreans' heliocentric view back to life.[28]

---

[25] Tolosani, *Compendio di sphera et machina del mondo nuovamente composto*, Firenze, Bernardo Zucchetta, 1514, p. 11r.

[26] Catholic believers were constantly reminded of the existence of hell and of the suffering that would be imposed on sinners by a great number of pictorial depictions and frescoes in places of worship. Some examples include: the chilling scenes in the Florence Baptistery (1260-1270) and in the Dome of Santa Maria del Fiore (1572-1579) by Giorgio Vasari and Federico Zuccari; Giotto's fresco cycle (c. 1306) in the Scrovegni Chapel in Padua; Buffalmacco's frescoes in the Monumental Cemetery (1336-1341) in Pisa, as well as Nardo di Cione's Strozzi Chapel (1351-1357) in Santa Maria Novella in Florence. Besides the flourishing of an infernal topography throughout the fifteenth and seventeenth century – such as Botticelli's *Chart of hell* (c. 1480-c. 1495), Giovanni Stradano's map (1587) – and the academic quarrel between Antonio Manetti (1506) and Alessandro Vellutello (1544) about Dante's hell, it is worth remembering the emergence of a specific literary genre: the descent into hell of the "chanson de geste" *Huon d'Auvergne*, Federico Frezzi's *Quadriregio* (c. 1395, printed in 1481), Matteo Palmieri's *Città di Vita* (written in 1464 c. but unpublished), Francesco Colonna's *Sogno di Polifilo* (1499), Orazio Romano's *Porcaria*, Astolfo's descent into hell in Ariosto's *Orlando furioso* (1532), and also Robert Greene's *Newes Both From Heaven and Hell* (1593).

[27] Juan de Pineda, *Commentariorum in Iob libri tredecim*, Hispali, In Collegio Hermenegildi eiusdem Societatis, 1598, chap. X, vers. 21-22, pp. 545-548.

[28] Id., *In Ecclesiasten Commentariorum Liber unus*, Antverpiae, apud Petrum et Ioannem Belleros, 1620, pp. 116-118 (quoted in Corrado Dollo, *Galileo e la cultura della tradizione*, Rubbettino, Soveria Mannelli, 2003, pp. 182-183).

Likewise, the Spanish Jesuit Benito Pereira also dealt with the issue of hell in a work that ran into many reprints and was considered to be the most authoritative commentary of the sixteenth century on Genesis – the *Commentaria et Disputationes in Genesim*.[29] In the eighth *Disputatio* Pereira examined the question of the different meanings of 'hell' in the Bible, namely whether it meant Jesus' sepulcher (which had been put underground), or whether it was the place in the depths of the earth to which Christ had actually descended. Pereira also pointed out that contemporary heretics interpreted the expression *descendit ad inferos* by referring to Christ's sepulcher, since they refused to admit that Christ's soul had descended into the infernal region after his crucifixion.[30] When translating the Jewish *Sheol* (kingdom of the dead) into the Latin *infernus*, Pereira sought to show that interpreters agreed on *inferos* as a place located in the bowels of the earth. He offered support for such an interpretation by quoting, among other things, *Quaestio* 126 of Augustine's *In Genesim*, where it referred to Jacob's words: «Descendam ad filium meum, lugens in infernum». More precisely, *infernus* would have meant the descent into Limbo, which was the highest level of hell.[31] Among a great number of biblical sentences, Pereira examined David's Psalm 138 «If I ascend into heaven, you are there: if I descend into hell, you are present», where the surface of the earth was thought to be the place in relation to which the opposite locations of heaven and hell could be defined.[32]

Several years later, Roberto Bellarmino devoted chapter X in IV book of his *De Christo* to the issue of hell: he carried out a thorough analysis of several biblical sentences and a meticulous comparison with Jewish and Greek terms, finally rejecting the identification of hell with Christ's sepulcher and placing the infernal realm in the core of the earth. Two aspects of Bellarmino's analysis are noteworthy: firstly, he referred to the cosmological-theological argument by specifying that if 'hell' had been a synonym of 'sepulcher', it would not have complied with the need to place it at the

---

29 Galileo owned the *Commentaria* and drew on Pereira's works in writing the so-called 'Copernican Letters': see ROMANO NANNI, *Per lo studio dei teologi di Galilei: le "regulae" del* Commentario in Genesim *di Benito Pereira*, in *I primi Lincei e il Sant'Uffizio: questioni di scienza e di fede*, Roma, Bardi, 2005, pp. 421-446.

30 BENITO PEREIRA, *Commentariorum in Genesim tomi quatuor*, Cologne, Ex officina Antonij Hierati, sub Monocerote, 1606, pp. 882 sq.

31 *Ibid.*, p. 883.

32 *Ibid.*, p. 884: «Ubi opponuntur inter se duo loca distantissima tum inter se, tum a superficie terrae ubi habitant homines. Unus locus est distantissimus secundum altitudinem, quod est coelum, alter locus est disiunctissimus secundum profunditatem qui est infernus».

maximum distance from the Empyrean;[33] secondly, he included this survey within a wider rebuttal of the Calvinistic interpretation that denied Christ's descent into a physical place lying in the depths of the earth.

For our purposes, it is extremely significant that among those who instead were inclined to identify hell with the sepulcher was Diego de Zuñiga in chapter 17 of *In Job Commentaria*[34] – a copy of which could be found in Galileo's own library: such an interpretation indeed paved the way for dismantling one of the anti-Copernican arguments on which the maximum opposition between earth and heaven was based.

The issue of the location of hell was vigorously debated since the Scriptures did not provide a clear answer, and the Church Fathers failed to agree unanimously. Thomas Aquinas, in the *Supplement* to his *Summa Theologiae*, qu. 97, had indeed quoted Augustine's words (*De civitate Dei* 20, 16) concerning the fact that nobody knew in which part of the world hell was located, except for those who had received a revelation from the Holy Spirit. He also clarified that, as Aristotle had claimed (*De caelo* 2, 13),[35] the Pythagoreans situated the place of punishment on a globe of fire, which they put at the center of the cosmos.[36] Nonetheless, Augustine, Gregorius and Thomas Aquinas himself were inclined to consider the localization of hell in the depths of the earth as the more probable option.

While during the sixteenth and seventeenth centuries the Catholic world tried to work out a justification for the localization of the infernal fire in the center of the earth by attempting to define its precise topographical dimensions and geography, the Protestant world did not share this interest as strongly. Contrary to the prevailing Catholic interpretation, Calvinist theologians were inclined to deny the physical descent of Christ in favour of a spiritual experience of universal pain: hell was not a place, rather a state of agony caused by the sins that people committed during their lives – or, in the case of Jesus, by the torments of humanity which he took on and which were symbolized by his agony on the Cross.[37] Although

---

33 See Roberto Bellarmino, *De Christo*, book 4, chaps. IX-X, in Id., *Opera omnia*, vol. I, Napoli, apud Josephum Giuliano, 1836, pp. 275-277; Id., *De Purgatorio*, book 2, chap. VI, in *Opera omnia*, BO III, pp. 109-112.

34 Diego de Zuniga, *In Job Commentaria*, Toleti, Ioannes Rodericus, 1584, chap. 17, pp. 419-421.

35 In the *De coelo* Aristotle only stated that Pythagoras had placed the Sun at the center of the earth's motion, without mentioning anything about hell, as instead St. Thomas stated.

36 Thomas Aquinas, *Supplementum* to the *Summa Theologiae*, qu. 97, art. 7.

37 See, for instance, Calvin's final addition to the *Institutio Christianae Religionis*, book II, chap. 16, § 12. Protestants, Anglicans and Puritans displayed a range of different interpreta-

this issue was only of minor importance among the various heated theological debates that the Counter-Reformation had been conducting in the face of a growing number of heresies, it is nevertheless significant to note that the heliocentric universe seemed to relate better to the reading put forward by the enemies of the Roman Church.

3. The Earth: Shining Star or Dreary Globe of Mortality?

> Reason, along with sense, teaches that the least noble, most impure and vile things are located in the most inferior and lowest places, and the most noble ones are in the most eminent and supreme: but the earth is the lowest of all natural bodies: therefore the earthly mass is in the lowest place in the World, that is, its center. That the earthly machine is such, who will be so devoid of eyes not to see any such thing? Is it perhaps that the Sun is inferior to the earth so that you, oh Copernicans, put it in the center?[38]

In his in-depth refutation of Galileo's *Nuncius*, Delle Colombe turned to Christoph Clavius' commentary on Sacrobosco's *De Sphaera*. After reviewing the traditional arguments that Sacrobosco and Ptolemy had put forward to uphold the centrality of the earth, Clavius added a fourth *ratio* concerning the Aristotelian cosmology: nature put the earth at the center of the celestial machine so as to separate this «vile and rude» body from the other parts of the heavens, which were instead «the most excellent».[39] Delle Colombe clearly understood Galileo's attempt to overturn the ontological hierarchy of the cosmos and to free the earth from its lowly status: in *Nuncius* the earth was no longer obscure, nor was it «a bilge of filth and of the

tions: see Dewey D. Wallace Jr., *Puritan and Anglican: The Interpretation of Christ's Descent into Hell in Elizabethan Theology*, «Archiv für Reformationsgeschichte», 69, 1978, pp. 248-286; David V.N. Bagchi, *Christ's Descent into Hell in Reformation Controversy*, «Studies in Church History», 45, 2009, pp. 228-247; Peter Marshall, *The Reformation of Hell? Protestant and Catholic Infernalisms in England, c. 1560-1640*, «Journal of Ecclesiastical History», 61, 2010, pp. 279-298; Catherine E. Laufer, *Hell's Destruction: An Exploration of Christ's Descent to the Dead*, London and New York, Routledge, 2016.

[38] Delle Colombe, *Contro il moto della Terra* (cit. note 21), p. 276.

[39] Christoph Clavius, *In sphaeram Ioannis de Sacro Bosco commentarius*, Lugduni, excudebat Guichardus Iullieron, 1593, cap. I, p. 160. See James M. Lattis, *Between Copernicus and Galileo: Christoph Clavius and the Collapse of Ptolemaic Cosmology*, Chicago, The University of Chicago press, 1994, pp. 118-120. This passage of Clavius' commentary is also echoed in Galileo's early *Tractatio de coelo*, which dated back to 1584-1590 and was included in a series of writings Galileo wrote while he was attending a course of Aristotelian philosophy: Galileo, *De motu antiquiora*, OG, I, pp. 48-49. See Wallace, *Galileo's Early Notebooks: the Physical Questions*, Notre Dame, University of Notre Dame Press, 1977, pp. 71-74.

world's dregs» (*non autem sordium mundanarumque fecum sentinam*)[40] or the «dregs of the world, defective, impure and very vile»;[41] conversely, it was brighter than the Moon and as luminous as the other planets, especially Jupiter, which was regarded as the 'noblest star'. Still, in the 1632 *Dialogue* Galileo expressed his disagreement with those – like the character Simplicio – who made a clear-cut distinction between the celestial bodies and the «vulnerable, transient and mortal» earth, which they as well described as the «dregs of the world», the «bilge of all waste».[42]

The despicable status of the earth, which was the consequence of the Aristotelian ontological distinction between the celestial and sublunary worlds and among the four elementary spheres, provided further grounds not only to argue for the presence of hell in the depths of the earth, but also to foster the theological debate over the descent and ascent of Christ, an issue that touched on the doctrine of faith.[43] For many Catholic theologians – even for those who played a crucial role in the 1616 affair, such as Roberto Bellarmino and Francesco Ingoli – agreeing with the Copernican system would have required a reappraisal of the traditional interpretation of Scripture passages concerning the ascension of Jesus to the heavens and equally his descent to hell. Specifically, earth – and hell – would have no longer been in the center of the cosmos, namely in the most distant place from the Empyrean, rather they would both have had to be moved into the heavens, where Christ would instead have gone up after his resurrection.

These were pivotal arguments, attested to as such in Foscarini's *Letter on the Opinion of the Pythagoreans and Copernicus*[44] as well: more precisely, the third category of objections raised against Copernicanism – which he was about to address – focused on Christ's descent onto the earth through

---

[40] GALILEO, *Sidereus Nuncius*, OG, III, p. 75. Hans Blumenberg picked out some alleged sources of that definition, such as Lucretius' *De rerum natura* and Plutarch's *De facie in orbe lunae*: HANS BLUMENBERG, *Paradigms for a Metaphorology*, Eng. trans. by R. Savage, Ithaca, Cornell University Press, 2010, pp. 109-110, note 27. See also ANDREA BATTISTINI (ed.) – GALILEO, *Sidereus Nuncius*, It. trans. by M.T. Cardini, Venezia, Marsilio, 1993, pp. 218-219, notes 206-207.

[41] Galileo to Gallanzoni, 16 July 1611, OG, XI, p. 148.

[42] GALILEO, *Dialogo sopra i due massimi sistemi*, OG, VII, p. 84. See also *ibid.*, p. 292. Galileo himself had used these same words in the *De motu*: see *infra* p. 47.

[43] See MELCHIOR INCHOFER, *Tractatus syllepticus, in quo quid de Terrae Solisque motu vel statione secundum Sa.am Scripturam et Sanctos Patres sentiendum*, Romae, excudebat Ludovicus Perignanus, 1633, chap. 13, p. 80, Eng. trans. by BLACKWELL, *Behind the scenes at Galileo's trial: including the first English translation of Melchior Inchofer's "Tractatus syllepticus"*, Notre Dame, University of Notre Dame press, 2006, p. 191.

[44] FOSCARINI, *Lettera sopra l'opinione de' pittagorici e del Copernico...*, Napoli, per Lazzaro Scoriggio, 1615, pp. 15-17, 52-56.

his Incarnation and his ascent to heaven after his resurrection; the fourth class of objections concerned hell. According to Foscarini, the debate that arose from it swung between two 'monstrous and absurd' options: on the one hand, the assumption that the maximum distance between hell and the Empyrean would have caused Lucifer's kingdom to be set on the Sun; on the other, the conflicting notion that hell was situated in the center of the earth would have implied that hell was continuously moving within the perfect and ethereal heavens. Foscarini rejected such readings by resorting to the argument of the ontological homogeneity of the cosmos, and by adhering to the stance that hell was located in the earth's center – since theologians themselves were generally agreed on this point. The only distinction Foscarini acknowledged was the one existing between the very thin air in which the planets moved, and the sublime sphere of the Empyrean, which became the reference point in defining the movements *sursum* and *deorsum*.

In May 1615 Piero Dini reported that the Peripatetics increasingly employed the argument of hell to refute the Copernicans and insisted on bringing into the sacristy a debate that instead should be kept far from the places of worship:

> I don't know that there is any news, if not some that might give rise to a continual explosion, so to speak, of those Aristotelians, who, when thinking about the other sect, say: "These put the Sun in hell, we put it in the third sky" and the like: all these things (although we don't say any of them in the way they do) can however greatly trouble the cause; but if they don't create anything else but length, it shall be of little importance.[45]

Just a few days after the anti-Copernican decree, Raffaello Gualterotti wrote to Galileo, who was still in Rome: «As for the motion of the earth, I surely would very gladly agree that it moves: and while someone claims that putting it in the fourth place is not possible, as we would put hell into the heavens, I would say that it is possible, and I would say many other very beautiful things».[46]

Francesco Ingoli's position was very different.[47] He regarded the issue of hell as the primary theological argument against Copernicanism. Based on Psalm 138, Isaiah 14 and Bellarmino's *De Christo* and *De Purgatorio*, the

---

[45] Piero Dini to Galileo, 2 May 1615, OG, XII, p. 175.

[46] Raffaello Gualterotti to Galileo, 3 April 1616, OG, XII, p. 253.

[47] See Francesco Ingoli, *De Situ et Quiete Terrae Contra Copernici Systema Disputatio ad Doctissimum D. Galilaeum Galilaeum* (1616), OG, V, chap. IV, pp. 407-408.

enormous distance that separated the angels and holy spirits on the one hand, and the demons and the damned on the other was grounds enough to invalidate the Copernican hypothesis. Not unlike Ingoli, in his 1616 *Peregrinatione coelesti* the Catholic astronomer Libert Froidmont excoriated one of the absurd outcomes that Copernicanism apparently led to, namely the presence of hell on the Sun.[48]

Even Tommaso Campanella addressed the issue of hell in 1616 in his *Apologia pro Galileo*, first by stressing that theology should not have gotten involved in the astronomical debate over geocentrism,[49] and then by showing the paradoxical conclusions reached by theologians. After summing up Saint Thomas' stance (especially by quoting the *Opusculum* XI, 21), Campanella presented an original interpretation of his own: by merging Copernicanism, an allegedly Pythagorean cosmology,[50] with Telesio's philosophy as well as elements of the current theological debates, he argued that those who place hell at the earth's center must acknowledge the earth's movement after all, since the heat generated by the infernal fire was the cause of its motion. Conversely, if they maintained that the earth was motionless, then they would need to put hell elsewhere. Campanella's interpretation met with strong criticism from Melchior Inchofer. In the chapter 13 of the *Tractatus Syllepticus* Inchofer challenged the foolish arguments Copernicans put forward in favour of the earth's movement, with veiled references to the pages written by Campanella, but without ever mentioning the author. The argument at hand was «hell is in the center of the earth, and in it fire torments the damned. Therefore it is totally necessary that the earth moves. [...] The consequent is proven by the fact that fire is a cause of motion. As a result Pythagoras, who according to Aristotle put the place of punishment in the center, thought that the earth is animated and is endowed with fire».[51]

More than fifteen years after the *Apologia* was published, the Anglican clergyman and philosopher John Wilkins focused on the topic of hell by

---

48 Libert Froidmont, *Saturnalitiae coenae, variatae Somnio, sive Peregrinatione caelesti*, Lovanii, Typis ae sumptibus Petri Sasseni & H. Nempaei, 1665^2, p. 48.

49 Tommaso Campanella, *Apologia pro Galileo...*, Frankfurt, G. Tampach, 1622; Eng. trans. *A Defense of Galileo*, ed. by R.J. Blackwell, Notre Dame, University of Notre Dame press, 1994, pp. 88-90, 95.

50 Campanella referred to Aristotle's *De Coelo* (II, 9, 293 a), according to which – Campanella claimed – Pythagoras was said to have founded the earth's movement on the presence of the infernal fire in it. Campanella provided a reading of this excerpt of the *De Coelo* that was completely at odds with St. Thomas' one: see note 36.

51 Inchofer, *Tractatus syllepticus* (cit. note 43), chap. 13, pp. 77-81, Eng. trans. by Blackwell, *Behind the scenes* (cit. note 43), pp. 188-192.

presenting a doxographical excursus. In proposition XIII of his *Discourse*, Wilkins wrote about those authors – beginning with Plutarch – who dragged hell into the medium region of the air and placed Paradise on the Moon. Although Augustine argued in his *De Civitate Dei* that the infernal place was difficult to define, many philosophers and theologians attempted to do so: among them Wilkins mentioned the Jesuit theologians Francis Ribera, author of *Comment on the Revelations*, and Leonardus Lessius, who wrote *De perfectionibus moribusque divinis*. Aligning himself with the Protestant interpretation, Wilkins was instead inclined to identify hell with a generic place of pain: «[...] the Place is not yet determined, but that hell is there where there is any Tormented Soul, which may be in the Regions of the Air, as well as in the Center; and therefore perhaps it is, that the Devil is stiled the Prince of the Air».[52] Wilkins continued his analysis by refuting Fromondus' interpretation at length. In the sixth proposition of his second book («That there is not any Argument from the Words of Scripture, Principles of Nature, of Observations in Astronomy, which can sufficiently evidence the earth to be in the Centre of the Universe»[53]) he made a distinction between theological matters (which concern scriptural commentary), natural philosophy and astronomical observations, clarifying that the debate over hell belonged to the first two fields. In the *Ant-Aristarchus* Fromondus had launched a well thought-out attack on Copernicanism; in it he included the text of the 1616 decree and went on to reaffirm the importance of placing hell in the center of the earth.[54] In the *Vesta* he addressed this same argument, summing it up with the following syllogism: hell is beneath the earth, hell must be in the furthest place from the Empyrean, therefore the earth is at the center of the empyrean sphere.[55] According to Wilkins, Fromondus' claims relied on three 'weak and doubtful' elements: firstly, that hell was at the center of the earth; secondly, the Empyrean needed to be concentric to the physical sky of planets and stars; and thirdly, the spatial distance reflected an ontological difference, which in turn implied different degrees of nobility within space.

One of Wilkins' opponents, the Scottish Alexander Ross, introduced in *The new planet no planet* an additional argument by interlacing astronomy

[52] John Wilkins, *A Discourse concerning a new world and another planet: in two bookes*, London, printed by John Norton and R. Hearne for Iohn Maynard, 1640, book I, p. 198.

[53] *Ibid.*, book II, pp. 106 sq.

[54] Froidmont, *Ant-Aristarchus, sive Orbis-Terrae immobilis liber unicus*, Antverpiae, ex Officina Plantiniana, 1631, chap. 12, p. 73.

[55] Id., *Vesta, seu Ant-Aristarchi vindex adversus Iac. Lansbergium*, Antverpiae, ex Officina Plantiniana Balthasaris Moreti, 1634, Tractatus VI, chap. 2, p. 150.

and theology: refusing the central position of the earth-hell in the cosmos would result in bedlam and open the way for the return of ancient beliefs and gnostic interpretations already refuted by Irenaeus in the second century.

For whosoever denied hell to be below [i.e., in the bowels of the earth], denied that there was any such place at all: as, *Pythagoras*, *Epicurus*, *Lucretius*, *Tully*, *Seneca*, *Lucian*, *Pliny*, and some others; to whom I may adde the Gnostickes, who held there was no other hell, but this world, whom Irenaeus resutes [*sic*]. [...] As hell must needs be in the earth below, so must heaven the place of the blessed, be above all these visible heavens.[56]

Notwithstanding his heartfelt defence of the Aristotelian cosmology, Ross foresaw that debasing the position of the earth in the cosmos might clash with the anthropocentrism underlying the religious interpretation of the Creation. He therefore resolved to trace a distinction between the cosmological and the religious interpretation of the world: he clarified that although the earth was «the most ignoble and most base element, in respect of its matter, and therefore the lowest»,[57] it lay in the center of the Creation as it was the house of the noblest creature.

The strong interlacing between astronomy and religious topics and the risk that Copernicanism might open the door to the gnostic heresy had also been mentioned by the Jesuit Melchior Inchofer. In the *Tractatus Syllepticus* he carried out an in-depth examination of the implications of Copernicanism for the notion of Christ's descent into hell, commenting on the theological debate that had arisen around a number of scriptural passages. Should Christ's descent be understood in a figurative sense, hence moving away from what the Catholic doctrine had largely established and giving new life to the Valentinian heresy? One year after Galileo's condemnation, Inchofer followed in Bellarmino's footsteps and disowned the notion of a clear demarcation between the fields of theology and natural philosophy, between astronomy and moral issues. The structure of the universe was 'a matter of faith' and Copernicanism had to be condemned as it overthrew the entire Catholic creed, even challenging the tenet of Christ's

56 Alexander Ross, *The new planet no planet: or, the earth no wandring star except in the wandring heads of Galileans... and Copernicus his opinion as erroneous, ridiculous and impious fully refuted*, London, printed by J. Young and are to be sold by Mercy Neighen and Gabriel Bedell, 1646, chap. 6, p. 59. See Irenaeus, *Against Heresies*, book 5, chapter 31. Ross had already refuted Copernicanism in his *Commentum de Terrae Motu circulari*, London, apud Thomam Harperum, 1634.

57 Id., *The new planet no planet* (cit. note 56).

death and resurrection and other topics that were crucial to the salvation of believers.[58]

For from this view [*i.e.* Copernicanism] it necessarily follows that one cannot accept in this proper sense what is said of Christ in the creed, namely, that he first descended into hell and then ascended into heaven. [...] Yet the defenders of the Copernican system want to take these words, and many others like them, to refer only to appearances. [...] And as a result they now invite their audience to revive the Valentinian heresy, which claims that many things did not really happen but are only apparent fantasies. Indeed would not the eternal Truth make a mockery of us if figures of speech were used in the established articles of faith and in what we need to know for salvation [...]?[59]

Christ's words addressed to the Scribes and Pharisees in John 8:23 «You are from below, I am from above; you are of this world, I am not of this world» (*Vos de deorsum estis, ego de supernis sum; vos de mundo hoc estis, ego non sum de hoc mundo*) might have been inverted by the Copernicans into «You are from above, I am from below; you are of this world»,[60] which was, as Inchofer pointed out, a claim worthy of heretics.

The same topic drew the attention of Jean-Baptiste Morin: undertaking a defence of geocentrism against Kepler and Philip Lansbergen, he devoted several pages to the analysis of Christ's motion after His crucifixion. The tenth chapter of *De Famosi et antiqui* (1631), which summed up the physical argument in support of geocentrism, addressed the relationship between the hierarchical structure of the universe and Christ's ascent,[61] whilst also denouncing the mistake and subterfuge used by the Copernicans, and principally by Kepler. The *Epitome Astronomiae Copernicanae* (1621) had clarified that the descending motion of Christ was to be thought of as directed towards the center of the earth – rather than towards the center of the cosmos – and, consequently, His ascent corresponded to the act of moving

58 See BLACKWELL, *Behind the scenes* (cit. note 43), p. 140.

59 INCHOFER, *Tractatus syllepticus* (cit. note 43), chap. 7, pp. 31-32 (quoted also in PESCE, *Gli ingegni senza limiti e il pericolo per la fede*, in JOSÉ MONTESINOS – CARLOS SOLÍS (eds.), *Largo campo di filosofare*, La Orotava, Fundación Canaria Orotava de historia de la ciencia, 2001, pp. 637-659: 656), Eng. trans. by BLACKWELL, *Behind the scenes* (cit. note 43), p. 139. Inchofer referred here to *Ef* 4,10: «He who descended is the one who also ascended far above all the heavens, that he might fill all things».

60 INCHOFER, *Tractatus syllepticus* (cit. note 43), p. 32: «Vos de sursum estis, ego de infernis sum, Vos de mundo hoc estis, (intellige de Mundo iuxta Systema Ptolemaei) Ego non sum de hoc mundo, sed de Mundo iuxta Systema Pythagoricum».

61 JEAN-BAPTISTE MORIN, *Famosi et antiqui problematis de Telluris motu vel quiete hactenus optata solutio*, Parisiis, apud authorem, 1631, pp. 76-79.

away from the earth, instead of being a motion that shifted from the center to the highest sphere.[62] Morin blamed such an interpretation: it would take advantage of simple and slow-witted minds as it failed to mention that the Scriptures refer to an 'ascent to heaven' and not merely to a generic 'ascent'. In addition, when defining motion – and, hence, when determining ascent and descent – it was always necessary to identify the points of departure (*a quo*) and arrival (*ad quem*). The Empyrean being the ultimate boundary of the world, every upward motion was compelled to tend towards it, whereas downward motion had to be defined in relation to the center of that sphere. Therefore, the moral and theological issue of ascent / descent required a precise astronomical structure.[63]

4. Galileo between Rashness and Fortitude. The Biblical Cosmos from *De motu* to *Sunspot Letters*

Galileo was fully aware that defending the Copernican system would entail breaking up the Christian cosmos. His appreciation of the fact that natural philosophy could not avoid facing several theological issues surfaced from one of the versions of the *De motu antiquiora*, whose second chapter opened with the following consideration:

> After the marvelous construction of the vast celestial sphere, the divine Creator pushed the refuse (*excrementa*) that remained into the center of that very sphere and hid it there. His purpose was to keep that refuse from impinging on the sight of the immortal and blessed spirits. But this dense and heavy matter did not with its own mass fill up the broad and spacious region that was left under the curved surface of the last sphere. [...] Thus the earth was left in the center. And similarly the denser bodies were placed nearer the earth.[64]

At that time Galileo still embraced the Ptolemaic system and did not forget to intertwine it with the biblical cosmos, by pointing out their perfect compatibility with the outcast status of the earth, as well as with the immense distance that divided the empyrean realm of the blessed and the

---

62 See Johannes Kepler, *Epitome Astronomiae Copernicanae*, KGW VII, p. 100.

63 Morin, *Famosi et antiqui* (cit. note 61), p. 79.

64 Galileo, *De motu antiquiora*, OG, I, p. 344; *On motion and on mechanics; comprising "De motu" (ca. 1590)*, Eng. trans. by S. Drake, Madison, The University of Wisconsin press, 1960, p. 116. On the Empyrean of theologians, see Galileo, *Iuvenilia*, OG, I, p. 41; Alimberto Mauri, *Considerazioni sopra alcuni luoghi del Discorso di Lodovico delle Colombe intorno alla stella apparita nel 1604*, Firenze, appresso Gio. Antonio Caneo, 1606, pp. 3v, 5v, 18v.

angels from the mortal place of the damned. It is no less significant that the lectures Galileo delivered at the Florentine Academy on the geography of hell date back to the same period (1588).[65] If as early as the 1590s Galileo sought to show the interplay between the geocentric structure of the cosmos and theological matters, it is clear that in his mind the abandonment of the Aristotelian-Ptolemaic model might call for the reappraisal of several doctrinal questions.

In the *Sidereus Nuncius* Galileo provided the first 'evidence' in support of the Copernican system: in addition to Jupiter's satellites, he focused on the parallelism between the earthly moon and the lunar earth, namely an earth that was as bright as the stars and took part in the dance of the heavens[66] – as he wrote, employing images that evoked neo-Platonic philosophy.[67] It is not insignificant that in the *Nuncius* he alluded to the idea of the earth that was sketched out earlier in *De motu* by dissociating himself from this model and by showing the shining and noble nature of our planet. In 1610 Galileo ruled out any reference to the Holy Scriptures by choosing to present the outcome of his research exclusively within the framework of natural philosophy. Again – and even more so – in the *Dialogue on the two chief world systems*, the character of Salviati did not trouble to answer Simplicio when the latter listed, among the cases of 'very solemn foolishness' ensuing from the adoption of Copernicanism, «that Christ, our Lord and Redeemer, went up to hell and went down to the heavens when he came closer to the Sun».[68]

Galileo's thinking prior to 1616 was characterized by a three-fold temerity (the first two of which he shared with Copernicus). The first was *epistemological*, namely the reckless attitude of astronomers who insisted on trying to grasp the true structure of the universe, thus defying the limits of human reason. The second was *cosmological* and *theological*, as it relied on the earth's ascent into heaven and on the ensuing doctrinal issues concern-

---

65 The Dominican friar Raffaello delle Colombe, who was Lodovico's brother, attacked Galileo for his survey on hell that, as every 'curious inquiry', should have been avoided. See Guerrini, *Galileo e la polemica anticopernicana a Firenze*, Firenze, Polistampa, 2009, pp. 62-64. On Raffaello Delle Colombe's sermons against Galileo, see *ibid.*, pp. 35-70; Id., *Echoes from the pulpit. A preacher against Galileo's astronomy*, «Journal for the History of Astronomy», 43, 2012, pp. 377-390.

66 See Natacha Fabbri, *The Moon as Another Earth: what Galileo owes to Plutarch*, «Galilæana», 9, 2012, pp. 103-135; Id., *Looking at an Earth-like Moon and Living on a Moon-like Earth in Renaissance and Early Modern Thought*, in Cecilia Muratori – Gianni Paganini (eds.), *Early Modern Philosophers and the Renaissance Legacy*, Dordrecht, Springer, 2016, pp. 135-151.

67 See Eugenio Garin, *Lo zodiaco della vita*, Roma-Bari, Laterza, 1982², p. 14.

68 Galileo, *Dialogo* (cit. note 42), p. 384.

ing hell and its opposition with the Empyrean – to which the debate over other inhabited worlds, which involved the doctrine of the Incarnation, could also be added. And finally, *theological* temerity, stemming from Galileo's boldness in interpreting Scriptural passages, which he attributed to something akin to fortitude, whereas most of his contemporaries regarded it as temerarious at the very least, if not actually heretical.

When drafting the *Sunspot Letters*,[69] Galileo in fact felt the need to seek advice on the theological implications of the earth's motion and the corruptibility of the heavens, and as is well known he addressed his inquiries to Cardinal Conti, whose answers echo in the first versions of the *Sunspot Letters*. Specifically, at a certain point Galileo described the Aristotelian statement on the incorruptibility of the heavens as 'repugnant to the Sacred Scriptures', repeating what Conti stated in the beginning of his letter of response, that «it was the common opinion of the Fathers that heaven was corruptible».[70]

Later Galileo asked Conti for further elucidations concerning scriptural passages that were in disaccord with Aristotle:

> On the Sacred Scriptures, I would like to know more in particular what Your Lordship seeks to know that does not support Aristotle: because if Your Lordship speaks about the corruptibility of heaven, there is no doubt that many places mention it; if he speaks about other dogmata, it is certain to be contrary to Aristotle, such as about the eternity and the government of the universe. But this has nothing to do with these observations.[71]

Conti ruled out that the Aristotelian doctrine on the eternity of the universe was relevant to Galileo's discussion on the corruptibility of heaven and on the earth's motion; meanwhile, he was willing to show him a collection of sources that supported Galileo's reading of the last two points.[72]

In arguing in defense of heliocentrism Galileo followed a multifaceted strategy. Besides refuting the physical and theological objections to Co-

---

[69] Thomas F. Mayer, *The Censoring of Galileo's* Sunspot Letters *and the First Phase of his Trial*, «Studies in History and Philosophy of Science», 42, 2011, pp. 1-10; Id., *The trial of Galileo, 1612-1633*, Toronto, University of Toronto press, 2012.

[70] Conti to Galileo, 7 July 1612, OG, XI, p. 354.

[71] Conti to Galileo, 18 August 1612, OG, XI, p. 376.

[72] It is a matter of fact that throughout the sixteenth and the beginning of the seventeenth century several Catholic theologians called the doctrine of quintessence into question and, consequently, the incorruptibility of heaven as well. See Campanella, *Apologia* (cit. note 49), pp. 86-88. Christoph Scheiner summed up this debate in the *Rosa Ursina*, Bracciani, apud Andream Phaeum, 1630, book IV, part II, chaps. XXIV-XXV. See Luigi Ingaliso, *Filosofia e cosmologia in Christopher Scheiner*, Catanzaro, Rubbettino Editore, 2005, pp. 235-238.

pernicanism, he also defended his apology of Copernicanism: from the *Nuncius* to the 1616 decree he strived to show a possible concurrence between the underpinnings of his philosophy and several assumptions from the Scholastic tradition – including some defended by Thomas Aquinas, whose teachings he had noted in his so-called *Iuvenilia*.

The first two versions of a paragraph in the *Sunspot Letters* (the third draft being almost identical to the final text)[73] help to shed further light on Galileo's rhetorical and theoretical strategy. The first version, which was later withdrawn due to censorship,[74] claimed:

> Now who will it be, who having seen, observed, and considered these things, will wish to persist longer in **an opinion that is not only false but erroneous and repugnant to the undoubted truths of the sacred letters**? which tell us that the heavens and all the world not only are generable and corruptible, but generated and dissolvable and transitory. See the divine goodness, which in order to draw us out of **such a great fallacy**, inspires in someone **necessary methods**.

In the foreground there is Galileo's aim to take on the role of upholder of the true theological-cosmological order of the universe: enlightened by divine goodness – i.e., speaking out thanks to the light of a natural reason that is not clouded by the senses, but is instead illuminated (albeit indirectly) by the Holy Spirit – he defines the theory of the incorruptibility of the heavens as «an opinion that is not only false but erroneous and repugnant to the undoubted truths of the sacred letters». In doing so, Galileo borrows the definition that had arisen from the debate on Copernicanism: by evoking the judgment on heliocentrism delivered by Juan de Pineda and quoted by Delle Colombe as well, Galileo aims to answer the latter by rebutting the accusation of rashness. More specifically, if – as the theologians declare – «something against the universal sentence of the Fathers, may be said to be temerarious» and if «a big mistake in philosophy is suspect to theology»,[75] the stubbornness of the upholders of incorruptibility who 'persist' in holding a wrong opinion is all the more

---

73 Galileo, *Istoria e dimostrazioni intorno alle macchie solari*, Biblioteca Nazionale dei Lincei, cod. Volpicelliano B, chart 61, already published in OG, V, pp. 138-139. Engl. trans. by Mayer, *The Censoring* (cit. note 69), pp. 5-6 (with some changes; bold is mine). On these passages, see Paolo Rossi, *Galileo Galilei e il libro dei salmi*, «Rivista di filosofia», 69, 1978, pp. 45-71: 47-49. See also Pesce, *L'ermeneutica biblica di Galileo e le due strade della teologia cristiana*, Rome, Edizioni di Storia e Letteratura, 2005, pp. 24-26.

74 In the letter of 10 November 1612 (OG, XI, pp. 428-429) Cesi informed Galileo of the censorship of the passages claiming that the incorruptibility of the heavens went against the Sacred Scriptures.

75 Delle Colombe, *Contro il moto* (cit. note 21), p. 290.

deplorable, given their refusal to follow the natural light that God has granted to man.

Galileo does not pretend to speak in the guise of a theologian – in point of fact, Lodovico delle Colombe was not a theologian either, but used the same definition – and does not venture into the arena of revelation: his conclusions are reached by resorting to the 'necessary methods' that men possess thanks to God's will and to the natural light which prevents the senses from making mistakes. The reading that Galileo suggests here with regard to the major role played by human reason appears to be brought back within the framework of Thomistic tradition.[76] As is clarified in the *Summa Theologiae*, divine illumination acts both through the ability given to man to access the intelligible, and through the survey that he could conduct using natural light. If the former might be the source of the well-known statement in Galileo's *Dialogue* concerning 'intensive knowledge' – according to which man can know several mathematical propositions with the same certainty as God[77] – the latter turns out to be very significant in Galileo's apology for his studies of nature. As Thomas Aquinas had explained, «Every truth is infused into anyone by the Holy Spirit who moves man to have intellectual understanding and to utter the truth. [...] Hence the natural light that is poured into the soul is God's enlightenment, whereby we are enlightened in order to know what pertains to natural cognition».[78]

The very same reference to God's illumination through man's natural reason is mentioned in the second version of this same passage in the *Sunspot Letters*, which underwent modification, including a long and telling addition:

> Now who will it be having seen, observed and considered these things, who will not embrace [...] the opinion so conformed to the undoubted truths of the sacred letters, which in so many places very openly and manifestly point out the unstable and fallen nature of celestial matter? *not, however, meanwhile despoiling of their deserved praise those sublime intellects which* ***with subtle speculations knew how to make the apparent discords of the physical discourses agree with sacred dogmas.*** *Now there is good reason that, the supreme theological authority also removed, they concede to the natural reason of other most serious authors and, more, to sense experience, to which I do not doubt that Aristotle himself should have approved [...]: whence if he argued the immutability of the heavens from there not being seen any sensible alteration in them*

[76] Such aim is taken to extremes and analyzed in depth by Campanella in his *Apology*: see Luca Bianchi, *La "Apologia pro Galileo" e la condanna di Tommaso d'Aquino*, «Intersezioni», 2, 1982, pp. 179-190.

[77] Galileo, *Dialogo* (cit. note 42), pp. 128-129.

[78] Thomas Aquinas, *Summa Theologiae*, I-II, qu. 109, art. 1.

over time, it is easily believable that when sense would have shown him what to us it makes manifest, he would have followed the contrary opinion, to which with such marvelous discoveries we are called. Here the divine goodness, which in order to remove from the mind **any ambiguity**, inspires in someone the necessary methods.[79]

In the first part of the text Galileo only mentions the corruptibility of the heavens, separating it from the debate on the eternity of the world – just as Cardinal Conti suggested that he do. The text that was added later can be broken down into two parts. The first section acknowledges – with caution – the worth of the great minds who looked for and found an agreement in the complicated debates on the physical structure of the universe and religious dogmata. Galileo then underscores the *accomodatio* practiced by the Church, whose aim is to silence philosophical discussion by attempting to identify or to work out an interpretative model of universe that is 'adapted' to the Sacred Scripture. This approach was so pervasive that even Lodovico delle Colombe, in his 1608 *Risposte piacevoli e curiose*, criticized those theologians who required philosophical doctrines to be 'accommodated' to Scriptural passages: «[...] it seems appropriate that we Catholics should have to accommodate philosophical subjects to Holy Scripture».[80] Such an endeavor is regarded as futile by Galileo, as Copernicanism corresponds to the meaning of Scriptures and does not need to be accommodated. In the second addition – which would be kept in the final edition – Galileo indeed outlines a feasible convergence of the views of the Peripatetics and the Catholic interlocutors on Copernicanism: the theological authority is 'removed' since the Fathers and Doctors of the Church agree on the corruptibility of the heavens; Aristotle as well would have supported this view if he had had the opportunity to study the sky through a telescope.

Compared to the second draft, the final sentence in the edited version is almost untouched, even though the syntagm «divine goodness» disappears and the preceding paragraph emphasizes the role played by «necessary methods» within the philosophical survey, thus contrasting them with «probable and apparent reasons»: «[...] we will philosophize better by giving our consent to conclusions that depend on manifest observations, rather than by persisting in opinions that are repugnant to sense itself, and only confirmed with probable and apparent reasons. [...] Here, by a higher

---

79 Galileo, *Istoria e dimostrazioni* (cit. note 5), pp. 138-139; Engl. trans. by Mayer, *The Censoring* (cit. note 69), p. 6 (with some changes; bold and italics are mine).

80 Delle Colombe, *Risposte piacevoli e curiose* (cit. note 19), p. 92r, also quoted in Guerrini, *Galileo e la polemica anticopernicana* (cit. note 65), p. 21.

virtue, in order to remove any ambiguity, necessary methods are inspired to someone».[81]

Indeed, it is always God who inspires the 'necessary methods', which in the case of Galileo were his improvements to the telescope and the *sensate esperienze* reported in the *Sidereus Nuncius* and in the *Sunspot Letters*. This reference also clarifies other statements to be found in Galileo's correspondence (in the so-called 'Copernican letters', as well as those written subsequently to the 1633 trial) and in the *Nuncius* with regard to the profound religious feeling that guided his observations: his «very religious and very devout mind»[82] led him to such discoveries as benefited from the enlightenment of divine grace. God did not reveal the message contained in the book of nature by means of the transcendent illumination of *lumen gloriae* – which enables the blessed to see the divine essence – nor by impressing the model of creation on the human mind, which would in this case be possessed innately by everybody.[83] God instead inspires only in someone the 'necessary methods' which enable the truth to be unveiled. Such statements did not come from an author with strong religious beliefs or a deep-rooted faith, nor from a priest-astronomer (as Kepler's model required): they were the outcome of a rhetorical strategy in which Galileo argued that his astronomical observations did not threaten the divine kingdom nor was it a blasphemous activity, but rather was a research carried out in obedience to God.

In advocating a theoretical viewpoint closely resembling that of his opponents, Galileo adapted Scholastic thought to contingent needs. He would do so again several years later: in the letter to Dini he presented himself as a champion of fortitude.

> It is not, therefore, my thought to embark on an undertaking so much greater than my force; although we must not doubt that the divine Goodness sometimes deigns to inspire some rays of his immense wisdom in humble intellects, and especially when they are at least adorned by sincere and holy fervor; and besides, when we have to accord sacred places with new and uncommon natural doctrines, it is necessary to have knowledge of such doctrines, as we cannot tune two chords together by listening to just one.[84]

---

81 Galileo, *Istoria e dimostrazioni* (cit. note 5), pp. 139-140.

82 Galileo, *Sidereus Nuncius* (cit. note 40), p. 56. See also Galileo to Peiresc, 22 February 1635, OG, XVI, p. 216. See also Galileo, *Frammenti al Discorso sui galleggianti*, OG, IV, p. 52.

83 The balanced relation between mathematical demonstrations and sensible knowledge is for instance clearly illustrated in Galileo's formulation of the principle of simplicity nature usually follows: Galileo, *Discorsi*, OG, VIII, p. 197.

84 Galileo to Dini, 3 March 1615, OG, XII, p. 300. Galileo owned a copy of a com-

When the debates that led to the 1616 decree were well under way, Galileo tried to defend himself from the charges of rashness: his undertaking did not go beyond the limits of *fortitudo*, nor did it come into conflict with the epistemological model widely accepted by Catholic theologians.

Galileo seemed confident that his rhetoric and epistemological defence would be successful. However, his efforts failed to stem the growing tide of dissent against him, and his measures turned out to be weak in the end. In the eyes of his contemporaries, it was evident that Galileo's natural philosophy did not bend to the role of the 'servant' (*ancilla*) of theology and therefore it maintained all the features of a reckless attitude. Piero Guicciardini's opinion is emblematic; he expressed his concern for Galileo's lack of moderation and his temerity, hinting at the ethical model of *fortitudo*: «He [Galileo] gets excited about his opinions, he has a great passion inside, and little fortitude and prudence to overcome it: so that he makes this sky of Rome very dangerous, utmostly in this century».[85]

5. Conclusion

As was foreseeable, following the anti-Copernican decree Galileo backtracked. This about-face is even more striking if the above-mentioned statements are compared with his 1624 reply[86] to Francesco Ingoli – who in the meantime had been appointed secretary of the Sacred Congregation of *Propaganda Fide*. In this letter Galileo passed over Ingoli's theological objections:[87] he claimed that he had abandoned any discussion of theological matters – among them the argument of the centrality of hell and Christ's descent and ascent, which Ingoli had given much space to – in order to focus solely on topics that were susceptible to 'confutation', namely to being

mentary to Aristotle's *Ethics* written by Lelio Pellegrini, a professor of moral philosophy in Rome – which was also in Federico Cesi's library (see Maria Teresa Biagetti, *La biblioteca di Federico Cesi*, Rome, Bulzoni 2008, p. 185): Lelio Pellegrini, *De moribus libri decem iis, qui Aristotelis ad Nicomachum inscribuntur, ordine perpetuo, atque sententia respondentes*, Romae, apud Aloysium Zannettum, 1600.

85 Piero Guicciardini to Cosimo II, 4 March 1616, OG, XII, p. 242.

86 On this letter, see Bucciantini, *Contro Galileo: alle origini dell'affaire*, Firenze, Olschki, 1995, pp. 149-168.

87 Even Guiducci suggested surrendering theological discussions: Mario Guiducci to Galileo, 21 June 1624, OG, XIII, p. 186.

examined on the basis of arguments concerned with the truth or falseness of statements.

Moreover, Galileo outlined a model of fortitude that was very different from the one described in his letter to Dini and adopted in the *Sunspot Letters* and Copernican letters. In this model he presented Copernicanism as a hypothesis, in addition to removing any defiance toward the Catholic cosmos and ruling out any attempt to free natural philosophy from its ancillary role.[88]

The idea of fortitude that surfaced here culminated in a model of religious zeal which was antithetical to the one he had embraced ten years earlier: in the main it distinguished the judicious attitude that Catholic natural philosophers were expected to maintain from the unwise behavior of heretical astronomers.

> I intend to deal with this topic rather extensively, and show them [i.e. heretics] that we Catholics hold on to the ancient certitude of the sacred authors not for lack of natural discourse, nor for having not seen as many reasons, experiences, observations and demonstrations as they saw, but because of the reverence we have for the Scriptures of our Fathers and because of the zeal of religion and of our faith.[89]

Notwithstanding his dissimulation, Galileo did not refrain from using the weapon of irony: it should not concern a true Catholic if «a heretic laughs at him because he places reverence and faith for the sacred authors before the many reasons and experiences that all astronomers and philosophers together have».[90] In other words, Galileo tacitly noted the inconsistency of the Catholic Church's demand: he claimed that the Church required philosophers to sacrifice the natural light of reason (which was a gift from God) on the altar of its Fathers' authority, and a good Catholic was expected to conform to this, even when he was aware that such a request was unfounded.

Galileo's caution therefore provides an insight into how dangerous the rashness of Copernican astronomers turned out to be for the 'Catholic Republic'. The temerarious ascent of the earth from being the dark prison

---

88 Galileo, *Risposta a Ingoli* (1624), OG, VI, p. 510: «[...] io non intraprendo questa impresa per pensiero o disegno ch'io abbia di sollevare e sostener per vera quella proposizione che già è stata dichiarata per sospetta e repugnante a quella dottrina la quale di maestà e autorità è superiore alle naturali e astronomiche discipline».

89 *Ibid.*, p. 511.

90 *Ibid.*

for the damned to becoming a bright planet, its outrageous approach to the Empyrean, as well as the temerity of those who intended to read the message written in heaven directly through a 'temerarious glass'[91] were actually important tiles in the mosaic of events that came together in the condemnation of Copernicanism.

[91] See note 12.

Franco Motta

NATURE, FAITH AND THE JUDGE OF FAITH
Some Considerations on the Historical-Political Context of Copernicus' Condemnation

1. The denunciation of Galileo and the first examination of Copernicanism by the Inquisition

The dossier against Galileo that reached the desk of the Prefect of the Index, Cardinal Sfondrati, in February 1615 contains several interesting charges. According to it, some «Galileists, who maintain that the Earth moves and the Sun stands still», are active in Florence; they have drafted a 'scripture' where it can be read that the Bible presents some unfit (*inconvenienti*) expressions, and that the Holy Writ should be the last text to be taken into consideration on questions of natural philosophy since it matters only in questions of faith and its exegetes often fall into error.[1] The author of the dossier, the Dominican friar Niccolò Lorini of Florence, enclosed a copy of this incriminating document with the «suspect or hazardous» sentences duly underlined. As we know, the 'scripture' or text in question was the letter of Galileo to Benedetto Castelli and the sentences brought to the attention of the censors all pertain to issues of biblical hermeneutics.[2]

---

1 Niccolò Lorini to Paolo Sfondrati, February 1615, in OG, XIX, pp. 298-299. For a deeper analysis of the inquiry that led to the condemnation of Copernicus in 1616 see, among the most recent studies, John L. Heilbron, *Galileo*, Oxford, Oxford University Press, 2010, pp. 200 sq.; Alfredo Damanti, *"Libertas philosophandi". Teologia e filosofia nella* Lettera alla granduchessa Cristina di Lorena *di Galileo Galilei*, Rome, Edizioni di storia e letteratura, 2010, pp. 49 sq.; Annibale Fantoli, *Galileo e la Chiesa. Una controversia ancora aperta*, Rome, Carocci, 2010, pp. 101 sq.; Michele Camerota, *Galileo Galilei e la cultura scientifica nell'età della Controriforma*, Rome, Salerno, 2004, pp. 260 sq.

2 *Processo di Galileo*, OG, XIX, pp. 299-304. The copy of the letter to Castelli delivered by Lorini to the Index diverges in some important passages from the more widespread version; according to Mauro Pesce, this is the effect of the rewriting of it by Galileo himself (Mauro Pesce, *L'ermeneutica biblica di Galileo e le due strade della teologia cristiana*, Rome, Edizioni di storia e letteratura, 2005, pp. 29 sq.).

Father Tommaso Caccini, who was the first to denounce Galileo, from the pulpit of Santa Maria Novella, confirmed these charges in a testimony released on March 20 at the Holy Office. According to him, it is «public knowledge» (the expression is borrowed from the terminology of criminal law of the time) that Galileo believes in the motion of the Earth and the immobility of the Sun. But his deposition cited further elements: that Galileo «is considered a good Catholic by many, but is suspect in matters of faith to others, since it is said that he is very close to that friar Paolo the Servite, so famous in Venice for his impieties»; that he has already attracted the attention of the Inquisition; and that he is the member of an «academy» which maintains contacts with «others, in Germany». Furthermore, his pupils would affirm that «God is in no way substance, but rather accident», that He is in some way provided with senses, and that «the miracles which are said to be performed by saints are not real miracles».[3]

Therefore, the picture taking shape before the eyes of Michelangelo Seghizzi, commissioner to the Holy Office, presented decidedly suspect features: Galileo was a close acquaintance of the Venetian intellectual Paolo Sarpi – for ten years among the most insidious enemies of the Apostolic See – and formed part of a cenacle with ties to Germany on various points of shared interest (in Rome «the matters of Germany» usually referred to anything relating to Protestantism). Assertions were attributed to Galileo's students that were apparently contrary to scholastic metaphysics and the cult of saints.

The investigation, which would stretch out over the following months, was assigned to Cardinal Millini, secretary to the Holy Office, since this congregation had exclusive competence in matters of faith. In his testimony Caccini cited Father Ferdinando Ximenes as the expert who had collected the compromising phrases *de auditu*, i.e. he had actually heard them spoken. Ximenes, when questioned on this, confirmed in large part Caccini's statements (except those regarding the cult of saints), mentioning in turn a presumed student of Galileo, Giannozzo Attavanti. When questioned a day later, Attavanti rejected the accusation *in toto*, attributing the charges to a misunderstanding on the part of Ximenes of their conversation.[4]

The case apparently finishes here. The friendship between Galileo and Sarpi; the suspicions that were circulating with regard to Galileo's position on «matters of faith»; the declarations ascribed to the «Galileists» concerning the substance and the senses of God or the origins of miracles – none

---

[3] *Processo di Galileo*, OG, XIX, pp. 307-311.

[4] *Ibid.*, pp. 316-320.

of this seemed to have had any consequences. Benedetto Castelli, the original recipient of the letter that Lorini had passed to the Holy Office, was not even interrogated. He merely received a paternal admonition from the Archbishop of Pisa, Monsignor Bonciani, who on March 12 warned him to abandon the «peculiar opinions» of Copernicus, since «beyond being a pack of nonsense, they were dangerous, scandalous, and reckless [*pericolose, scandalose e temerarie*], being in outright opposition to the Sacred Scriptures».[5]

The adjectives 'dangerous', 'scandalous' and 'reckless' were technical terms drawn directly from the terminology used by the qualificators and consultants to the Inquisition when defining the degree of deviation of a proposition from orthodox doctrine. This meant that even before the testimony of Caccini issued on March 20, based on the denunciation of Lorini alone, heliocentrism had already raised the suspicions of the authorities responsible for the appraisal and censoring of ideas. Indeed, during a plenary session of the Holy Office held on March 19, Pope Paul V ordered Caccini to be interrogated regarding the «errors of the aforesaid Galileo», thus branding them as such even before any theological examination had been conducted.[6]

One year later, notwithstanding a written denunciation and two testimonies focusing not on Copernicanism itself, but rather on its implications for the interpretation of the Bible, on the hierarchy between philosophy and theology in the explanation of natural phenomena, and on other theses concerning theological and metaphysical questions, the decision of the Holy Office was to submit to the censors two simple sentences limited to the strictly astronomical aspect of Copernicus' astronomy and leaving aside any philosophical implications: «That the Sun is the center of the world, and consequently motionless in its place [*immobile di moto locale*]. That the Earth is neither the center of the world nor motionless, but moves around itself, also with a daily motion».[7] Such was the text submitted to the consultants on 19 February 1616, exactly one year after Lorini's denunciation – two sentences drawn almost word for word from the testimony of Caccini («The Earth moves around itself, also with daily motion; the Sun

---

5 Benedetto Castelli to Galileo, 12 March 1615, OG, XII, pp. 153-154. It must be remarked that, from Bellarmine's correspondence preserved in the Jesuit archives in Rome (see below, notes 12 and 21), Monsignor Bonciani appears to have close and stable relationships with the Cardinal.

6 *Processo di Galileo*, OG, XIX, p. 276.

7 *Ibid.*, p. 320.

is motionless»), as if no one had taken it upon himself to actually read Copernicus' contested book.

In other words, the Inquisitors seemed to be suffering from a certain lack of zeal, which is quite surprising, especially since in the middle of February 1615 another anonymous consultant to the Holy Office had been given a copy of Galileo's letter to Castelli, and he pointed out three further dubious assertions dealing with the question of the truth of Scripture, Galileo having used unwary expressions such as «false propositions», «abstain» and «pervert» in relation to the actual text of the Bible. In the consultant's report these assertions were qualified as «*male sonans*» ('evil sounding').[8]

Now the use of the term *male sonans* when referring to a proposition under theological evaluation is not to be taken as a vague or generic accusation. It belonged to the legal lexicon of the Inquisition, which defined to a nicety the different possible degrees of deviation from Catholic truth: 'evil sounding' denoted the fourth degree of error after those of 'heretical', 'erroneous' and 'tasting of heresy' (*sapiens haeresim*).

We can find a detailed discussion of this in an important work that laid out the methodology for the Counter-Reformation scholastic theology, the *Loci theologici* (*Theological Commonplaces*) (1563) by the Salamancan theologian Melchor Cano, in particular in book 12, "De locorum usu in scholastica disputatione" ("On using commonplaces in a scholastic controversy"). Cano's tome was not intended as a handbook for inquisitors, unlike frequently cited texts such as the 14th century Inquisitor General of Aragon Nicholas Eymerich's *Directorium Inquisitorum* (republished by Francisco Peña in 1578) or Eliseo Masini's *Sacro Arsenale, ovvero Pratica dell'Uffizio della Santa Inquisizione* (1621), which provided generations of inquisitors with the rules for trials in cases of faith. Cano's treatise was rather intended to train theologians in the correct use of the sources of faith and in the construction of a well-established theological discourse. In this sense, it can be considered a broadly shared set of standards on Catholic theology.[9]

---

[8] *Ibid.*, p. 305. The first two censored assertions can be found already underlined by Lorini in the copy of the letter transmitted to the Index, while the third is autonomously identified by the anonymous consultant. If not otherwise specified, all the original texts are translated by myself.

[9] The importance of Cano in the Second scholasticism is stressed by MELQUIADES ANDRÉS, *El método teológico en la época moderna*, in *El método en teología*, Valencia, Facultad de Teología San Vicente Ferrer, 1981, pp. 201-233, 223 sq., and IGNACIO JERICÓ BERMEJO, Regula fidei et veritas fidei catholica. *El artículo de fe según Melchor Cano*, «Scriptorium victoriense», 30, 1983, pp. 277-315. Cano's consideration of the doctrinal notes is analysed by BRUNO NEVEU, *L'erreur et son juge. Remarques sur les censures doctrinales à l'époque moderne*, Naples, Bibliopolis, 1993, pp. 272 sq.

In a technical sense (*specialiter*), according to Cano «We properly term evil sounding, and offensive to pious ears those propositions where not a manifest error against faith, but rather something out-of-tune and absurd can be noted, appearing shameful to pious and religious ears».[10]

Obviously, ascribing to a suspect sentence this lesser and more subtle degree of incorrectness could be a delicate task and Cano reminded his readers: «It is foolhardy insolence to regard as a crime what the Church has never considered as such». Declaring that the Blessed Virgin was tainted with the original sin of Adam would sound bad «to the ears of the people», even if theologically it was a fully legitimate sentence (the Dominican order, to which Cano belonged, was rigidly hostile to the doctrine of the Immaculate Conception), and the same would be found even for some passages in the Gospel, «should we let ourselves be led by the mob». So we may properly deem as *male sonantes* those sentences that «although not tasting of heresy, nevertheless show an absurd and extraneous sound, disagreeing with the sound teaching and the authentic and well-founded way of speaking of the Church».[11]

Cano added some examples of evil-sounding statements: for example, the declaration that the emperor Constantine and Pope Sylvester had made a mistake when they endowed the Church with temporal possessions, or that the Roman Church is the synagogue of Satan, and so on. It can be seen that these assertions were not strictly heretical since they had never been formally condemned by a council, but they were nonetheless openly opposite to tradition. This was the degree of incorrectness that the anonymous consultant to the Holy Office attributed to three passages in Galileo's letter to Castelli.

There is no doubt that an assessment of this kind must have created some alarm among the cardinals of the Congregation of the Inquisition (indeed, during this period revised copies of Galileo's letter were being circulated on the scientist's behalf in power circles in Rome by his friend, the prelate Monsignor Piero Dini of Florence, reaching the cardinals Barberini, Del Monte and Bellarmine). Furthermore, at the beginning of March 1615 Foscarini published his *Lettera sopra l'opinione de' Pittagorici*, which was entirely devoted to establishing an unprecedented theological concordance between Copernicanism and the Bible. And no later than early June the manuscript letter written by Galileo to the Grand Duchess Christina was

---

10 Melchor Cano, *Locorum theologicorum libri XII*, here in the 1734 edition, Patavii, typis Seminarii, XII, 7, p. 391.

11 *Ibid.*, p. 392.

beginning to circulate; in it the scientist expanded on some of the considerations that he had touched upon in his letter to Castelli, advancing his new hermeneutical proposal of a separation between the scientific and the religious spheres and the pre-eminence of the experimental method in explaining natural phenomena.[12]

2. The crisis of the Sacred College and the "solitude" of Cardinal Bellarmine

To what I've summed up so far it must be added that in 1615 the whole matter still revolved around a "missing guest" – that is, Nicolaus Copernicus, whose book *De revolutionibus orbium coelestium*, which he dedicated to Pope Paul III, had appeared in print more than seventy years before, remaining since then a rare volume (it was reprinted only once, in Basel, in 1566), and circulating only among professional astronomers.[13] On the other hand, Galileo's two recent booklets showing strong evidence of the tenability of the Copernican system, the *Sidereus nuncius* (1610) and the *Istoria e dimostrazioni intorno alle macchie solari* (1613), had gained a Europe-wide reputation, and the author was even praised in 1611 for his heavenly discoveries by the renowned Jesuit astronomers of the Roman College.

We must also keep in mind that the Copernican question remained a marginal problem for the Roman Curia when compared with the much larger issues which were at stake at the time, involving the fundamental relationship between temporal and spiritual power and the struggle for hegemony between the Apostolic See with its historic prerogatives and the growing claims of secular rulers. In this very context, however, interesting analogies can be found between the debate over heliocentrism and the theological-political issues of the period – analogies that, as I will attempt to show, should be taken into consideration in assessing the reasons that led the Congregation of the Index (*Congregatio pro Indice Librorum Prohibitorum*, created in 1571 by Pius V) to issue their prohibition of Copernicanism *ut thesis* ('as a thesis') in 1616.

As we know, the responsibility for resolving the question raised by Lorini and Caccini and countered by the alternative proposals of Foscarini

---

12 That Bellarmine must have known Galileo's letter to Christina is proved by the Cardinal's autograph note in Archivum Romanum Societatis Iesu, *Opp. NN. 245*, c. 43 (see Damanti, *"Libertas philosophandi"*, cit. note 1, pp. 365-366).

13 The rareness of the *De revolutionibus orbium coelestium*, at the time, is remarked by Michel-Pierre Lerner, *L'hérésie heliocentrique: du soupçon à la condamnation*, in *Sciences et religions de Copernic à Galilée (1540-1610)*, Rome, École Française de Rome, 1999, pp. 69-91, 75.

and Galileo fell to the Jesuit cardinal and theologian Robert Bellarmine. A trusted advisor to the papacy, he served as the unofficial mediator on the case for the Holy Office and undoubtedly for Paul V himself. On such a complex issue Bellarmine probably seemed the natural candidate; he was universally esteemed for his learning and had earned a reputation for diplomatic tact after serving in the summer of 1591 on the special commission established to correct the many errors introduced by Pope Sixtus V in the revised edition of the Vulgate – an embarrassing incident that had shown the Sacred College how delicate questions touching on the letter of the Bible could be.[14]

Bellarmine accompanied, so to speak, from start to finish what has been called the 'first trial' of Galileo. As we know, in April 1611 he had requested the opinion of the mathematicians of the Roman College on the veracity of the new astronomical discoveries: notwithstanding the full confirmation which he received, his initial suspicion regarding the scientist, as far as we know, dates from that circumstance.[15] In a conversation with Prince Federico Cesi (founder of the Accademia dei Lincei) presumably dating to early January 1615 and therefore preceding Lorini's denunciation, he declared, in Cesi's words, that: «As for Copernicus' opinion [...] he holds it to be heretical, and that the motion of the Earth, without any doubt, is against Scripture».[16] He was the advisor whom Paul V summoned to a meeting on 24 February 1616 following a harsh confrontation with Cardinal Alessandro Orsini, at the end of which the pope concluded that «Galileo's opinion» must be considered «erroneous and heretical».[17] Finally, Bellarmine is the one who notifies the scientist on February 26 of the report handed down two days before by the consultants to the Holy Office and

[14] The incident is briefly retraced in Franco Motta, *Bellarmino. Una teologia politica della Controriforma*, Brescia, Morcelliana, 2005, pp. 586 sq.

[15] «Among the others, Bellarmine told me [...] that if [Galileo] would have stayed here too long, they couldn't do without asking him some justifications of his case»: Piero Guicciardini to Curzio Picchena, 5 December, 1615, recalling Galileo's visit to Rome four years earlier, OG, XII, pp. 206-207; Bellarmine's letters to the Fathers of the Roman College and their response, OG, XI, pp. 87-88. On the cardinal's views in astronomical matters see Ugo Baldini, *L'astronomia del cardinale* and *Bellarmino tra vecchia e nuova scienza*, in Id., Legem impone subactis. *Studi su filosofia e scienza dei gesuiti in Italia, 1540-1632*, Rome, Bulzoni, 1992, pp. 285-303, 305-344. Baldini's studies have raised some critics, centered not on the depth of his inquiry but rather on his interpretive approach, emphasizing the 'modernity' of the perspective of Bellarmine and, more generally, of the Jesuits about the relation between faith and science: see, for instance, Maurizio Torrini, *Da Galileo a Kircher: percorsi della scienza gesuitica*, «Galilæana», 2, 2005, pp. 3-17.

[16] Cesi to Galileo, 12 January 1615, OG, XII, pp. 128-130, 129.

[17] Piero Guicciardini to Cosimo II, 4 March 1616, OG, XII, pp. 241-243, 242.

implicitly approved in the plenary session of the *Congregazione dell'Indice* the day before – for which we have no minutes – whereby Copernicanism is condemned as erroneous and heretical.[18]

Based on the hints in letters received by Galileo from Tuscans living in Rome and members of the Florentine court (all supporters of the scientist, but each of whom had his own views on his chances: Monsignor Dini was optimistic, while Prince Cesi was more cautious and the ambassador of the grand duke, Guicciardini, was impatient with Galileo's eagerness to stand up for his ideas), Cardinal Bellarmine's stance during the fourteenth months preceding the official censure of Copernicanism was somewhat vacillating. While, as we have seen, in January 1615 he declared «the opinion of Copernicus» and in February 1616 «the opinion of Galileo» to be heretical (although in the latter case he may have been referring to the Copernican cosmology combined with the hermeneutical theses contained in Galileo's letters to Castelli and Christina), in the interval between these two uncompromising statements he showed some willingness to engage in dialogue. On 7 March 1615 Monsignor Dini informed Galileo of a conversation that he had had with the cardinal: «As for Copernicus [...] he does not believe that it is going to be prohibited, and the worst that could happen is that some glosses will be added, explaining that his doctrine was introduced to save appearances, or something of the kind».[19] And in fact this would prove to be the substance of the decree issued one year later, not prohibiting but rather «suspending» the *De revolutionibus orbium coelestium* until it could be amended. At the same time Father Christoph Grienberger of the Roman College assured Dini that Bellarmine was well disposed toward the receipt of any further mathematical evidence of heliocentrism that Galileo might provide before approaching the biblical question.[20]

It must be added that Bellarmine's behavior was probably influenced by serious diplomatic considerations – on one side the deference due to the House of Medici of which he was a 'natural' subject, having been born in Montepulciano, and on the other his close dealings with Christina of Lor-

---

[18] *Processo a Galileo*, OG, XIX, pp. 321-322. As is well known, the genuineness of the Bellarmine's notification to Galileo has been strongly debated. The hypothesis that the document is a forgery drafted in Rome, during the 1633 trial, has been advanced through a well-researched analysis by Vittorio Frajese, *Il processo a Galileo Galilei. Il falso e la sua prova*, Brescia, Morcelliana, 2010. Much criticism about Frajese's hypothesis has then been expressed by Sergio Pagano, *Il precetto del cardinale Bellarmino a Galileo: un 'falso'?, con una parentesi sul radio, Madame Curie e i documenti galileiani*, «Galilæana», 7, 2010, pp. 143-203.

[19] Dini to Galileo, March 7, 1615, OG, XII, pp. 151-152.

[20] *Ibid.*, p. 151.

raine as an intermediary with the Florentine court – as he protected the interests both of his own family (Bellarmine presided over a large patronage network comprising his brothers, sisters, cousins and nephews) and the diocese of Montepulciano, for which he held the role of 'protector' in Rome.[21] It was a member of the House of Medici – Mary, the widow of Henry IV of France and cousin of the grand duke – who prevented the burning of Bellarmine's *De potestate Summi pontificis in rebus temporalibus* ('On the power of the pope in temporal matters') in November 1610 by suspending the sentence issued by the Parliament of Paris. The book had been condemned for damaging the monarchical prerogative and directly inspiring the theories on regicide that had materialized so menacingly a few months before with the assassination of Henri IV.

There is no clear evidence regarding the differing viewpoints held within the Sacred College, especially among the theologians of the Holy Office and the Index, with respect to Copernicanism. The cardinal of Aracoeli, Agostino Galamini – a zealous Dominican friar, and formerly inquisitor in Brescia, Genoa and Milan as well as Master of the Sacred Palace – appears to have been Caccini's contact and the promoter behind the scenes of the trial.[22] Felice Centini – cardinal of Ascoli, member of the Friars Minor Conventual, and a docile *creatura* of Pope Paul V with staunch pro-Spanish leanings – probably shared his position.[23] In contrast Alessandro Orsini, Francesco Maria del Monte, and above all Bonifacio Caetani and Maffeo Barberini assumed a different posture; they were more prudent, even overtly favorable to Galileo and to the legitimacy of a heliocentric exegesis of the Bible.[24]

Faced with such a constellation of opinions, Bellarmine appears to have adopted the position of the mediator, made possible by his personal authority and by his close proximity to the pope as his most influential advisor on theological matters. On the other hand, the fact that he was entrusted to

---

21 Bellarmine's correspondence with the Grand-Duchess Mother Christina, the Grand-Duke and the Grand-Duchess, stored in the Archivio di Stato of Florence and the Archivum Romanum Societatis Iesu, is also available in the typewritten copies drafted in the early 20th century by Fathers Xavier-Marie Le Bachelet and Sebastiaan Tromp (*Epistolae Bellarmini*), in the Archivio della Pontificia Università Gregoriana in Rome (Apug 1601-36).

22 Massimo Bucciantini, *Contro Galileo: alle origini dell'*affaire, Firenze, Olschki, 1995, pp. 38 sq.

23 On Galamini see *Dizionario biografico degli italiani*, vol. 51, pp. 325-326 (S. Rivabene); on Centini *ibid.*, vol. 23, pp. 593-597 (G. Benzoni).

24 On the opposition of Caetani and Barberini toward a condemnation of Copernicanism for the charge of heresy see the 1633 memo of Giovanfrancesco Buonamici, OG, XV, p. 111 and OG, XIX, pp. 407-411, mentioned by Camerota, *Galileo Galilei* (cit. note 1), p. 320.

handle Copernicus' case could be read as a symptom of the overall weakness of the Sacred College, and more specifically of the Congregations of the Index and of the Inquisition in the second decade of the 17th century.

Cardinal Bellarmine was born in 1542 (one year before the death of Copernicus and the publication of his work) and by 1615 he was an elderly man. In January of that year he published his celebrated *De ascensione mentis in Deum per scalas rerum creatarum* ('On the ascension of the mind to God through the ladder of created things'), which testifies to his allegorical vision of nature and hence his intellectual distance from any realist approach to natural philosophy. This book was also the first in a series of devotional pamphlets that would appear annually until 1620, as if in anticipation of the canonization process that the Society of Jesus would undertake after his death in 1621.[25]

In other words, despite his tireless dedication to the Roman congregations of which he was a member (the Holy Office and the Index, but also the Congregation of Rites which was responsible for the processes of beatification and canonization, matters where his theological expertise would have been invaluable), Bellarmine seemed to be preparing to retire from his institutional commitments and withdraw into a life of ascesis and spirituality. He had produced the magnum opus of his intellectual career, the *Disputationes de controversiis christianae fidei* ('Disputations about controversies on the Christian faith'), some thirty years before and his later contributions to the theological and political debates of the time – such as the *De potestate Summi pontificis in rebus temporalibus* or his confutations of the arguments of James I and the theologians of Venice against the supreme authority of the pope – were little more than a reworking of themes already dealt with in it.

In 1615 Bellarmine therefore was one of the last remaining protagonists in a crucial period in the Church's history that was coming to a close. He belonged to the generation of theologians and prelates trained in the 16th century and called upon to interpret and enforce the decrees of the Coun-

---

25 On Bellarmine's epistemology in the light of the *De ascensione mentis in Deum* see *ibid.*, pp. 289-290, and more recently FRANCESCO BARRECA, *Cosmologia ed ermeneutica biblica nel* De ascensione mentis in Deum *di Roberto Bellarmino*, «Galilæana», 10, 2013, pp. 119-136. A comparison between his allegorical concept and the emblematic approach to natural knowledge, which was peculiar to the Jesuits' Aristotelism – most famously represented by Athanasius Kircher –, has been first advanced by WILLIAM B. ASHWORTH, Jr., *Catholicism and Early Modern Science*, in DANIEL C. LINDBERG – RONALD L. NUMBERS (eds.), *God and Nature. Historical Essays on the Encounter between Christianity and Science*, Berkeley, University of California Press, 1986, pp. 136-166, and RIVKA FELDHAY – MICHAEL HEYD, *The Discourse of Pious Science*, «Science in Context», 3/1, 1989, pp. 102-142.

cil of Trent under the pontiffs Pius V, Gregory XIII and Sixtus V, thereby contributing to the rise of the Counter-Reformation.

During the pontificate of Paul V the Roman Curia was gradually depleted of the best of this generation of prelates, leading to the erosion of its spiritual and intellectual standing and to a situation of «crisis of the Counter-Reformation», as Delio Cantimori described it in his essay on the trial of Galileo.[26]

In my opinion the key to understanding the difficulty of Bellarmine's role in the episode of Galileo's 'first trial' lies here; that is, in the entire absence of any official interlocutor in the Congregations of the Index and the Holy Office, one capable of drawing up a strategy to cope with an emerging new order of knowledge that was gaining support across the Catholic world.

Here it will suffice to note a series of prosopographical facts. The *sommo inquisitore*, Giulio Antonio Santori, last heir to the uncompromising line adopted by Paul IV and Pius V, a man of the utmost severity who kept the reins of power at the Holy Office firmly in his hands for more than twenty years, died in 1602. He was followed the next year by one of his most influential opponents, Silvio Antoniano. The father of Catholic pedagogy and the most gifted Latinist in the Curia, Antoniano had pressed for a milder policy with regard to the censorship of books and therefore often found himself at loggerheads with the Congregation of the Inquisition. In 1606 Cardinal Agostino Valier passed away; a humanist and a 'reforming bishop', one of the founders of the Tridentine rhetorical style, and mediator between Rome and the Republic of Venice, he himself advocated a certain degree of tolerance (the partial lifting of the ban on the works of Erasmus' works began with him). The scholarly Caesar Baronius, prefect of the Vatican Library and author of the founding text of Catholic historiography, the *Annales ecclesiastici*, died in 1607, and Ascanio Colonna, one of the most learned scholars and manuscript collectors in Rome, in 1608.

Some even more influential cardinals had passed away at the end of the preceding century: the powerful Giovanni Ludovico Madruzzo (1600), a key figure in relations between Rome and the Hapsburg imperial court; Marcantonio Colonna (1597), a member (together with Bellarmine) of the special commission on the Sixtine Vulgate, Prefect of the Index, and himself favorable to a policy of expurgation rather than the outright prohibition of condemned books; and Francisco de Toledo (1596), the first

---

26 Delio Cantimori, *Galileo e la crisi della Controriforma*, in Cantimori, *Storici e storia*, Turin, Einaudi, 1971, pp. 657-674.

Jesuit to be appointed a cardinal, distinguished professor of philosophy and theology at the Roman College, and a key player in the strategy of rapprochement between the Apostolic See and the new French king, Henry of Navarre, the former leader of the Huguenots.[27]

Most of these prelates (apart from Santori) supported the policy of a partial opening up of the rights of the book market; this led to the Index of 1596, a measure decided by Pope Clement VIII to moderate the harshness of the previous Sixtine Index. Thus, formerly prohibited works were for a brief period allowed to circulate after undergoing a process of expurgation – books such as the Latin Bibles published in Protestant countries, and astrological works that had been condemned by Sixtus V in his *Coeli et terrae creator*.[28] This period of relative tolerance ended with the death of Clement VIII (1605), but the decision in favour of the expurgation of Copernicus' work rather than prohibiting it outright seems to echo this precedent.

Moreover, these cardinals – like other prominent members of the papal entourage such as Antonio Possevino, the author of the *Bibliotheca selecta* (d. 1611) – were historians, writers, scholars, theologians and humanists of outstanding intellectual stature, architects of the Tridentine cultural strategy that, by employing profane knowledge for apologetic ends, provided the Church with suitable tools to oppose the Reformation and create a new hegemony in Catholic countries.

However, this exceptional generation left no real successors and in 1615-1616 only Bellarmine and the celebrated Cardinal Federico Borromeo, founder of the Biblioteca Ambrosiana in Milan, remained. The scion of an influential family, Borromeo became a prelate of exemplary virtue and scholarship, but he was based in Milan, having been appointed archbishop there in 1595. These two cardinals found themselves having to deal with the Congregations of the Index and the Holy Office, which were now dominated by lower level officials with a purely bureaucratic approach to censorship.

---

[27] On Santoro, Antoniano, Valier, Baronio and Marcantonio Colonna, from the perspective of their role in the Roman congregations, see the entries in Adriano Prosperi (ed.), *Dizionario storico dell'Inquisizione*, Pisa, Edizioni della Normale, 2010, respectively III, pp. 1370-1377 (S. Ricci); I, pp. 70-71 (M.T. Fattori); III, pp. 1638-1639 (M.T. Fattori); I, pp. 138-139 (M.T. Fattori); I, p. 350 (P. Mazur). On Madruzzo and Ascanio Colonna see *Dizionario biografico degli italiani*, respectively vol. 67, pp. 181-186 (R. Becker); vol. 27, pp. 275-278 (F. Petrucci). On Toledo, see *Diccionario histórico de la Compañía de Jesús. Biográfico-temático*, Rome-Madrid, Institutum historicum SI-Universidad pontificia Comillas, IV, pp. 3807-3808 (J.P. Donnelly).

[28] Frajese, *Nascita dell'Indice. La censura ecclesiastica dal Rinascimento alla Controriforma*, Brescia, Morcelliana, 2006, pp. 151 sq.

It is not surprising that in the face of the unprecedented challenge posed to the conception of biblical truth by the implications of heliocentrism and heliocentric exegesis, the Jesuit cardinal Bellarmine found himself having to cope with a general sense of uncertainty widespread in the Roman Curia.

3. An extrajudicial solution: Bellarmine's letter to Foscarini

It is perhaps partly on these grounds that Bellarmine's response to Galileo and Foscarini took the shape of a private admonition, as if to avoid an official pronouncement from the Holy Office. Indeed, Bellarmine's solution was a political rather than a judicial one, conveyed in the well-known 'confidential letter' – which had in fact been conceived and written as a public letter – to Foscarini dated 12 April 1615.

Detailed analyses of this document can be found elsewhere in the ample historiographic literature on Galileo and Copernicanism.[29] It is sufficient here to remind the three main points under which – as was his custom – the cardinal developed his arguments. To begin, he posits that the merits of heliocentrism can be discussed in an equitable manner if the theory is treated as a hypothesis («*ex suppositione*, and not absolutely») rather than as an objective fact since it overtly contradicts Scripture. Second, a Copernican interpretation of Scripture does not accord with the consensus of the Church Fathers and later commentators, and as such it is contrary to the statements on doctrine issued by the Council of Trent. Third, the orbital motion of the Earth and the immobility of Sun has not yet been demonstrated by any physical evidence, and it is unlikely that it ever will be. Therefore, there is no reason to abandon the long-established empirical observation that the Sun moves in the third heaven.[30]

It is clear that Bellarmine chose to conclude that the notion of heliocentrism must be rejected based on disciplinary rather than theological or scientific grounds, based on the precept that the first obligation of believers is obedience to the decrees of the Church. He cited «the shared consensus of the Church Fathers» and «the modern commentaries on Genesis» without explicitly mentioning any authors; after all, even the most recent of these,

---

29 On the historiographical interpretations of Bellarmine's letter to Foscarini see Motta, *Epistemologie cardinalizie. Ipotesi, verità, apologia*, in Gian Mario Bravo – Vincenzo Ferrone (eds.), *Il processo a Galilei e la questione galileiana*, Rome, Edizioni di storia e letteratura, 2010, pp. 93-103.

30 OG, XII, pp. 171-172.

such as the influential Jesuit theologians Jean Lorin (Joannes Lorinus) and Nicolaus Serarius, had written their works before Galileo's discoveries.[31] Instead, he bowed to the authority of the Council of Trent, considered at the time to be the definitive repository of Catholic dogma.

As for Galileo's argument that astronomical matters could be excluded from the binding jurisdiction of the Bible because they did not pertain to faith, Bellarmine appealed to a strict biblical literalism when he ascribed the value of truth to Geocentrism using evidence presented *ex parte dicentis* ('according to who is speaking'). Yet he was perfectly aware that, according to the Tridentine decrees, *pars dicens* ('the part which is speaking'), i.e. the source of truth, lies not only in Scripture (as stated in the Reformed principle of the self-sufficiency of the Bible), but also in the continuity between Scripture and tradition guided by the Church's interpretation. Or, to cite the Council of Trent, «the meaning which the Holy Mother Church, which is entitled to pass judgment on the real sense and the interpretation of Scripture, has believed and currently believes».[32] Here it is in fact the possibility of a change in the Church's attitude towards the connection between biblical truth and natural philosophy that is under discussion, at least according to Galileo's proposal.

The position of Bellarmine in the *Controversiae* – as well as that of the entire body of Catholic controversialist theologians in the age of confessions going back to Luther's first opponent, Johannes Eck – was founded on the premise that «Scripture is not so clear in itself as to be sufficient, without further explanation, to resolve controversies of faith» – where such controversies also, and indeed primarily, include where to draw the dividing line between what pertains to faith and what does not.[33]

Within the doctrinal context of Catholic scholasticism, which was characterized by an objective rather than a subjective conception of faith – that

---

31 Special attention has been devoted by scholars to the Jesuits Juan de Pineda, *Commentariorum in Job libri tredecim*, 1598 and 1602, Jean Lorin, *Acta apostolorum commentaria*, 1605, Nicolaus Serarius, *Iosue, ab utero ad ipsum usque tumulum*, 1609, who rely on Clavius in ruling out the plausibility of a heliocentric exegesis of the Bible. VOLKER R. REMMERT, *"Sonne steh still über Gibeon"*, «Zeitschrift für historische Forschung», 28, 2001, pp. 539-580: 559 sq. See also IRVING A. KELTER, *The Refusal to Accommodate: Jesuit Exegetes and the Copernican System*, «Sixteenth Century Journal», 26/2, 1995, pp. 273-283.

32 *Concilium Tridentinum*, ed. Societas Goerresiana, V, *Actorum pars altera*, sess. IV, April 8, 1546, decr. II on the authenticity of the Vulgate, p. 92.

33 *Disputationes de controversiis christianae fidei adversus huius temporis haereticos*, here in the Venetian edition of Bellarmine's *Opera omnia*, apud Ioannem Malachinum, 1721-28, I, I controversy *De Verbo Dei*, b. 3 *De Verbi Dei interpretatione*, ch. 1, *Scripturam non esse tam apertam per se, ut sine explicatione sufficiat ad controversias fidei terminandas*, p. 64.

is, faith conceived as a complex of dogmatic articles rather than the moving of the individual conscience towards God – the importance to the religious sphere of a statement made in the profane (that is, merely secular or philosophical) sphere depended not so much on its content as on whether the Church considered it to be binding. In other words, it was not actually the subject of a proposition that decided whether it was to be considered *de fide*, rather that the *magisterium* or Church tradition had taken a position on the matter (or were thought to have, as in case of Geocentrism according to Bellarmine). What was *de fide* simply had to be believed in 'on faith', whereas what was not *de fide* could be left to the discretion of the faithful or – better – could be debated among the different theological schools.

Against this complex backdrop, several hotly debated issues with far more theological weight than the motion of the Earth – and with explicit religious significance today – were not considered to be *de fide* at the time: the doctrine of the Immaculate Conception; the question of the action of divine grace on the human disposition toward God (which lay at the heart of the controversy *De auxiliis* between the Dominicans and the Jesuits, and which Paul V – on the advice of Bellarmine – brought to an end in August 1607 by imposing the silence on the two parties); and one of the most controversial issues in the history of the early modern Church – the hierarchy of power between the papacy and the council: indeed, this last fuelled a subterranean conflict between Rome and the French episcopacy that rumbled on until the dogma of papal infallibility was declared in 1871.

In contrast, an implicit indulgence was perceptible in the Church's attitude toward the theses in the *De revolutionibus orbium coelestium*. Copernicus' book, with its dedicatory letter to Paul III, had circulated for decades without opposition, even during the papacies of the notoriously unbending Paul IV and Pius V. This detail did not escape the attentive eye of Paolo Sarpi; in the advisory opinion that he prepared for the Venetian Senate regarding the censorship of Copernicanism he wrote: «The suspension of the book cannot but have the effect of causing amazement at the unprecedented act of suspending an old book already seen by the entire world and never censored in the past, whether by the Council of Trent or in Rome».[34]

In other words, Bellarmine's letter to Foscarini advancing an extrajudicial compromise – «It seems to me that Y[our] P[aternity] and Mr. Galileo would act wisely if they content themselves with speaking *ex suppositione*,

---

[34] PAOLO SARPI, *Sopra un decreto della congregazione in Roma in stampa presentato per l'illustrissimo signor conte del Zaffo a 5 maggio 1616. 7 maggio 1616*, in SARPI, *Opere*, ed. by G. and L. Cozzi, Milan-Naples, Ricciardi, 1969, pp. 602-605, 603.

and not absolutely» – was hardly couched in the tones of an authoritative statement intended to settle a dangerously contentious issue. Indeed, it sounded weak and dilatory – «I will not believe that this evidence [of the motion of the Earth] exists, until it will be shown to me» –, betraying the absence of a clear position on the matter within the Roman Curia.

At least this is how it was perceived by its addressees. Benedetto Castelli, writing to Galileo on May 6, regarded it as a victory for their side. Piero Dini noted a few days earlier that it represented «a point now gained, that is, the possibility to address issues as a mathematician, hypothetically», and on May 16 urged his friend to «wind up that piece of writing [i.e. the Letter to Christina] that you told me you have drafted», adding that «as for the letter of the Carmelite Friar, Sir Prince [Cesi] tells me that soon it will be reprinted with the addition of further authorities, to have his interpretation made more clear».[35] Indeed, Foscarini was even then working on a «full and exhaustive treatise in the Latin language» provided, Dini writes, with «many passages from the holy Fathers, whereby he strengthens himself».[36]

The feeling in Rome was that the informal intervention of Bellarmine had brought the issue to an honorable conclusion; on April 18 Cardinal Barberini informed Dini that «I hear no more talk about the affairs of Mr. Galileo»; two days later an unnamed Jesuit Father also congratulated Dini that «Mr. Galileo's affairs have been sorted».[37] As we know, this perception would be proved wrong in the following autumn, as the inquiry was resumed after the interrogation of Ximenes and Attavanti. Galileo's departure for Rome «to account for himself concerning some charges, or rather calumnies» – as Cosimo II wrote in a letter recommending his mathematician to no less a personage than the Cardinal-nephew Scipione Borghese – marked the turning point of the whole case.[38]

4. The literal meaning of the Scriptures

Galileo's stay in Rome from December 1615 to June 1616 has been generally interpreted as a spontaneous move, implicitly falling in with the agreement that had been proposed by Bellarmine in the hope of persuad-

---

35 Castelli to Galileo, 6 May 1615, OG, XII, pp. 177-178; Dini to Galileo, 2 May and 16 May 1615, OG, XII, pp. 175-176, 181.

36 Cesi to Galileo, 20 June 1615, OG, XII, pp. 189-190.

37 Dini to Galileo, 18 April and 20 April 1615, OG, XII, pp. 173-175.

38 Cosimo II to Scipione Borghese, 2 December 1615, OG, XII, pp. 205-206.

ing the Church authorities of the veracity of the heliocentric model by employing «the tongue instead of the pen».[39]

Various clues suggest, however, that Galileo came to Rome at the specific request of the Holy Office rather than voluntarily. In his correspondence from that period he maintained the most rigorous discretion regarding the matters discussed at various meetings, as was the custom in inquisitorial proceedings; he wrote about his direct or indirect dealings with the cardinals of the Holy Office («those most distinguished personalities handling such matters»); he hints at discreet, albeit in-depth inquiries into his behavior and ideas («kind access and a hearing were granted me, and the freedom to reassure them as to my every act, saying, thought, opinion and teaching»).[40]

Again the most plausible hypothesis is that the tribunal on issues of faith was trying to find an extrajudicial solution to a case that had serious diplomatic implications, that had been reopened on the basis of the contradictory testimony of two witnesses, and that was further fuelled in Rome by the charges of certain «enemies», who can be identified as Caccini and, probably, the Dominican friars of the Church of the Minerva.

The concurrence in time between the interrogations of Father Ximenes and Galileo's student Attavanti and the initiation of the inquiry in Rome is confirmed by some veiled references by Galileo to «so many and such grievous slanders» that have been brought against him by his accusers, who have «plotted against my reputation», and regarding which, nonetheless, he has managed to justify himself. It seems, therefore, that we are dealing with charges that were not based on written testimony, and it is not difficult to identify in such charges the alleged theses regarding the «substance» and the «senses» of God that we find in the denunciations by Lorini and Caccini that led to the opening of the inquiry.[41]

However, in the overall context I believe that it is important to underscore two details. The first detail is that this informal inquiry represented the first time that the pope's inquisitors focused on both the theses of Galileo – or those ascribed to him – and Copernicanism itself as a cosmological theory:

---

[39] Galileo to Dini, May 1615, OG, XII, pp. 183-185. See Heilbron, *Galileo* (cit. note 1), pp. 215 sq., Damanti, *"Libertas philosophandi"* (cit. note 1), pp. 107 sq., Fantoli, *Galileo e la Chiesa* (cit. note 1), p. 104, Camerota, *Galileo Galilei* (cit. note 1), pp. 303-304. Bucciantini, *Contro Galileo* (cit. note 22), pp. 69 sq., on the contrary, is more inclined toward the hypothesis of a convocation of Galileo. See Querenghi to Alessandro d'Este, 1 January 1616, OG, XII, p. 220.

[40] See Galileo's letters to Picchena, especially the ones of 8 January and 6 February 1616, OG, XII, pp. 222-223, 230-232.

[41] Galileo to Picchena, 8 January 1616 (see above).

I say to you – Galileo wrote to Picchena on February 6 – that my affair has been wholly concluded as for the part concerning my own person; [...] But [...] a charge is annexed to my cause, not concerning my own person, rather the entire whole of those who have been or still are endorsing in printed works, or private writings, or in public demonstrations and lectures, or even in personal speeches a certain doctrine and opinion that is not unknown to Y[our] Most Illustrious L[ordship], the decision regarding which is presently being discussed.[42]

The *Discorso del flusso e reflusso del mare* ('Discourse over the flow and reflow of the sea') written by Galileo in Rome and dedicated to his patron Cardinal Orsini, is dated 8 January 1616. Despite its (laconic) profession of hypotheticism («Taking, therefore, *ex hypothesi* the motion of the Earth...»), the text is intended to prove the reality of the Copernican system through the evidence of the tides, so as to offer Bellarmine the «demonstration» of the Earth's motion that the latter, however skeptically, had referred to in his letter to Foscarini.[43]

The *Disputatio de situ et quiete terrae* ('Disputation on the place and immobility of the Earth') by Francesco Ingoli, a prelate in the household of Cardinal Caetani, dates to these same weeks. As is well known, Ingoli was soon to be appointed by the Congregation of the Index as the consultant charged with the task of emending Copernicus' text. The *Disputatio* is a short refutation of heliocentrism, the drafting of which was very likely entrusted to Ingoli by the Holy Office itself. It was founded upon *argumenta mathematica, physica* and *theologica*, and appears to be the first official statement of the Church's position with regard to Copernican cosmology (apart from the Dominican theologian Giovanni Maria Tolosani's *De coelo supremo immobili*, a work which was written, however, some seventy years earlier).[44] In other words, early in the year 1616 the Holy Office was beginning to take a more direct interest in the Copernican theory, in its physical and astronomical perspectives, and in its doctrinal consequences.

The second detail relates to the fact that Galileo came to his confidential interrogations armed with documents that he presented to his interlocutors for examination. «I thank God – he wrote to Picchena after the

42 *Ibid.*, p. 230.

43 OG, V, pp. 377-395: 381. The importance of this text is underscored by Damanti, *"Libertas philosophandi"* (cit. note 1), pp. 109-110, albeit stressing too much, in my opinion, Galileo's alleged hypotheticist approach.

44 Bucciantini, *Contro Galileo* (cit. note 22), pp. 84 sq. On Tolosani see Miguel Angel Granada, *Giovanni Maria Tolosani e la prima reazione romana di fronte al* De revolutionibus*: la critica di Copernico nell'opuscolo* De coelo et elementis, in Bucciantini – Torrini (eds.), *La diffusione del copernicanesimo in Italia 1543-1610*, Florence, Olschki, 1997, pp. 11-35.

conclusion of the affair – that what I said has been always proved by me with writings, copies of which are kept by myself». A few months later this fact was confirmed by Cardinal Del Monte, who assured the Grand Duke that «to his person [i.e. to Galileo] there is not the least flaw to be ascribed, and he will be able to answer for himself and to refute the calumnies of his persecutors, since all that he has been requested to provide is proven by writings».[45]

Among those «writings» was perhaps the *Discorso del flusso e reflusso del mare* even if, in my opinion, this hypothesis is not highly plausible. It is more likely that Galileo included the *Istoria e dimostrazioni intorno alle macchie solari* ('History and demonstrations concerning sunspots'), which the Holy Office had in fact ordered to be examined in the very period coinciding with Galileo's departure for Rome.[46] Almost certainly there would have been the letter to Castelli – which Monsignor Dini had been circulating on behalf of Galileo among certain cardinals of the Holy Office – and the letter to Christina of Lorraine.

This means that very probably between December and January, before beginning an evaluation of the Copernican system, the Holy Office carried out an informal inquiry on the suspect assertions ascribed to Galileo but then quickly clarified by him, and on the methodological principles laid out in his public letters: i.e., the Bible may sometimes speak inaccurately to settle the opinions of uncultivated people; its exegetes may be subject to error; it has binding authority only in matters concerning faith; and that therefore on questions of natural philosophy it is less trustworthy than mathematical and philosophical inferences.

As we know, these premises formed the foundation of Galileo's hermeneutical proposal. They were described and discussed as such in the written denunciation sent to Rome by Lorini, as already seen at the beginning of this paper, and in Bellarmine's letter to Foscarini. In other words they must be considered as the conceptual core around which the entire Galileo *affair* of 1615-1616 revolved.

Notwithstanding this, after the interrogation of Galileo behind closed doors, those assertions did not become the object of any formal censure, nor was he notified of any obligation to renounce them. Indeed the conclusion of the trial centered entirely upon the mathematical aspect of the

---

45 Galileo to Picchena, 26 March 1616, OG, XII, p. 251; Cardinal Del Monte to Cosimo II, 4 June 1616, OG, XII, p. 264.

46 *Processo a Galileo*, OG, XIX, p. 320: «Die 25 9mbris 1615. Videantur quaedam litterae Gallilei, editae Romae cum inscriptione *Delle macchie solari*».

heliocentric theory, which was judged to be false and contrary to Scripture. It was on this basis that Foscarini's *Lettera sopra l'opinione de' Pittagorici* was prohibited and the commentary on the Book of Job by Diego de Zuñiga was sentenced to expurgation, since they now figured as attempts to harmonize the letter of the Bible with a condemned thesis.

In fact, the hermeneutical theses advanced by Galileo contained nothing that was intrinsically incorrect. First, the idea that when disputing issues of natural philosophy the Bible should not be contrasted with «the experiences and reasons of philosophy» (*experimentis et rationibus philosophiae*), to use the words of Father Benito Pereyra – professor of Sacred Scripture at the Roman College and one of Galileo's sources for documents in his own defense – was widely shared. It meant that on such issues the standard arguments should be drawn from physics (Aristotelian physics, in Pereyra's case) and the exegetes had to conform with them.

In his *Commentarii et disputationes in Genesim* (1591-99) Pereyra gives several examples of what he means. He mentions ancient authors who committed macroscopic errors concerning the structure of the world and the universe; some of these had traditionally been viewed with suspicion, like the early Christian theologians Origen and Lactantius, but there was also John Chrysostom, an influential Father of the Church. Adhering strictly to the letter of the Scriptures, they argued that the heavens are not spherical in shape, that they are motionless, and that the planets move freely through them «like fishes in the water, and birds through the air»; they also rejected the existence of the antipodes and asserted that the level of the oceans could exceed the highest mountain ranges. Statements like these, adds Pereyra, «now appear to be false, in the light of evident experiences and necessary demonstrations».[47]

Obviously, I am not arguing that Pereyra anticipated Galilean physics; as shown by his rejection of the free motion of heavenly bodies, he remained attached to the Ptolemaic vision of the solidity of the planetary spheres, a vision that had already been challenged in the late 16th century and that Bellarmine himself dismissed in his Louvain lectures. Nonetheless, when arguing for the stability of the Earth at the center of the universe, Pereyra based his reasoning on the Aristotelian principle of the heaviness

---

[47] Benito Pereyra, *Commentarii et disputationes in Genesim*, Lyon, ex officina Iuntarum, 1594, I, p. 25. On Pereyra hermeneutical theses see Corrado Dollo, *Galileo Galilei e la cultura della tradizione*, ed. by G. Bentivegna, S. Burgio, G. Magnano San Lio, Soveria Mannelli, Rubbettino, 2003, pp. 170 sq.; Remmert, *"Sonne steh still über Gibeon"* (cit. note 31), p. 550.

of this element, so contradicting the literalism of Justin and the Second Epistle of Peter which, following Ex. 20 and Ps. 24 [23] and 136 [135], had affirmed that the Earth was founded upon the waters. In the same way, when commenting on the two *luminaria magna* of Gen. 1,16, he made it clear that the passage must be interpreted as referring to how these phenomena commonly appeared to the senses, given that the Moon is actually much smaller than the Sun and all the other heavenly bodies, apart from Mercury.[48] In other words, setting aside his loyalty to Ptolemaic cosmology and Aristotelian physics, what Pereyra was showing was a notable example of an exegesis of biblical passages based on philosophical and astronomical evidence.

So, from a hermeneutical perspective the adapting of scriptural interpretation to the findings of natural philosophy was not in itself subject to censorship, at least not in the early 17th century. A more rigid form of biblical literalism arose soon thereafter, especially under the tutelage of Cornelius a Lapide, who was appointed professor of Sacred Scripture at the Roman College in 1616. While this literalism would blur over time in Catholic culture, it continued to influence the more orthodox strands of Lutheran and Calvinist scholarship.[49]

Secondly, it was acknowledged that the Bible could contain what at first sight might appear to be misleading sentences, that its text could sometimes be unclear or appear to indulge popular beliefs, and that therefore its exegetes were subject to errors of interpretation. Indeed, this was the premise upon which the entire system of Catholic controversialist theology rested, since the uncertainty and obscurity of the Biblical text justified the spiritual authority of the Roman pontiff as the ultimate judge on controversies over matters of faith.

Bellarmine's *Disputationes de controversiis* make the principle of the pope's authority clear at the very beginning, as they deal with the «interpretation of the Word of God» in the chapter mentioned above (see above, note 33), reminding the reader that there are «countless sources of diffi-

---

48 Pereyra, *Commentarii* (cit. note 47), I, pp. 56-57, 142-143. Bellarmine's views on the fluidity of the heavens have been highlighted and discussed by Baldini – George V. Coyne, *The Louvain Lectures (*Lectiones Lovanienses*) of Bellarmine and the Autograph Copy of his 1616 Declaration to Galileo*, Vatican City, Specola Vaticana, 1984. Remmert, *Im Zeichen des Konsenses. Bibelexegese und mathematische Wissenschaften in der Gesellschaft Jesu um 1600*, «Zeitschrift für historische Forschung», 33/1, 2006, pp. 33-66, 50 sq., mentions further Jesuit authors that stressed the importance of philosophical and mathematical learning for the purpose of a correct biblical exegesis, in particular José de Acosta, Jerónimo de Prado and Juan Bautista Villalpando.

49 François Laplanche, *Herméneutique biblique et cosmologie mosaïque*, in Olivier Fatio (ed.), *Les Églises face aux sciences du Moyen Age au XXe siècle*, Geneva, Droz, 1991, pp. 29-51, pp. 34 sq.

culty» in the text, «words and expressions that are ambiguous», «imperfect», or «inappropriate», as well as «phrases that are unique to the Jews».[50] Another influential theologian of the Society of Jesus, Father Jean Lorin (who succeeded Pereyra in the chair of Sacred Scripture at the Roman College), when explaining the use of the term *firmamentum* in the Vulgate to indicate the heavens of the planets, resorted to the will of Moses to «accommodate, when writing, the impressions of the common people».[51]

Concord among authors and loyalty to Catholic tradition were of course a guarantee of correct interpretation, but Bellarmine himself – a young man still at the beginning of his teaching career – had overlooked that concord in his commentaries on the *Prima* of Aquinas' *Summa theologiae* delivered in his Louvain lectures between 1570 and 1572. At the time, with reference to the story of the Creation, he indeed resorted to Scripture to reject the theory of the incorruptibility and inalterability of the heavens posited by orthodox Aristotelians, as well as by Aquinas and «many scholastic doctors».[52]

5. A hypothesis on the reasons behind Copernicus' condemnation

Hence, if we look at the case from a purely doctrinal perspective, it seems unlikely that the conditions for a condemnation of Galileo's hermeneutical theses, as explicated in his letters to Castelli and Christina of Lorraine, were present. Instead, things change if we adopt a historical perspective.

At that particular time, during the reign of Paul V, those principles must have appeared extremely suspect to the Holy Office, and to the Roman Curia in general. First of all, they were expounded in quite a singular form.

---

[50] I controversy *De Verbo Dei*, b. 3, ch. 1 (see above, n. 33), p. 65. On these passages see also Remmert, *Im Zeichen des Konsenses* (cit. note 48), pp. 40-41, who indicates comparable assertions in the *Prolegomena in sancta evangelia* (1598) of Father Alonso Salmerón, one of the major experts in biblical exegesis of the first Jesuit generation. On the correlation between biblical hermeneutics and the power structure within the Church see Laplanche, *Bible, sciences et pouvoirs au XVIIe siècle*, Naples, Bibliopolis, 1997, pp. 24 sq.

[51] Jean Lorin, *Commentarii in librum Psalmorum*, Venice, apud Iacobum Tomasinum, 1737 (orig. 1612), p. 247 (in Ps 18 [19]). On Lorin see also Dollo, *Galileo Galilei* (cit. note 47), pp. 140-141.

[52] *The Louvain Lectures*, *An coelum natura sua sit corruptibile*; *An coelum de facto corrumpetur*, pp. 9-11. Bellarmine deems also «more probable» the autonomous motion of the stars than their fixity to the heavenly spheres, though adding that «if there will ever come out an evidence [of this latter theory] it will be necessary to correct the comprehension of the Scripture, so that they will not conflict with a manifest truth»: *ibid.*, p. 21.

The statements about the pre-eminence of the mathematical and physical sciences in the explanation of natural phenomena, as well as the emphasis on the risks connected with a literalist exegesis of the Bible, were set forth in Galileo's writings as the foundations of a precise and explicit hermeneutical argumentation. On the contrary, the biblical commentaries of Pereyra and Lorin were scholarly canonical works intended for the higher education of the clergy – and moreover their authors had already been subjected to the pre-emptive internal censorship of the Society of Jesus.

Secondly, over and above this, even more sensitive topics can be found in Galileo's letters, especially in his letter to Christina; for instance, the appeal for *libertas philosophandi*, the 'freedom to philosophize', or the distinction that he made between «the debatable and the demonstrative disciplines», with the significant «great difference [...] between giving orders to a mathematician or a philosopher, and ruling over a merchant, or a lawyer».[53] Even his continual references to Augustine could raise suspicions, since the controversy *De auxiliis* had shown how difficult, if not impossible, it was to bend this particular Church Father's writings to the disciplinary demands and logic of the scholastic theology embraced by the regular orders.

These topics must have been perceived as elements of a possible new knowledge system that was exempt from and even a rival to the Church's teachings, at least in the sphere of biblical hermeneutics and with regard to the description of natural phenomena.

At the time, the interpretation of the Bible was the methodological framework within which not only the categories of religious and moral discourse, but also the legitimacy of the temporal and spiritual powers were elaborated. As such, in the first quarter of the 17th century it was the sphere in which the harsh theological-political debate between the Roman Church and other 'centers' of power such as the English monarchy and the Gallican Church unfolded.[54]

I believe that one of the main reasons that led to the condemnation of Copernicanism by the Catholic Church can be identified if we approach

53 Galilei, *Lettera a Cristina di Lorena sull'uso della Bibbia nelle argomentazioni scientifiche*, ed. by F. Motta, Genoa, Marietti 1820, 2000, p. 114. I rely on this critical edition since, as is well known, the version of the letter to Christina included in OG has been reconstructed by the editor, Antonio Favaro, by blending together texts taken from different witnesses. See also the more recent critical edition by Bucciantini – Camerota in *Scienza e religione. Scritti copernicani*, Rome, Donzelli, 2009, pp. 33-84.

54 According to Laplanche, *La Bible en France entre mythe et critique. XVIe-XIXe siècle*, Paris, Albin Michel, 1994, p. 13, in the decades between the Council of Trent and the Thirty Years' War the biblical issue emerged as the leading issue around which claims for truth and the use of violence were acted out: «De l'une et de l'autre, la solution est cherchée dans le Livre».

the whole affair from this historical perspective. The hermeneutical theories delineated in Galileo's writings, as I noted before, were scarcely censurable in themselves, if viewed as abstract proposals of methodology, but their significance changed if they were seen as the premises for substantial, unorthodox conclusions. In such cases, an exegesis of the Bible aimed at legitimating assertions contrary to the official doctrine would have been wholly unacceptable. The same distinction between issues that were *de fide* and adiaphoric ones would have made no sense, since such issues, if judged to be in conflict with the sources of faith (Scripture and tradition), would have been considered as pertaining to faith.

This is exactly what happened to Copernicus' heliocentric theory. Having ascertained Galileo's «candidness and integrity» in early February, the supreme tribunal of faith moved on to the examination of «a certain doctrine and opinion [...], the resolution concerning which is now under discussion».[55] On February 19 a brief description of the Copernican system – based, as I noted above, on two sentences extracted from Caccini's denunciation – was distributed to the consultants of the Holy Office; a copy of Ingoli's *Disputatio* was probably made available to them as well.

That the two sentences submitted to the censors lacked any astronomical or theological contextualization appears to be in line with inquisitorial practice. This ordinarily involved the extrapolation of particular sentences from the text under examination using an analytical method in which such sentences might not even be direct citations from the text, but paraphrases deemed to represent its theses. The most well known case is that of the five propositions ascribed to the *Augustinus* of Cornelius Jansen and condemned by Pope Innocent X with the bull *Cum occasione* in 1653, which sparked the huge Jansenist controversy.[56]

The unanimous judgment of the consultants on Copernicanism was recorded on paper in a meeting held on 24 February 1616. Heliocentrism was evaluated not only from a theological, but also from a philosophical point of view, a fact that might appear surprising at first glance since profane knowledge did not pertain, strictly speaking, to the jurisdiction of the Holy Office. However, it reflects the increasing focus of Church censorship on natural philosophy in the late 16th century (as evidenced by the inclusion of Telesio and Patrizi in the Index), based on the principle of a concordance between truths of reason and truths of faith proclaimed in the bull

55 Galileo to Picchena, 6 February 1616 (as in note 40).

56 *Dictionnaire de théologie catholique*, 8/1, pp. 474-496 (J. Carrèire). Neveu, *L'erreur et son juge* (cit. note 9), pp. 512 sq.

Apostolici regimini (1513) of the Fifth Lateran Council.[57] In any event, the double judgment issued in 1616 effectively blocked the possibility that the heliocentric theory, now defined as being contrary to Scripture, could be considered as compliant with philosophical truth.

The assertion that the Sun is the center of the universe and does not orbit around the Earth was defined as «foolish and absurd in philosophy» and from a theological viewpoint «formally heretical»; the theory of the motion of the Earth received the same censure on philosophical grounds, while theologically it was judged to be «erroneous», that is, just one degree below heresy in the scale of deviation from orthodoxy.[58]

The document was signed by eleven theologians of the Congregation, among them the commissioner of the Holy Office, Seghizzi, and Father Benedetto Giustiniani of the Society of Jesus, the same priest who, when entrusted some twenty years earlier with the examination of Francesco Patrizi's *Nova de universis philosophia*, did not raise any objections to the author's thesis of a daily rotation of the Earth in a geocentric system.[59] Giustiniani also served as rector of the Roman College between 1599 and 1602, and in 1607 actively participated in the debate over the Venetian Interdict, writing two pamphlets, one of them a defense of Bellarmine's statements on the jurisdictional privileges of the Church against the attacks of the theologians of the Republic.[60]

As for Galileo, the inquiry closed as it began, quite informally. He was privately informed by Bellarmine on behalf of the pope of the Holy Office's censorship of the heliocentric theory – as ostensibly decided on February 25 by the congregation in a plenary session, but of which, as far as we know, no memorandum exists –, with the warning to «abandon it». If he refused, then any further research on, or the teaching or defense of heliocentric ideas would be prohibited.[61]

---

57 As Maria Pia Donato remarks, the *Apostolici regiminis* «acted as the ideological and juridical framework of theological censorship against philosophical-scientific thought in the early modern age» (*Scienze della natura*, in *Dizionario storico dell'Inquisizione*, III, pp. 1394-1398, 1394).

58 *Processo a Galileo*, OG, XIX, pp. 320-321.

59 Lerner, *L'hérésie heliocentrique* (cit. note 13), p. 77.

60 *Diccionario histórico de la Compañía de Jesús*, II, p. 1740 (J.P. Donnelly). It is worth reminding at least two other prestigious members of the commission: the Augustinian Gregório Nuñez Coronel, former preacher to the Duke of Savoy and secretary of the *De auxiliis* commission, and the Dominican Tomás de Lemos, himself involved in the proceedings of the *De auxiliis* as attorney of his own order. S. *Lexikon für Theologie und Kirche*, III, p. 61, and VI, p. 942.

61 *Processo a Galileo*, OG, XIX, p. 321.

The story of the relationship between Galilean thought and the Bible, with its remarkably modern ring, ends here; the works that followed did not touch upon topics of biblical hermeneutics. The publishing history of the letter to Christina of Lorraine unfolded entirely within the world of the Reformation and of Catholic anti-Roman dissent, from the Latin translation published in Strasbourg in 1636 by Elia Diodati (*Nov-antiqua sanctissimorum Patrum, ac probatorum theologorum doctrina de Sacrae Scripturae testimoniis in conclusionibus mere naturalibus*) to the Italian edition printed secretly in Naples by Lorenzo Ciccarelli in 1710.[62]

From this perspective the extrajudicial solution chosen by the Holy Office, probably on the suggestion of Cardinal Bellarmine, proved to be effective: Galileo was silenced with regard to his hermeneutical proposals, but the official mathematician and philosopher of the Grand Duke of Tuscany was saved from any possible judicial consequences. The attempt to legitimize a heliocentric scriptural exegesis was forestalled with the prohibition of Foscarini's *Lettera*. This was, so to speak, a pre-emptive use of doctrinal censorship, whose effect was to avert the possible opening up of an area of uncertainty on such a thorny issue as the monopoly over the interpretation of the Bible and Catholic tradition.

As we know, the fate of Copernicus' work was somewhat different from that envisaged by the consultants to the Holy Office. The concise memorandum of the session of the Index held on March 1st in Bellarmine's residence hints of a prolonged debate in which some cardinals of the Congregation managed to change the accusation of formal heresy against heliocentrism to the more vague charge of 'falsity' and 'opposition to Scripture'. Accordingly, Copernicus' *De revolutionibus* was not prohibited, but suspended until it could be 'corrected' – a provision that would have been regarded as far too mild for a book written by a heresiarch, as Copernicus was in all respects after the judgment handed down by the consultants of the Inquisition. The corrected version, as we know, was prepared with surprising rapidity; the decree of emendation, which adopted the proposals advanced by Ingoli, to whom the question had been entrusted, dates to May 1620.[63]

As I noted above, this revision to the doctrinal censure of the heliocentric theory was ascribed at the time to the intervention of Cardinal Caetani and Cardinal Barberini (although it is possible that pressure had also been

---

62 Motta, *Nota al testo*, in *Lettera a Cristina di Lorena sull'uso della Bibbia* (cit. note 53), pp. 67-80.

63 Pierre-Noël Mayaud, *La condamnation des livres coperniciens et sa révocation à la lumière de documents inédits des Congrégations de l'Index et de l'Inquisition*, Rome, Editrice Pontificia Università Gregoriana, 1997, pp. 37-42, 69-71, 77-79.

applied by Cardinal Orsini on Paul V).[64] This decision was taken during a session of the Index, and therefore not under the authority of the Congregation of the Holy Office, which was formally responsible for decisions concerning the nature of a suspect proposition. The Congregation, however, took official note of the change in a session held on March 3 *coram Sanctissimo* 'in the presence of the Most Holy', that is, the pope himself.[65] The decree of the Index, dated 5 March 1616, made public the text of the condemnation drafted four days earlier, whereby the «Pythagorean doctrine» was declared *falsa* and *Sacrae Scripturae adversans*.

Diplomatic considerations must have provided the main reason for such a change of opinion, with the evident wish to avoid giving offence to the Grand Duke.[66] Perhaps it is no coincidence that on February 13 the Cardinal-nephew Scipione Borghese wrote to Cosimo II personally to assure him that special consideration would be accorded to Galileo. The scientist himself, writing to the Florentine Secretary of State in the aftermath of the censure decree, referred – not without reason, although his tone perhaps betrayed an excessive degree of optimism – to a danger that had finally been averted, since the Church had dismissed the accusation that the heliocentric theory was «against the faith and heretical», merely declaring that «such opinion does not agree with the Sacred Scripture».[67] All the same, the real significance of the fact that the grave accusation of heresy and error hovered over the thesis of heliocentrism – even if for just a few days – still must be assessed.

In introducing my assessments, I would like to make it clear that I do not wish to re-open the question concerning the doctrinal notes relating to Copernicus issued by the consultants to the Congregation of the Inquisition, i.e. whether the notion of heresy they applied had a dogmatic or a mere technical nature – a question that primarily touches on the 'second trial' of Galileo and his condemnation in 1633, rather than on the 1615-1616 affair.

---

64 See above, note 24. In a memory written during his own pontificate, and preserved in the Biblioteca Apostolica Vaticana, Barb. Lat. 4900, Maffeo Barberini reminds that the solution had been approved by Bellarmine: Francesco Beretta, *Urbain VIII Barberini protagoniste de la condamnation de Galilée*, in José Montesinos – Carlos Solís (eds.), *Largo campo di filosofare*, Eurosymposium Galileo 2001, La Orotava, Fundación canaria orotava de historia de la ciencia, 2001, pp. 549-574, 557, note 37.

65 *Processo a Galileo*, OG, XIX, p. 278.

66 Camerota, *Galileo Galilei* (cit. note 1), pp. 320-321.

67 Galileo to Curzio Picchena, 6 March 1616, OG, XII, pp. 243-245.

As is well known, the need to distinguish between a 'theological' and an 'inquisitorial' heresy was introduced in Galilean historiography more than a century ago – although with very little critical reaction at the time – by the *abbé* Léon Garzend with his weighty treatise *L'Inquisition et l'hérésie* (1912), which was recently brought back to the attention of historians by Bruno Neveu during the first conference on the history of the Holy Office held in Vatican City in 2003.[68] Since then we have had the stimulating considerations of Francesco Beretta, who points out that the abjuration by Galileo in 1633 was an act endowed with intrinsic magisterial value by the Holy See, and the equally well documented reappraisals by Neveu and by Father Pierre-Noël Mayaud.[69]

For my part, I believe that the idea of enclosing the complex and often contradictory facts of historical dynamics within the abstract categories of theology and canon law is always unsafe, particularly when dealing with the governing mechanisms of the Old Regime, with its juridical instruments, political concerns, and the will of those in positions of power perpetually intertwining without actual discontinuity.

6. Heresy as a mirror of Church authority

What I would like to underscore here is just one among the many possible interpretations of the historical, and especially the theological-political reasons – meaning by this term the permanent and reciprocal mirroring between theology and politics that formed the background against which the Christian confessions and the temporal powers of the time maneuvered – that led the Roman Church to censure Copernicanism in 1616, even coming close to condemning it outright as heresy.

If we turn back to the theological categories envisaged by Melchor Cano, we can see how the two higher degrees of deviance from orthodox faith are described in great detail. Basing himself on the writings of the eminent theologians Thomas Aquinas and Cardinal Cajetan, Cano insisted

---

[68] Neveu, *Y a-t-il une hérésie inquisitoriale?*, in Agostino Borromeo (ed.), *L'Inquisizione. Atti del Simposio internazionale*, Vatican City, Biblioteca Apostolica Vaticana, 2003, pp. 469-491.

[69] Beretta, *Le procès de Galilée et les archives du Saint-Office. Aspects judiciaires et théologiques d'une condamnation célèbre*, «Révue des sciences philosophiques et théologiques», 83, 1999, pp. 441-490; Neveu – Mayaud, *L'affaire Galilée et la tentation inflationniste. À propos des notions d'hérésie et de magistère impliquées dans l'affaire*, «Gregorianum», 83/2, 2002, pp. 287-311; Beretta, *L'affaire Galilée et l'impasse apologétique. Réponse à une censure*, «Gregorianum», 84/1, 2003, 169-192; Neveu – Mayaud, *Impasse apologétique ou tentation inflationniste? À propos de l'affaire Galilée*, «Archivum historiae pontificiae», 42, 2004, pp. 205-209.

that no heresy can be established in the absence of the violation of an article of faith proclaimed as such by the Church:

> As long as a question is uncertain, and the Church has not spoken yet in favour of one of the two parties in a controversy, there is no judgment about the concordance of one opinion or the other with Catholic faith. We can understand from this that there does not exist any heresy unless someone takes a position against a Catholic truth that the Church has prescribed with certainty. We must add to this that it is very far in itself from the name of Catholic truth, that someone calls Catholic truth [a doctrine] that is shared by everyone as a probable opinion, but that neither the Church has ever requested to be believed nor the faithful believe out of instilled faith.[70]

Cano ascribed the quality of erroneousness to three different kinds of statements. The first is the class of sentences that «shake», but «do not overturn» faith, since they oppose «Catholic and universal truths in which the universal Church believes» but that are not truths of faith; for example, that the mendicant orders must live on charity rather than manual labor is true, but is not a matter of faith. The second class of statement includes propositions that contradict «a truth that, according to the strong [*vehemens*] conviction of the experts, is a truth of faith, despite it not having yet been declared as such by the Church nor proved with indisputable arguments». Finally, a third class of erroneous statement regards a proposition «that is certainly in opposition to Catholic faith, although not in a manifest way, but rather on the basis of the very probable and nearly necessary opinion of all the experts».[71]

We may note that on the basis of this theological taxonomy a charge of heresy against the heliocentric cosmology would have required as a necessary precondition the clear acknowledgment of Geocentrism as a revealed truth by the Church, a pronouncement that historically had never been made. Instead Cano's second, more flexible designation of erroneousness would seem to apply to this issue, referring to theses that may contradict a truth of faith that has not yet been officially proclaimed as such but is believed to be true with a high degree of certainty by expert theologians. And yet, in this case as well the fact remains that neither a council nor the *Cathedra Petri*, the teaching of the Bishop of Rome, had ever explicitly ad-

---

[70] Cano, *Locorum theologicorum libri XII* (cit. note 10), p. 370.

[71] *Ibid.*, p. 388. A detailed account on the discussion of the notes of error and heresy by the other prominent theologians of Spanish scholasticism, such as Domingo de Soto, Bartolomé Carranza and the *Salmanticenses*, can be found in Neveu, *L'erreur et son juge* (cit. note 9), pp. 251 sq.

dressed the problem, nor had any clear and firm opinion been issued by scholarly theologians regarding whether Geocentrism was a truth of faith. It was not by chance that Diego de Zuñiga's *Commentarii* passed unscathed the examination of the *Suprema*, the Spanish Inquisition, when they were first published in 1584 (and were eventually reprinted without any changes in Rome in 1591).[72]

Now, I consider it to be worth emphasizing that, at the time of the first informal 'trial' of Galileo, the theological tradition of which Melchor Cano was a representative – and, with him, other outstanding authors of mid-16th-century Spanish scholasticism such as Domingo de Soto, and the teacher of both Cano and de Soto, Francisco de Vitoria – had been progressively weakened by the competition of other elements. Most important among those was the process of centralization of the governing powers of the Church in the hands of the papacy, a process that had been accomplished not only in the institutional sphere with the reform of the Roman congregations by Sixtus V and the diplomatic efforts to implement the Tridentine decrees throughout Catholic Europe, but also on a strictly doctrinal level with the production of a rich corpus of theological treatises, written primarily by Jesuits who had been educated at the Roman College or one of the Order's other centers of higher education, like its colleges in Ingolstadt and Cologne. The aim of these theologians was to construct a full model of Church authority based on the papal office of *iudex fidei*, 'judge of the faith'.

In this corpus of works – made up variously of pamphlets, treatises of controversialist theology, and commentaries on the *Summa theologiae* – the exclusive privilege of the Apostolic See to define dogma was emphasized within the framework of a Church model in which the Pope's spiritual sovereignty over the Christian world found legitimacy in his prerogative as the ultimate judge in cases of faith. This argument identified in the Chair of Saint Peter the superordinate source of doctrine, since the Roman pontiff and the bodies appointed by him, such as the Holy Office and the Index, summarized the functions of dogmatic legislation – properly the domain of the Council – with the instruments of doctrinal censorship.

From this perspective censorship operated as a tool of legitimization and as an expression of the spiritual sovereignty of the papacy, hence as an instrument of political action, assessed and administered on a case by case basis, sometimes even to the detriment of the stringent logics formally envisaged by canon law.

---

72 Mayaud, *La condamnation des livres coperniciens* (cit. note 63), p. 44.

This sort of practice was lucidly denounced by Sarpi in one of the most significant of the reports that he prepared for the Venetian Senate, dating to February 1609. Here the theologian pointed out the gap that formed during the Interdict between the doctrinal censures issued by the Roman Curia and their actual consequences. According to Sarpi, Roman censures were often wanting in doctrinal substance and appeared to reflect political pressures above all else:

> [In Rome] they are well aware that the doctrine [of Venice] is well established in divine Scripture, in the canons of the universal Church, in the laws of the Empire and of other Catholic kingdoms, and in the writings of the Holy Fathers. For that reason, they deem it necessary to escape the objections that may be raised against them with such powerful weapons, so they talk in generalities and assert that [this doctrine] contains reckless, slanderous, heretical, erroneous, and scandalous things, respectively. With those names they frighten the world, and, not specifying further details, they don't give anyone the opportunity to protest.[73]

Cano had served as the official theologian representing the emperor Charles V at the Council of Trent in 1551-1552. His *Loci theologici* were conceived as a response to the need of the controversialists to reshape the methods of theology in the aftermath of the council's asserted hierarchy of the sources of faith. From a historical perspective he shared – together with the other authors of the School of Salamanca – the vision of a balance of powers in the Church that acknowledged the primacy of the Roman See within a framework of harmony with the council and the main theological faculties.

At the turn of the century, in the intense and decisive years that witnessed the absolution of Henry IV, the death of Philip II, and the crowning of James I in England, the historical context had decisively changed. From this perspective, at that time the ecclesiological concepts of the School of Salamanca, although still central in the system of higher education of the clergy in the late 16th century, were no longer the prevailing standard – at least not in Rome.

One example of this may be found in the controversy over grace, where the Spanish Dominicans (who referred to themselves as *thomistae*, 'followers

[73] *Consulto 50, Scrittura in diffesa delle opere scritte a favore della Serenissima Repubblica nella controversia col Sommo pontefice*, February 25, 1609, in Sarpi, *Consulti*, I/1, *I Consulti dell'Interdetto (1606-1607)*, ed. by C. Pin, Pisa-Rome, Istituti editoriali e poligrafici internazionali, 2001, pp. 672-695, 677-678. On the different perspectives of the jurisdictional conflict between Rome and Venice, see Frajese, *Sarpi scettico. Stato e Chiesa a Venezia tra Cinque e Seicento*, Bologna, il Mulino, 1994, pp. 383 sq., and the studies collected in Marie Viallon (ed.), *Paolo Sarpi. Politique et religion en Europe*, Paris, Garnier, 2010, pp. 28-54.

of Thomas Aquinas') – represented first and foremost by Domingo Bañez, a pupil and then a colleague of Cano – failed to obtain the condemnation of the Molinist theses upheld by the Society of Jesus, which were stigmatized as 'new' and 'Pelagian' by the majority of observers at the time. It must also be remembered that in the summer of 1601 Clement VIII sought to resolve the controversy himself, based on his papal authority and his prerogative of *iudex fidei*. This effort to define the correct thesis regarding the relationship between human will and divine grace on the basis of the vast amount of Augustinian sources available and the endless paperwork produced by the *De auxiliis* commission eventually came to nothing with the death of the pope, but it provides nonetheless a significant example of the doctrinal power claimed by the Chair of Saint Peter.

It is worth adding that on this particular issue Cardinal Bellarmine, the leading theorist on the question of the spiritual sovereignty of the pope, had cautioned Clement VIII against the solution of acting by himself invoking his role of *iudex fidei*, even submitting an opinion suggesting that a council be convened to settle the question, in a remarkable reversal of roles that tells us a great deal about the insufficiency of a purely doctrinal reading of the intertwining tendencies within the Roman Curia.[74]

7. The Jesuits and the judge of controversies

There is now a vast bibliography available on the role played by the Society of Jesus in the events that culminated in the censures of 1616 and, more generally, their contribution to the germination of experimental science. In particular, a unique feature of Jesuit culture that has been identified by historians of early modern science is their longstanding, steadfast effort to transplant observational and experimental practices into the soil of Aristotelian epistemology.[75]

---

[74] Motta, *Bellarmino* (cit. note 14), pp. 582 sq.

[75] Among the many titles, see Rivka Feldhay, *Galileo and the Church. Political Inquisition or Critical Dialogue*, Cambridge, Cambridge University Press, 1995, p. 203; Id., *The Jesuits. Transmitters of the New Science*, in Bucciantini – Camerota – Franco Giudice (eds.), *Il caso Galileo. Una rilettura storica, filosofica, teologica*, Florence, Olschki, 2011, pp. 47-74; Richard J. Blackwell, *Galileo, Bellarmine, and the Bible. Including a Translation of Foscarini's* Letter on the Motion of the Earth, Notre Dame, University of Notre Dame Press, 1991, pp. 135 sq.; Kelter, *The Refusal to Accommodate* (cit. note 31); Michael J. Gorman, *Molinist Theology and Natural Knowledge in the Society of Jesus 1580-1610*, in *Sciences et religions de Copernic à Galilée* (cit. note 13), pp. 235-254. A detailed bibliographical essay in Sheila J. Rabin, *Early Modern Jesuit Science. A Historiographical Essay*, «Journal of Jesuit Studies», 1, 2014, pp. 88-104.

Regarding the question under discussion here, I feel that it is important to highlight another aspect of the historical identity and role of the Jesuits in this period – an aspect forged by political conflict, first and foremost the design which they championed of affirming the unlimited sovereignty of the pontiff over the individual consciences of believers.

The decades between the last quarter of the 16th century and the outbreak of the Thirty Years War witnessed the height of religious conflict in Europe and the momentum of the theological-political developments of the early modern period. At that time the Jesuits underwent a fundamental process which – in the eyes of their enemies – reshaped their identity from that of an essentially Hispanic order to a multifaceted body devoted to the establishment of papal tyranny.[76]

Expelled from France in 1594 after the attempted murder of Henry IV by a former Jesuit student, Jean Chastel; expelled from Venice in 1606 during the Interdict; expelled from Bohemia in the aftermath of the insurrection against Hapsburg rule, the Jesuits were perceived as a powerful presence dominating the area where the civil and the religious spheres, the temporal and spiritual powers, intersected. In this period Jesuit hegemony over the consciences of the princes of the *Ancien Régime* gained momentum, with many powerful Catholic rulers from the dukes of Bavaria and Savoy to Henry of France and the imperial dynasty choosing Jesuit priests as their spiritual advisors, from Johannes Buslidius to Guillaume Lamormaini, and Pierre Coton, the most renowned and most detested of them all.[77]

The ideal model of the court confessor was outlined by Bellarmine in one of the devotional booklets that he produced during the latter part of his life, the *De officio principis christiani* of 1619. The confessor should not be intimidated by the majesty of his penitent, but he rather sounds out his conscience with even greater religious zeal and devotion, since the private sins of the man could pollute the public person of the sovereign. The Jesuit confessor embodied the hierarchical relationship between the temporal

---

76 Eric Nelson, *The Jesuit legend: superstition and myth-making*, in Helen Parish – William G. Naphy (eds.), *Religion and Superstition in Reformation Europe*, Manchester-New York, Manchester University Press, 2002, pp. 94-115, 109 sq.; José Martínez Millán, *La trasformazione della monarchia* hispana *alla fine del XVI secolo. Dal modello cattolico castigliano al paradigma universale cattolico-romano*, in Paolo Broggio – Francesca Cantù – Pierre-Antoine Fabre – Antonella Romano (eds.), *I gesuiti ai tempi di Claudio Acquaviva. Strategie politiche, religiose e culturali tra Cinque e Seicento*, Brescia, Morcelliana, 2007, pp. 19-53.

77 Robert Bireley, *The Jesuits and the Thirty Yeas War. Kings, Courts, and Confessors*, Cambridge, Cambridge University Press, 2003; Harro Höpfl, *Jesuit Political Thought. The Society of Jesus and the State, c. 1540-1630*, Cambridge, Cambridge University Press, 2004, pp. 20 sq.

and sacred powers: he remained seated wearing his *biretta* while hearing the confession, while the prince was on his knees, bareheaded.[78]

This hierarchical relationship between the director of conscience and the sovereign *quoad peccata*, with regard to sins and their absolution, was in a sense the iconic representation of the relationship between spiritual and temporal power outlined in Bellarmine's theory of the indirect power of the pope. According to Bellarmine, as is well known, the prerogative of judging and sanctioning the acts of rule of a prince fell to the Roman See. This comprised acts pertaining to temporal issues if they affected the preservation of faith (one textbook case was the granting of freedom of worship to heretic subjects) and even, if necessary, the excommunication of an errant sovereign and the absolution of his subjects from their bonds of fealty.

The fact that the bond of obedience between subjects and their prince could be broken by excommunication as an act that separated an individual from the community of Christians (an act that was not without precedent, going back to Gregory VII's excommunication of Henry IV in 1076), as opposed to considering this bond to be fixed and indissoluble because it is a tie of nature to the paternal authority of one's sovereign, as established by God at the Creation (the declared position of the English king, James I, and of the French Parliaments) constituted one of the most controversial issues in the political debate across Europe in those years.

What appears interesting to me is that we have here two diverging theories regarding the social pact, one identifying a juridical-political sphere ruled by its own laws stemming from nature and hence not constrained by spiritual power, and the other denying that the sphere of secular power was exempt from religious authority, based on the priority of faith and of the Catholic Church.

To a certain extent an analogy can be drawn between this struggle for authority between the Church and the secular powers and the emerging confrontation between the Church and experimental science, between the hermeneutics of the Letter to Christina, with its notion of the autonomous jurisdiction of natural philosophy, and the hermeneutics of the theologians of the Holy Office, who designated the Copernican theory as false, erroneous and heretical.[79] It must be noted that in both cases the Apostolic See

---

[78] NICOLE REINHARDT, *Voices of Conscience. Royal Confessors and Political Counsel in Seventeenth-Century Spain and France*, Oxford, Oxford University Press, 2016, pp. 67 sq.

[79] Analogous considerations can be found in VOLKER REINHARDT, *Il concilio di Trento e le scienze naturali: la controversia fra Bellarmino e Galilei come paradigma*, in PAOLO PRODI – WOLFGANG REINHARD (eds.), *Il concilio di Trento e il moderno*, Bologna, il Mulino, 1996, pp. 485-501.

exercised its powers with recourse to an act of judgment, be it with regard to the conscience of the prince or to the lawfulness of heliocentrism.

After the failed attempt by Sixtus V to have Bellarmine's *Disputationes* placed on the Index in 1590 because of his rebuttal of the idea of an immediate papal dominance over the kingdoms of the earth, the theory of the indirect power of the pope on temporal issues, though never explicitly recognized as dogma, became the pillar of the theological system substantiating the hegemonic strategies of the Roman Church. It is against this theory, as we will see below, that in 1606 James I forced the suspect Recusants to take an oath of allegiance to the English Crown and in the same year Paolo Sarpi sought to develop a full-fledged theory of the State in opposition to the political views of Jesuit scholasticism.[80]

According to Sarpi, the claim of the *totato* – the all-encompassing authority of the Apostolic See, in his sarcastic words – was to raise to the level of dogma theological concepts that were actually devoid of any real foundation in Scripture and ancient tradition, concepts that had emerged over time with the evolution of the Church. «The imposition of the right to license or to block from kingship, to appoint, or to remove kings – Sarpi wrote in the above-mentioned account – is something that has been attempted for five hundred years, and it has never had success or been put into execution».[81] Later Sarpi wrote to the Gallican magistrate Jacques Leschassier: «Bellarmine condemns [our opinions] as heretical, and establishes as an article of faith that the pope can excommunicate kings and release subjects from their oath, and he unceasingly argues [...] that the excommunicated, as long as they remain such, cannot be given any obedience without committing a sin».[82]

In his political considerations the Servite friar and theologian Fulgenzio Micanzio, another well-known correspondent of Galileo (and biographer of Sarpi), grasped the enormity of the theory of indirect power: «Even eating, drinking, sleeping and everything must be subordinated to the spiritual end and salvation; therefore [the Apostolic See] has authority over everything»; and furthermore: «Every habit that is not directed toward God's glory and toward salvation is an abuse, hence all things, none excepted, would fall under the authority of the ecclesiastical court».[83]

---

80 This is opportunely noted by Corrado Vivanti, *I* due governi del mondo *negli scritti di Sarpi*, in Viallon (ed.), *Paolo Sarpi. Politique et religion en Europe* (cit. note 73), pp. 28-54.

81 *Consulto 50*, p. 680.

82 22 December 1609, in Sarpi, *Lettere ai gallicani*, ed. by B. Ulianich, Wiesbaden, Franz Steiner, 1961, pp. 61-66: 64.

83 Fulgenzio Micanzio, *Annotazioni e pensieri*, in Gino Benzoni – Tiziano Zanato (eds.),

In 1616 consultors and cardinals of the Index and the Holy Office did indeed resolve that the theory of the motion of the Earth and the stillness of the Sun was «subject to the authority of the ecclesiastical court» since it conflicted with the Biblical image of the universe which, though never doctrinally codified, was implicitly received by tradition. The same could be said about the hierarchy of the spiritual and temporal powers, which was never recognized as dogma by a council nor did it even formally receive the consensus of scholastic theologians, but which had been firmly and unwaveringly upheld by the acts of a long sequence of popes from Gregory VII to Pius V and also, it must be said, had clearly been envisaged as a thesis by Aquinas.[84]

The point is that explicit recourse to the *potestas indirecta* – notwithstanding the opposition of a widespread, resolutely anti-Roman wing of the Catholic Church – was supported by a vast amount of apologetic scholarship centered around controversialist theology, with the theory of the pope as *iudex controversiarum*, 'judge of controversies in the realm of faith', as the conceptual foundation for his indirect power. Not by chance, the central thesis of the most renowned and disputed of Bellarmine's *Controversiae*, concerning the spiritual power of the pope, was argumented as – much of it word for word – a long annotation entitled *De iudice controversiarum disputatio* ('Disputation on the judge of controversies'), written by him in the margin of his lecture on Aquinas' *Secunda secundae* that was delivered in Louvain in 1574.[85]

In the area of controversialist theology, during the decades under consideration here, the Jesuits – who were far more prepared, due to their rigorous education and scholarship, than the members of other orders for the debate with Protestants and with the advocates of State authority – were the architects of the pope's sovereignty based on his position as the highest judge in matters of faith.

The polemical works of the greatest authorities on controversialist matters in the Society of Jesus – such as Martin Becanus, Adam Tanner, Grégorio de Valencia, Edmund Campion, and François Veron, who were active primarily along the confessional border between Catholic and Reformed Europe, i.e., in Germany, the Low Countries, and the Austrian He-

---

Storici e politici veneti del Cinquecento e del Seicento, Milan-Naples, Ricciardi, 1982, pp. 733-863: 760, 761.

[84] THOMAS AQUINAS, *Summa Th.* II-II, q. 12, *De apostasia*, a. 2, *Utrum princeps, propter apostasiam a fide, amittat dominium in subditos, ita quod ei obedire non teneantur.*

[85] MOTTA, *Bellarmino* (cit. note 14), pp. 363 sq.

reditary Lands, as well as in France after the Edict of Nantes and in the underground circles of English Catholicism – invariably included a treatise *De iudice controversiarum*.

The argumentation in their treatises was usually based on the linking of two elements. The first was hermeneutical: since Scripture was not the transparent, perspicuous text proclaimed by Luther and the Reformation, an impartial judge was needed to resolve the virtually limitless points of conflict that could arise between different interpretations of what was often a rather opaque and obscure text. The second element was analogical, and derived from the quintessentially Tridentine principle of the visibility of the Church and its nature as a *societas perfecta*, a society provided with all the necessary means to reach its own divine ends. «Just as in political and civil matters – Becanus wrote in his commentary to the *Secunda secundae* in 1615 – quarrels and disputes often arise, requiring the presence of a judge issuing a judgment over the parts, the same holds in matters of faith and religion». Moreover, just as in regular tribunals a distinction is made between judge, law and jurisprudence, «in the same way, in controversies on faith we have the judge, the Scripture of the two Testaments, and tradition».[86]

8. The theological-political conflict in the age of Paul V

Niccolò Lorini's denunciation and Galileo's letters to Benedetto Castelli and Christina of Lorraine came to the attention of the Holy Office at the very moment when the hermeneutical and political importance of the debate over who should be recognized as the ultimate judge in faith controversies and biblical interpretation had reached its climax. The ruling élite in the Roman Curia had just come through a bruising doctrinal confrontation over the nature and boundaries of ecclesiastical power – a battle that had mobilized governments and learned milieus from all over Europe – when it was asked to pass judgment on Galileo's theses regarding the distinction between the spheres of natural philosophy and divine Revelation, and the daring proposal for a heliocentric exegesis advanced by Foscarini.

As a matter of fact, at that historical juncture not only Galileo, but also far more powerful political players emerging in the form of secular States – with their own political doctrine and their own apologists who

[86] Becanus, *De iudice controversiarum*, in *Opera omnia aucta revisa et in duos tomos distributa*, here in the 1649 edition, Mainz, impensis Ioan[n]is Godefredi Schönwetteri, II, pp. 1235-1252, 1247.

were becoming ever more practiced in theological argumentation – were breaking the monopoly on the interpretation of Scripture traditionally entrusted, in accordance with the Catholic system of learning, to the schools of the regular orders.

In 1609 the Letter of Majesty, an edict emitted by Emperor Rudolf II that guaranteed religious tolerance for his subjects in Bohemia, had been harshly criticized in Rome. Thereafter, for the ensuing decade up to the opening stages of the Thirty Years War, the efforts of Paul V and the Jesuits – his strategists in the campaign to re-Catholicize the Hapsburg lands – had been directed at averting the danger that the Austrian, Hungarian, and Bohemian *Stände* might manipulate the imperial succession to their advantage.

Rome's goal was to prevent a dynastic crisis in the House of Hapsburg after Rudolf II died, since the new emperor Matthias had no direct heirs, a situation that could have led to the assigning of provisional powers to the Elector Palatine Frederick V, a prominent representative of the aggressive Calvinist front. The imperial election of Ferdinand of Hapsburg in 1619, a model pupil of the Jesuits and known for his religious devotion and his zealous campaign against the heretics, was an undeniable victory for the Apostolic See. In this and a handful of other cases the notion that the papacy had the right to exercise surveillance over State affairs when these impacted on the interests of the Church assumed blatantly concrete form.[87]

Also in 1609 a book *De potestate papae* ('On the power of the pope') was published in London. Its author, William Barclay, was a well-respected Catholic jurist of Scottish origin who taught for thirty years in France, and King James I saw in him a precious ally in the war of documents that he was waging against the papacy. Barclay laboured over his book for years, during the darkest period of the French religious wars, during the reign of Henry III. The debate between Rome and London about the Oath of Allegiance and the Interdict of Venice finally persuaded him to prepare a definitive version of his work, which his son had published after his death.

Barclay's aim was to advance a theological-political theory marked by an idea that had characterized the French *Politiques* – the division of power between different spheres of interest:

---

[87] On the political dynamics in the Holy Roman Empire during the 1610s and 1620s with regard to the foreign policy of the papacy, see JAN PAUL NIEDERKORN, *Papst, Kaiser und Reich während der letzten Regierungsjahre Kaiser Rudolfs II. (1605-1612)*, in ALEXANDER KOLLER (ed.), *Die Aussenbeziehungen der römischen Kurie unter Paul V. Borghese (1605-1621)*, Tübingen, Max Niemeyer, 2008, pp. 83-99, and KOLLER, *Papst, Kaiser und Reich am Vorabend des Dreissigjährigen Krieges (1612-1621). Die Sicherung der Sukzession Ferdinands von Innerösterreich*, *ibid.*, pp. 101-120.

> We must know that, since the beginning, those two powers by means of which the world is bound to its duties, that is the political and the ecclesiastical, are so distinct and separate between them, that (though they both originate from God) each of them is closed within its own limits, and cannot by its own right cross the boundaries of the other, and neither of them exercises power over the other.[88]

The authority over the civil commonwealth that was claimed, even if only in indirect form, by the popes was an usurpation of power dating no further back than Gregory VII. Had God actually ordered it, the Church Fathers would have surely have advised the emperors about it.[89] This was what Barclay argued, and there was no doubt at the time that he had on his side the composite Catholic world which Rome had distanced herself from since the Council of Trent.

The *De potestate papae* was placed on the Index in 1609 together with other titles, among them the *Vindiciae contra tyrannos*, a cornerstone of Calvinist political theory written by Iunius Brutus (Philippe de Duplessis-Mornay) some thirty years earlier, in 1579. Bellarmine refuted Barclay's positions in his last published work on controversialist and theological-political issues, the *De potestate Summi pontificis in rebus temporalibus adversus Gulielmum Barclaium* (1610). In it Bellarmine presented well-known theses already elucidated in the *Disputationes* and widely taught in Jesuit schools; nevertheless, at the time it was received as an official intervention of the Holy See, intended to reaffirm the doctrinal nature (as an article of faith) of the *potestas indirecta* of the pope, against the Gallican tradition that prevailed in France.

The moment chosen could not have been worse: the deadly attack on Henry IV in May of that year made the country sink again into the fear of religious war. As I have already noted, the *De potestate Summi pontificis* was condemned in November by the Paris Parliament. Just a few months earlier the Parliament had condemned the *De rege* of Father Juan de Mariana, the author who had come to be widely regarded as the Jesuit theorist of regicide.[90]

The death of Henry IV marked the climax of a historical stage in which the theoretical structure of papal sovereignty proposed by the Ro-

---

88 WILLIAM BARCLAY, *De potestate papae, an et quatenus in reges et principes seculares ius et imperium habeat*, Pont-à-Mousson, apud Iacobum Garnich, 1610, p. 9.

89 *Ibid.*, p. 40.

90 SYLVIO H. DE FRANCESCHI, *La crise théologico-politique du premier âge baroque. Antiromanisme doctrinale, pouvoir pastoral et raison du prince: le Saint-Siège face au prisme français (1607-1627)*, Rome, École française de Rome, 2009, pp. 203 sq.; STEFANIA TUTINO, *Empire of Souls. Robert Bellarmine and the Christian Commonwealth*, Oxford, Oxford University Press, 2010, pp. 160 sq.

man Church in the second half of the 16th century reached a critical point within the context of open conflict with other institutional players who were seeking to enforce their absolutist monarchical powers, as in the case of England and France, or to reinforce or defend their customary rights, as in the case of the Republic of Venice. What these players were laying claim to was full autonomy in the realm of the *saeculum* – the political sphere with its own forms of sovereignty and its own ties of obedience and control between governing bodies and the governed, between the ruler and the ruled – and its separation from the realm of spiritual matters and the authority of the Church.

These players perceived the doctrine supporting the interference of the Roman Church in the legitimate structures of sovereign rule and in the jurisdiction of the lay courts as a completely new and unjustified theory, fashioned in the theological workshops of the papacy and overtly going against Christian tradition. In his writings on the Venetian Interdict, Sarpi never tired of reaffirming that the reasons of the Republic were founded on custom and the shared consensus of the universal Church, while the pretensions of Rome in matters pertaining to the judiciary and benefices originated out of distorting modern trends on the part of the Holy See: «A new doctrine seeking boundless and limitless power», aimed at handing over the rights and jurisdictions of the community of the faithful – abbots, bishops, chapters of long-standing Venetian loyalty – to the agencies of the Roman Curia.[91]

At the time, it must be said, clear boundaries dividing the spiritual and the temporal spheres did not exist, and could not even be defined due to the entangled rules in matters of mixed jurisdiction (such as astrology, witchcraft, polygamy, and blasphemy), and because both spheres sought legitimacy in the authority of Scripture, and hence in the interpretation of it.

William Barclay supported his arguments on the separation of powers with numerous passages taken from the Old and New Testaments and the Church Fathers, as did Bellarmine in the treatise written to counter him. The difference lay in the fact that Bellarmine not only could counterpose interpretation to interpretation, biblical passage to biblical passage; he was also able to claim that *his* was the ultimate interpretation as it rested on the pope's authority as the highest judge on matters of controversy.

A theological argumentation would also be used by Galileo in his second letter on sunspots, composed in 1612 and therefore shortly before his 'Copernican writings'. Here the scientist declared that the thesis of the in-

---

91 Frajese, *Sarpi scettico* (cit. note 73), pp. 303 sq.

corruptibility of the heavens was linked to the 'Aristotelian error' of the eternity of the world, and hence was «an opinion that was not only false, but also erroneous, repulsing the unquestionable truths of the Sacred Writings». Here Galileo was deploying expressions and judgments borrowed from the domain of the Inquisition. Not by chance the Roman censorship intervened, requiring that this passage be erased from the text.[92]

The adoption of the figure of the *iudex fidei* by the controversialists meant that every proposition founded on Scriptural or, more generally, theological arguments fell under the jurisdiction of the Chair of Saint Peter, even if only to establish whether or not something should be considered as pertaining to faith, as in the case of Barclay's *De potestate papae*. But the breadth and diffusion of the religious discourse, as well as the prominence of the Bible as a virtually infinite repository of ethical, political, and even scientific models, meant that the pope's authority extended over an enormous area. Within a few years both the jurisdictional debates between the Apostolic See and its secular rivals, and the debate on the character of heliocentric theory would test the boundaries of that jurisdiction.

It must be added that in the view of the Roman Curia the distinction between the field of the jurisdictional claims made by temporal powers and that of real heresy was extremely nuanced.

France was a country of two confessions, and in the Parliaments where debates over papal claims and Jesuit plots were being conducted, Protestant judges were also seated by right of law. As for Venice, Paolo Sarpi was using all of his influence and the English embassy was acting feverishly behind the scenes to bring about an official conversion of the Republic's ruling class to Protestantism, a possibility that was causing great concern to Pope Paul V.

The Anglican controversy itself – which raged between 1606 and 1610 and sealed the minority status of English Catholics well into the 19th century – was played out in the vague borderline area dividing political obedience and membership in a religion.

With the Oath of Allegiance imposed in 1606 on suspect recusants in the aftermath of the Gunpowder Plot, King James sought to shed light on the distinction between 'loyal' subjects and the followers of the *potestas indirecta* – which, in his view, was tantamount to a theological justification of

---

92 Galileo, *Istoria e dimostrazioni intorno alle macchie solari*, OG, V, pp. 138-140, n. 6. See Philippe Boulier, *L'inaltérabilité du ciel pose un problème théologique*, «Galilæana», 10, 2013, pp. 41-71: 53-54. According to the author, these expressions would be suggested by Castelli, in the perspective of reconciling Galilean physics and Christian metaphysics.

regicide. Indeed, the oath was formulated to include not only a declaration of allegiance to the king but also the rejection of the doctrine that the pope «hath any power or authoritie to depose the king, [...] or to authorize any forraigne prince, to invade or annoy him, or his countreys, or to discharge any of his subiects of their allegiance, and obedience to his Maiestie».[93]

According to King James the oath was merely a political measure – and, for the purposes of my analysis, a further example of a secular sovereign's claim of distinct and autonomous jurisdictions. In the view of the Roman Curia it was instead a wholly unacceptable document, one containing heretical statements, and as such it was condemned by the Holy Office in the spring of 1606.

There followed a long, drawn-out controversy with repeated interventions on both sides. It arose simultaneously with the Interdict and would continue for several years as a Europe-wide debate between the *regnum* and the *sacerdotium* (an issue later summarized in Hobbes' *Leviathan*). In 1609 King James addressed an *Apology for the Oath of Allegiance* (*Apologia pro iuramento fidelitatis*) to all the princes of Europe, an appeal to unite in defense of the rights of the temporal powers (the papal nuncio in France, Monsignor Ubaldini, read between its lines «the style and opinions of Fra Paolo»).[94]

English Catholics were forbidden to take the oath by Paul V's brief *Magno animi moerore* issued in September 1606, «seeing it conteines many things which are flat contrary to faith, and salvation».[95] In one of his contributions to the debate, the *Apologia Matthaei Torti* (1608), Bellarmine confirmed the dogmatic nature of the indirect power of the pope, and hence the impossibility of separating membership in the Catholic Church and the obligation to obey the pope as one's spiritual sovereign, i.e. higher in authority than any temporal prince:

> This oath not only entails civil obedience in merely temporal matters, [...] but also the denial of pontifical power, which is not a merely temporal matter, but a sacred matter, established by Heaven, which no mortal can abrogate, or diminish.[96]

---

[93] The text of the oath may be found in James I, *Triplici nodo triplex cuneus. Or an apologie for the Oath of allegiance, against the two breves of Pope Paulus Quintus, and the late letter of Cardinal Bellarmine to G. Blackwel the Archpriest*, London, by Robert Barker, 1607, p. 48. See on the issue Motta, *Bellarmino* (cit. note 14), pp. 385 sq.; De Franceschi, *La crise théologico-politique* (cit. note 90), pp. 61 sq.; Tutino, *Law and Conscience. Catholicism in Early Modern England, 1570-1625*, Aldershot, Ashgate, 2007, pp. 117 sq.

[94] De Franceschi, *La crise théologico-politique* (cit. note 90), p. 213.

[95] James I, *Triplici nodo triplex cuneus* (cit. note 93), p. 49.

[96] Tutino, *Law and Conscience* (cit. note 93), p. 165.

This overlapping of spheres was indeed implicit in the text of the Oath: in a further passage, James' subjects were asked to «abhorre, detest and abiure as impious and hereticall» the doctrine of the dissolution of the duty of obedience to an excommunicated king, and to believe «in conscience» that the pope could not abolish the binding nature of the oath itself.[97]

In all respects James was fully entitled, as the head of the Church of England, to stigmatize a doctrine as impious and heretical, but of course in the eyes of Rome (notwithstanding his diplomatic expressions of deference toward the prestige of the Apostolic See), this could not but emphasize his heretical position. Even worse, his attempt to impose on his subjects a persuasion «in conscience» meant that temporal power was encroaching on the private sphere of the individual conscience, a jurisdiction reserved for the Church and upon which it exerted rigid control by sacrament.[98]

In conclusion, I would like to highlight once again that the hermeneutical issue raised by Copernican cosmology occurred at a moment in history when the Roman Church was in deep crisis caused by what one might describe as an «excess of sovereignty» in the person of the pope, as had been envisaged by Catholic controversialist theology in the late 16th century on the basis of the theological-political theory of the judge of controversies. This claim to sovereignty triggered a reaction from the defenders of temporal power, who were becoming increasingly accustomed to using theological arguments to legitimize their own jurisdictional claims.

The heart of the crisis lay in the impossibility of clearly dividing the spiritual and temporal jurisdictions in a context where the Bible still represented the norm and the reference text, but in Rome the very idea of the separation and respective autonomy of the two forms of power was perceived as a challenge to the full spiritual sovereignty of the pope.

I therefore believe it is plausible to argue that this perception reflected – at least by analogy – on Galileo's hermeneutical proposals, causing a reaction that was aimed at preventing another sphere of jurisdiction (in this case the relationship between nature and Scripture) from being subtracted from the highest judge of faith.

---

97 JAMES I, *Triplici nodo triplex cuneus* (cit. note 93), p. 114.

98 This aspect of the Oath of allegiance has been underscored by TUTINO, *Law and Conscience* (cit. note 93), pp. 133 sq. On the different interpretations of the document, see JOHANN P. SOMMERVILLE, *Papalist Political Thought and the Controversy over the Jacobean Oath of Allegiance*, in ETHAN H. SHAGAN (ed.), *Catholics and the 'Protestant Nation'. Religious Politics and Identity in Early Modern England*, Manchester, Manchester University Press, 2005, pp. 162-184.

Luigi Guerrini

THE ARCHBISHOP AND ASTRONOMY
Alessandro Marzimedici and the 1604 Supernova

An undertaking that might shed new light on our knowledge of the spread of observational astronomy in the Tuscan milieu in which Galileo operated is a detailed reconstruction of the early seventeenth-century theological and literary culture in Florence. In particular, increasing and enhancing what we know about Archbishop Alessandro Marzimedici's attitude towards scientific studies of the sky could improve our understanding of the Florentine discussions over the 1604 supernova (1604-1606) and the issue of the earth's movements (1610-1615). Bearing in mind that Galileo and Marzimedici were coeval and met each other under many circumstances, we know that at the very beginning of his prestigious ecclesiastical career Marzimedici held the rank of Canon of the metropolitan church of Florence, and then was appointed Auditor of the Apostolic Nunciature. He began to climb the ladder of the higher clerical hierarchy, in 1595 becoming Bishop of Fiesole. Ten years later he was elected Archbishop of Florence by Pope Paul V and in this conspicuous new capacity he became Moderator and Cesarean Chancellor of the Florentine Theological Faculty. He took control of the activities of this ancient institution, alongside overseeing the discussions that took place within city's monastic schools. Such a position offered him the opportunity to regulate and limit the theological debate, easily having cultural dissidents and opponents eating out of his hands.[1]

---

[1] I would like to express my own sincere gratitude to Dr. Catherine Karnitis for her precious help and valuable assistance. On the activity of the Florentine theological faculty in the early seventeenth century and on the connection between religion and science in the capital of the Grand Duke of Tuscany, see: Luca Giuseppe Cerracchini, *Fasti teologali ovvero notizie istoriche del Collegio dei teologi della sacra Università fiorentina dalla sua fondazione sino all'anno 1738 [...]*, Florence, F. Moücke, 1738; Michele Cioni, *I documenti galileiani del S. Uffizio di Firenze*, Florence, Pagnini, 1908; Arnaldo D'Addario, *Aspetti della Controriforma a Firenze*, Rome, Edizioni degli Archivi di Stato, 1972; Adriano Prosperi, *L'inquisizione fiorentina al tempo di Galileo*,

In accordance with a long-standing tradition followed by many authorities in public life in Renaissance Florence, Marzimedici also gathered and chaired a circle («cenacolo») of intellectuals that remained active up until his death in 1630. This *cenacolo* involved not only theologians and clerics, but also literary men, natural philosophers, and even mathematicians. Among those admitted to its meetings were Ludovico delle Colombe, the natural philosopher who dedicated his 1606 *Discorso sopra la nuova stella* to the Archbishop,[2] and the Dominican friar Tommaso Caccini, who later would publicly denounce Galileo's sympathies for mathematics and for the Copernican cosmology as heresy and unorthodox beliefs.[3]

Marzimedici had a deep appreciation of the importance of astronomy and grasped its wider intellectual connection with theology. He discussed his ideas about how to survey celestial phenomena with his circle of experts and friends. We can say that Marzimedici was especially interested in the appearance of the supernova in 1604, which he realized was an event bound to bring huge consequences for both theology and cosmology. He publicly debated the topic, together with that of the corruptibility and fluidity of the sky, and sought to devise some kind of solution to settle the contentions about the disciplinary boundaries and the right hierarchical order connecting astronomy and theology. Starting in 1610 he also followed the debate over the double movement of the earth and the Copernican cosmology that was unfolding in Florence and watched with growing concern Galileo's rise in favor at the court of Tuscany. Marzimedici did not really try to address astronomical problems using mathematics, and perhaps he did not grasp the inner unity of astronomy, mathematics and physics implied in the works of Copernicus and Galileo. However, he certainly did have a clear understanding of the far-reaching consequences of the new astronomy on traditional cosmology and was filled with increasing anxiety

in Paolo Galluzzi (ed.), *Novità celesti e crisi del sapere. Atti del convegno internazionale di studi galileiani*, Florence, Giunti Barbèra, 1984, pp. 315-325; Luigi Guerrini, *Galileo e le origini della polemica anticopernicana a Firenze*, Florence, Polistampa, 2009, *passim*. Even though it does not deal with Marzimedici, see also Kathleen M. Comerford, *"The care of souls is a very grave burden for the pastor": Professionalization of clergy in early modern Florence, Lucca, and Arezzo*, in Wim Jansen – Barbara Pitkin (eds.), *The formation of clerical and confessional identities in early modern Europe*, Leiden-Boston, Brill, 2006, pp. 349-367.

2 Ludovico delle Colombe, *Discorso nel quale si dimostra che la nuova stella apparita l'ottobre passato 1604 nel Sagittario non è cometa né stella generata o creata di nuovo né apparente, ma una di quelle che furono da principio, e ciò esser conforme alla vera Filosofia, Teologia e Astronomiche demostrazioni*, Florence, Giunti, 1606.

3 See Sergio Pagano (ed.), *I documenti vaticani del processo di Galileo Galilei (1611-1741)*, Vatican City, Archivio Segreto Vaticano, 2009, p. 23.

in the face of any kind of cultural transformation that might pose a threat to the supremacy of Catholic theology.

Marzimedici was convinced that theology and astronomy were allied sciences and that learning about Aristotelian cosmology and Ptolemaic astronomy could effectively help Christian believers to rekindle their faith. He saw the study of the traditional universe as a powerful tool for worship and it was for this reason that he never accepted Galileo's astronomical revolution and constantly tried to counter the spread of the heliocentric theory.

I will begin my paper with an analysis of the involvement of the archbishop in astronomical studies, and in particular his interest in the 1604 supernova, by considering the discussions this phenomenon triggered within the Medici court. For this task, I present carefully selected passages from two sources containing an account in some form of these discussions, namely Raffaello Gualterotti's *Discorso sopra l'apparizione de la nuova stella* (1605),[4] and Michelangelo Buonarroti's tragic opera *Il natal d'Ercole* (1605).[5] I will then direct my attention towards Ludovico delle Colombe's *Discorso sopra la nuova stella* and Carlo Marucelli's *Canzoni*, both published in 1605 and both dedicated to the archbishop.[6] These latter two works directly stem from the discussions held in the intellectual circle of Marzimedici and provide valuable information about his personal assessment of the latest developments in astronomy, in particular those connected with the fluidity of the sky and the appearance of new stars. After comparing the discussions held at the court and at the palace of the archbishop, my general assessment will be that among traditionalist thinkers a substantial disagreement emerged about the ways in which the threatening idea of the corruptibility of the skies and the disturbing issue of the nature and position of the new star ought to be countered.

Dealing with these kinds of problems was a tricky and dangerous matter, which required a clever and artful strategy involving a great deal of dissimulation. The new astronomy of Tycho, Kepler and Galileo was rejected utterly and unanimously by theologians and natural philosophers. Some-

---

4 Raffaele Gualterotti, *Discorso sopra l'apparizione de la nuova stella e sopra le tre oscurazioni del sole e de la luna nel anno 1605. Con alquanto di lume del arte del oro*, Florence, Giunti, 1605.

5 Michelangelo Buonarroti the Younger, *Il natal d'Ercole. Favola rappresentata al Serenissimo Signor don Alfonso da Este principe di Modana e all'Eccellentissimo Signor don Luigi suo fratello, nella venuta loro a Firenze da Madama Serenissima di Toscana nel palazzo dell'Eccellentissimo Signore D. Antonio de' Medici*, Florence, Giunti, 1605.

6 Carlo Marucelli, *Canzoni all'Illustrissimo e Reverendissimo Monsignor Alessandro Marzimedici*, Florence, Tosi, 1605.

thing effective had to be done, however, to reaffirm that the true account of the origin of the universe was to be found in the book of *Genesis*, and to reinforce the idea that the sky's immutability had been definitively asserted by Aristotle in *De caelo*. However, unlike those who later denounced Galileo to the Inquisition, neither Marzimedici nor his closest cultural advisors thought that the most effective way to prevent or curtail the spread of the new astronomy in Florence was to defame and publicly accuse its upholders, let alone denounce them to the Holy Office in order to punish them for their alleged heresy and arrogance. Therefore, a rift slowly but steadily opened up between those who called for more rigorous repressive methods (first and foremost some of the Dominican friars at the convents of Santa Maria Novella and San Marco) and those who were convinced that the stronger the effort at repression, the greater the resistance that would be produced as a result of it. Additionally, the inevitable clashes caused great annoyance to the political powers in Florence, who sought to protect Galileo against various vagaries as the controversy grew.

In short, for Marzimedici the choice was not simple, because he did not want to appear as a mere referee, but did not wish to engage himself in a public dispute either. Fundamentally, he believed that the faith in the Aristotelian-Ptolemaic universe that had been closely integrated into the Biblical account of Creation during the previous centuries (roughly since the time of Thomas of Aquinas) had to be reignited and strengthened through the enhancement of the Church's communication and propaganda campaigns, using speeches, lessons, and any form of written work that would be able to take on the enemy dialectically. Nevertheless, this should be done without waging a holy war against heresy, carefully avoiding any sort of political disagreement with the authorities who in one way or another supported the development of the sciences within the Grand Duchy of Tuscany.

At the outset, those most involved in the discussion deemed the archbishop's strategy to be well crafted and sufficiently effective to deal with the issues both of the 1604 supernova and the corruptibility of the sky. However, after the publication of *Sidereus Nuncius* in 1610, Marzimedici's attitude began to be seen by some as in fact creating a chink in the traditionalists' armor rather than offering a truly effective course of action against their opponents. Hence, the most determined among Galileo's foes – Dominicans such as Tommaso Caccini and Raffaello delle Colombe, as well as some staunch natural philosophers such as the already mentioned Ludovico delle Colombe, who was ceaselessly mocked by Galileo in his works – stepped into the fray, leaving Marzimedici and the Florentine clergy behind them. Addressing themselves to the highest authorities of the Dominican order in Rome, they decided to strike a blow against Galileo by denouncing his heliocentrism to the Roman Inquisition.

One may wonder whether this decision really grew out of the conviction that the Church hierarchy in Florence would never take decisive action to put a stop to Galileo's campaign or if some more personal motivations may have played a role. It is my observation that within traditional historiographical strands individual grudges and personal obsessions have been overemphasized. As my study will show, it was chiefly the feeling that Marzimedici was being over-cautious during the discussions in 1604 over the new star that eventually led to the decision by this group of Galileo's most stubborn adversaries that the time was ripe to spread their own wings.

Contrary to the Florentine Dominicans, at the end of the discussions on the 1604 supernova Marzimedici took no public action against the forward-looking astronomers. He did not react publicly either to the publication of Galileo's *Sidereus Nuncius* in 1610 nor to the success of the book on sunspots in 1613. This does not mean that from the very beginning of the controversy he did not grasp the significance and the possible consequences of the new astronomical discoveries for religious and theological culture. However, even when the maelstrom broke loose, he recoiled and did not publicly join the altercation, avoiding any kind of direct confrontation with Galileo's powerful patrons. From a historical point of view, this is something that can be fully understood only after a more accurate reconstruction and a clearer picture of the Florentine debate over the 1604 supernova has been provided.[7]

Archbishop Alessandro Marzimedici's Coat of Arms.

---

[7] See Stillman Drake, *Galileo against the philosophers in his Dialogue of Cecco di Ronchitti (1605) and Consideration of Alimberto Mauri (1606) in English translations*, Los Angeles, Zeitlin and Ver Brugge, 1976; *Firenze e la Toscana dei Medici nell'Europa del Cinquecento. La corte, il mare,*

1. 'Favole' and 'Discorso'. Representing and decoding the New Star

Two alternative intellectual perspectives and even two artistic traditions faced each other during the discussions triggered in Florence by the appearance of the 1604 supernova, and the Medici court was the center around which both revolved. In 1605 two notable works were published in the capital of the Grand Duchy of Tuscany and they both directly and indirectly dealt with the issue of the new star that had appeared in the sky the previous year. Their authors, the non-Aristotelian natural philosopher Raffaello Gualterotti and the much sought-after poet Michelangelo Buonarroti the Younger (the great-nephew of the celebrated artist), were important figures at the Medici court, to which they were often summoned and even paid to write about and discuss artistic and scientific issues.[8]

The origin of Gualterotti's explanation of the new star and the conformity of his arguments with the Stoic model of condensation and rarefaction have been highlighted by Eileen Reeves in her 1997 book *Painting the Heavens. Art and Science in the Age of Galileo*, in which it is also suggested that Gualterotti and Galileo actually met in Florence in the summer of 1605 to discuss the contents of Gualterotti's *Discorso sopra l'apparizione de la nuova stella*.[9] Reeves delves into this work, looking for similarities and affinities

i mercanti. La rinascita della scienza. Editoria e società. Astrologia, magia e alchimia, Milan, Electa, 1980; Galluzzi, *Il mecenatismo mediceo e le scienze*, in Christian Bec *et al.* (eds.), *Idee, istituzioni, scienza ed arti nella Firenze dei Medici*, Florence, Giunti, 1980, pp. 189-215; Michael H. Shank, *Setting the stage: Galileo in Tuscany, the Veneto and Rome*, in Ernan McMullin (ed.), *The Church and Galileo*, Notre Dame, University of Notre Dame Press, 2005 pp. 57-87.

8 Kepler's was the first description of the nova. See Johannes Kepler, *A thorough description of an extraordinary new star which first appeared in October of this year 1604* (Eng. trans. by Judith Veronica Field and Anton Postl), «Vistas in Astronomy», 20, 1977, pp. 333-339. On which, Miguel A. Granada, *Kepler vs. Roeslin on the Interpretation of Kepler's Nova: (1) 1604-1606*, «Journal for the History of Astronomy», 36, 3, 2005, pp. 299-319 and Patrick J. Boner, *Kepler's Copernican campaign and the new star of 1604*, in Boner (ed.), *Change and continuity in early modern cosmology*, Dordrecht, Springer, 2011, pp. 93-114. See also David H. Clark – Richard F. Stephenson, *Which historical new stars were supernovae?*, «The Quarterly Journal of the Royal Astronomical Society», 17, 1976, pp. 290-302 and Id., *The historical supernovae*, Oxford, Pergamon Press, 1977. For more details, Boner, *Kepler's cosmological synthesis: Astrology, mechanism and the soul*, Leiden, Brill, 2013, pp. 69-103.

9 Eileen Reeves, *Painting the Heavens. Art and Science in the Age of Galileo*, Princeton, Princeton University Press, 1997, pp. 78-81. See also Massimo Bucciantini, *Galileo e la nova del 1604*, in Bucciantini – Maurizio Torrini (eds.), *La diffusione del copernicanesimo in Italia: 1543-1610*, Florence, Olschki, 1997, pp. 237-248; William R. Shea, *Galileo and the supernova of 1604*, in Massimo Turatto (ed.), *1604-2004: Supernovae as cosmological lighthouses. Proceedings of a meeting held in Palazzo del Bo, Università degli studi di Padova, Padua, Italy 15-19 June 2004*, San Francisco, Astronomical Society of the Pacific, 2005, pp. 13-20. For further more detailed studies, but

with Plutarch's and Seneca's hypotheses on the causes that bring about the appearance of new stars:

> Gualterotti thought that the new star might have produced by various means, and these hypotheses are recognizably Stoic in origin. He believed, for instance, that the fluid heavens might have become so compressed as to have formed the nova, a possibility associated with Stoicism in Plutarch's *On the Face of the Moon*, and briefly described there as pertaining to the creation of the fixed stars. In conformity with the Stoic model of condensation and rarefaction, Gualterotti also maintained that earth contained, or as he put it, "participated with", water, water with air, and air with the heavens, this last being composed of a pure fire identical with ether, and he insisted, as did Seneca in his discussion of underground rivers and caverns in Book Three of the *Natural Questions*, that each element contained all the other ones.[10]

Gualterotti, who was also an accomplished astrologer, alchemist and poet, certainly must have been well versed in the classical literature. However, he was a proficient practitioner of observational astronomy as well.[11] In Florence he was famed for the accuracy of his astronomical observa-

---

outside the Florentine borders, see Ugo Baldini, *La nova del 1604 e i matematici e filosofi del Collegio romano. Note su un testo inedito*, «Annali dell'Istituto e Museo di storia della scienza di Firenze», 6, 1981, pp. 63-98; Saverio Ricci, *Federico Cesi e la Nova del 1604. La teoria della fluidità del cielo e un opuscolo dimenticato di Joannes van Heeck*, «Atti dell'Accademia Nazionale dei Lincei. Rendiconti. Classe di scienze morali storiche e filosofiche», 43, 1988 (1990), pp. 111-133; Granada, *Johannes Kepler and David Fabricius: Their discussion on the nova of 1604*, in Boner (ed.), *Change and continuity* (cit. n. 8), pp. 67-92 and Id., *After the nova of 1604: Roeslin and Kepler's discussion on the significance of the celestial novelties, 1607-1613*, «Journal for the history of astronomy», 42, 2011, pp. 353-390.

10 Reeves, *Painting the Heavens* (cit. note 9), p. 79.

11 In 1605, he also arranged an optical device for examining stars' features and later contested the fact that the telescope was invented by a Dutch: «A friend of both Galileo and Sarpi, Raffaello Gualterotti, for instance, contested the glory of the anonymous Dutch inventor, in part because he felt that such credit naturally belonged to a Florentine, but also because he saw both his own use of the cerbottana as a means of observing stars by day in 1605 and his alleged development of another optical instrument "for jousting and for war" around 1598 as worthy precursors of the device from the Netherlands. Even though Gualterotti himself called that invention "a puny thing", he associated it and the cerbottana with his early experimentation, around 1580, with a pinhole aperture in a camera obscura. Gualterotti's guide in this experiment was almost certainly the first version of della Porta's *Natural Magic*, where the reader was invited to descend into "the deepest of wells" in order to see stars by day, and in the following chapter offered, as if by default, a much more manageable procedure, that of making a pinhole aperture in a dark room. The fact that the pits of wells, the cerbottana, the camera obscura without a lens, and the eventual telescope all involved restricted visual rays and diminished circumambient light, in other words, persuaded some to see these earlier devices as clear antecedents of the Dutch invention» (in Reeves, *Galileo's Glassworks. The Telescope and the Mirror*, Cambridge (Mass.), Harvard University Press, 2008, pp. 85-86).

tions, which he usually conducted in the *Galleria Medicea* of the Grand Duke's palace.[12] With regard to the new star, the point I would like to emphasize is that Gualterotti chose to rely on a direct and in some sense radical approach to describe both its nature and its position, one which reflected his own scientific mindset and his own certainty about the penetrability and mutability of the skies. He did not rely solely on classical texts and the accounts of others and, having a great deal of confidence in his own abilities, mustered the courage to work out his own answer. Thus, he came to the conclusion that, while the birth of the star ought to be considered a perfectly natural event, condensation alone did not provide a sufficient explanation. In addition to the simple process of solidification, time was required for this unusual type of phenomenon to unfold. Therefore, he argued that «[...] to create a new star, heaven needs do no more than what it ordinarily does; that is, to condense itself, but to go just a little further and prolong this condensation».[13]

As a natural philosopher, Gualterotti thought that any kind of natural event could be understood through observation and experiments («*Perché il cielo è penetrabile e questo lo insegna l'esperienza*»),[14] and he believed that the human mind was perfectly equipped to solve the riddle of the star spotted in the sky over Florence. Moreover, he was a bold thinker and was convinced that every scientist had the obligation to set forth what he thought, even if his conclusions were openly at odds with the Aristotelian view of the natural world:

> I believe in large part that this star newly appeared in the eighth heaven was created... And according to generally held Aristotelian opinion, i.e. that the sky is impenetrable, inalterable, and eternal, never a body in such a form could be generated in the sky; but I do not retain that this opinion is actually correct.[15]

Gualterotti's conclusion was audacious, blunt, and resolute, and rested on the prolonged observation of celestial condensation, whose length and duration he was able to confirm. He did not fear either the anger or the negative reactions of those natural philosophers who were committed to Aristotle's thought and doctrines. Addressing himself directly to the Grand Duke Ferdinando de' Medici, Gualterotti was keen to show his own

---

12 Antonio Favaro, *Raffaello Gualterotti*, in Galluzzi (ed.), *Amici e corrispondenti di Galileo*, Florence, Salimbeni, 1983, pp. 647-669: 658.

13 Gualterotti, *Discorso* (cit. note 4), p. 20.

14 *Ibid.*, p. 12.

15 *Ibid.*, p. 11.

resolution in the face of the appearance of the new star, which rested first and foremost on an unshakeable belief in the reliability of the senses, in particular vision. He regarded vision as the most trustworthy source of evidence and as the main tool whereby celestial phenomena and planetary movements could be explained.

However, at the Florentine court the appearance of the new star was not just a matter for professional natural philosophers. The unexpected celestial phenomenon also drew the attention of members of the court, and the consideration of artists and poets, who evaluated the event differently and drew divergent conclusions. When the duke of Modena, Alfonso III d'Este, along with his brother Luigi d'Este, arrived in the capital of the grand duchy in the early days of October, a tragic opera («melodramma») in five acts entitled *Il natal d'Ercole* was performed in the courtly theater at the residence of Don Antonio de' Medici. Its author, Michelangelo Buonarroti the Younger, arranged the entire festival and dedicated the *libretto* to Alfonso d'Este on October 22, 1605.

Buonarroti the Younger was born in Florence on November 4, 1568. In 1585, having received the most complete education he could dream of, he became a member of the *Accademia Fiorentina*. In 1596, after his graduation from the university in Pisa, he was appointed president (*arciconsolo*) of the Accademia della Crusca, where he shared his passion for poetry and the vernacular with fellow men of letters such as Giovanni and Pietro de' Bardi, Iacopo Soldani, and Maffeo Barberini, the future Pope Urban VIII. By 1605 Buonarroti the Younger undoubtedly had become one of the most prominent figures in Florentine literary circles, and at the Medici court he belonged to the group surrounding the Grand Duchess, Cristina of Lorena, for whom he composed a number of festival books in which events such as ceremonial entries into cities, weddings, christenings, and ambassadorial visits were recorded and lauded.[16]

The performance of *Il natal d'Ercole* held in honor of the Estense family has a dramatic plot crammed with references to astronomical phenomena, including the star discovered in 1604. Among other things, it testifies to both the enjoyment and the awe in observing the skies shared by the members of the Medici family and by most of the courtiers who frequented the

[16] Among the most recent contributions, see Janie Cole, *Se di fuori è dorata, dentro è d'oro: Maffeo Barberini, Michelangelo il Giovane e Galileo*, «Belfagor», 60, 355, 2005, pp. 1-26 and Id., *Cultural Clientelism and Brokerage in Early Modern Florence and Rome: New Correspondence between the Barberini and Michelangelo Buonarroti the Younger*, «Renaissance Quarterly», 60, 2007, pp. 729-788. See also Massimiliano Rossi, *La Crusca nell'occhio: l'Empoli tra Galileo e Michelangelo il Giovane*, «Galilæana», 4, 2007, pp. 189-209.

aristocratic circles in Florence in the early years of the seventeenth century. In this theatrical piece one can find references to the Milky Way, as well as a number of remarks on the new star. The latter clearly implied that the sky could change its – once allegedly eternal – image and undergo any kind of alteration. The clairvoyant Apollonian prophet, Tiresias, plays an active part in the story, especially in the fourth act, where we hear him refer to the new star:

Di nova luce m'allumò la mente,
E del futuro il cor presago rese.
Ond'io se vere son queste sembianze,
Questi novi prodigi, e gli altri segni,
Quai voi narrate or palesarsi in cielo;
Dirò, non senza de gli Iddei disdegno
Lassù cangiarsi il manto a' lumi alteri;
Poscia ch'a mezzo dì sparso di stelle
Risplende il cielo, e di novello cinto,
Che voi dite sembrar di neve un fiume,
Tutto si fregia il puro azzurro eterno.
[...]
Stelle infinte, e de' più degni aspetti,
Ch'immortalmente in Cielo han fisso il nido.[17]
[...]
O d'infinite stelle insieme accolte
Luce più folta, e di maggior sembianza.
O di simil cagioni opra palese.
Ma forma nuova che lassù si spanda.[18]

In these lines Buonarroti the Younger seems to be admitting the existence of an infinite number of stars in the sky, and shows no hesitation in acknowledging that something new was unfolding above human heads. Unlike Gualterotti, though, he claims that no man can fathom the natural reasons shrouded behind the birth of a new star. For this enigmatic phenomenon, any kind of logical or natural inquiry or explanation seems pointless. Scientific research was inconsequential, since human beings ought to restrict themselves to enjoying the sublime brightness of their light without searching for the causes that brought these stars into existence. Tiresias broaches this subject by suggesting that any kind of inquiry in this direction is basically useless and bound to fail:

---

17 Buonarroti the Younger, *Il natal d'Ercole* (cit. note 5), p. 27.

18 *Ibid.*, p. 29.

Qual anima presaga,
Qual di divino spirto uman pensiero
De' celesti segreti effetti, ed opre
Immaginar ne può ch'al vero aggiunga.
Come dianzi fallace il parlar nostro
Vaneggiava in cercar l'alta avventura,
Di cui lieto ogg'il ciel si fa più bello.
Vano il pensiero, e stolto ogni consiglio
E' de' mortali in terra,
Ove l'opra immortal si tesse in cielo,
Che delle stelle sol goder la luce
N'è dato a noi: ma chi le guida, e regge
Sol ne intende il voler della sua legge.[19]
[...]
Prodigi e segni nuovi, che dal Cielo
Si fan palesi per felice mostra,
Non son se non di bene, e di valore
Sembianze illustri, e sovra il natio figlio
Presaghi segni di fatal trionfi.[20]

Taking the gist of Buonarroti the Younger's 1605 poem, we can conclude that only God actually knows the way in which nature is arranged. Natural laws can perhaps be understood by men on earth with regard to the measures that control life in the corruptible beings scattered throughout the lower part of the universe, but knowledge of the eternal rules overseeing the skies is restricted to the creative mind who designed and wove the web that connected together all celestial phenomenon. Naturally, this also implies that theologians, rather than natural philosophers or mathematicians, are allowed to wisely speculate on God's decisions about the best arrangement of the natural world beyond the sphere of the Moon, and that all the novelties in the sky – real natural events though they were – should be seen as signs of the Christ's triumphal second advent.

Some partial conclusions can be drawn from what has been said so far. Between 1604 and 1605 the Florentine court was disturbed by the appearance of the new star, and personages such as Gualterotti and Buonarroti the Younger, both of whom were in the service of the Medici family, lent expression to the diffuse feeling of bewilderment. They tried to reassure

[19] *Ibid.*, p. 34.

[20] *Ibid.*, p. 45.

IL NATAL
D'ERCOLE
DI MICHELAGNOLO
BVONARROTI.
Fauola rappresentata al Serenissimo Signor Don
ALFONSO DA ESTE
Principe di Modana,
E all'Eccellentissimo Signor Don LVIGI suo fratello,
Nella venuta loro a Firenze.
DA MADAMA SERENISSIMA
DI TOSCANA.
Nel palazzo dell'Eccellentiss. Sig. D. ANTONIO MEDICI.
IN FIRE'
Nella Stamperia de' GI'

Michelangelo Buonarroti the Younger, *Il natal d'Ercole* (Florence, 1605).

their noble readers and listeners, each coming up with his own explanation. Gualterotti stated that the new celestial event could be assessed and explained by natural philosophers through observation and reason. Buonarroti the Younger reminded his readers that it could be admired without going into or rejecting its deeper implications, but stated his conviction that any attempt to conduct a focused inquiry into its primary causes would prove useless and ineffectual. Many in Florence shared one or the other of these positions, including probably some who did not belong to the Medici court. Nonetheless, both were utterly dismissed in the theological circles of the city, especially the one led by Archbishop Marzimedici.

2. Astronomy and Theology in Alessandro Marzimedici's Intellectual Circle

A few days before Christmas in 1605, a booklet dedicated to Archbishop Alessandro Marzimedici came out in Florence. It contained three *canzoni* that gave poetical expression to the moral and religious beliefs derived from the study of the sky. They lauded the traditional sacred mysteries of the Christian faith together with the earth-centered universe surrounded by celestial orbs made up of incorruptible quintessential matter and moved by an immobile divinity who was constantly worshipped by intelligences (or angels) and blessed souls. The heavens encircling the elements of the sublunary realm were believed to be eternal and incorruptible, as well as nobler than those of the disparaged nether realm. These *canzoni* were written by Carlo Marucelli, the elderly brother of the renowned polymath Francesco Marucelli. Carlo was a talented and popular poet, and a good friend of Alessandro Adimari. He was also known as the most dogged enemy of the lyricist Francesco Ruspoli, a composer of flowery exuberance whose work was greatly admired by his patron Don Antonio de' Medici.[21]

The first *canzone* describes the ascending order of the spheres that carry the planets and stars, enriched by additional spheres above the firmament in accordance with the Christian tradition. It praises the geocentric order of the universe, whose spiritual dimension is based on the presence of angels whose ardor keeps the spheres turning. Orbs are defined as God's «eternal offspring» and «potency in action» to which a unique circular motion has been transmitted:

E negl'Orbi celesti ordin distinto,
Nel Sacro Dì che più non alse, ed arse
Misto, e confuso l'onda, e 'l fuoco allato;
Prodotto all'alta cura
Dall'Ente, al moto, al ben'ovrare accinto
Fu qualunque di questi have il beato
Regno, liberi d'atto e di natura
Lor merto, altrui mercede
Confermar parte in Dio l'eterno piede.
Potenze in atto, al moto
D'eterno genitore eterna prole,
S'apriro, e grazia illuminante ad esse
Che le fece Alme dell'Eterea mole
Quando questi, ed il tutto ei mosse immoto.[22]

The investigation of this cosmological scheme will allow the human mind to worship the Divine through His natural creation and the spiritual and physical movement caused by the angels. The universe as a whole is an ideal structure unifying the natural and supernatural worlds. With study

[21] On Marucelli, see Giulio Negri, *Istoria degli scrittori fiorentini*, Ferrara, Pomatelli, 1722, pp. 198 and 218; Francesco Saverio Quadrio, *Della storia e della ragione d'ogni poesia, nel quale tutto ciò che alla narrativa o melica s'appartiene, è ordinatamente mostrato*, Milano, Agnelli, 1741, p. 490. On the Marucelli family, see Massimiliano Albanese, *Marucelli Francesco*, in *Dizionario Biografico degli Italiani*, Roma, Istituto dell'Enciclopedia Treccani, 71, 2008, pp. 376-379.

[22] Marucelli, *Canzoni* (cit. note 6), p. 8.

and contemplation, the human mind should grasp the universe as an ordered whole, harmoniously organized, without lingering on its particularities and its less estimable features. If the mind reflects in itself this universal harmony and sees the cosmos as a «divine sphaera», no conflict can arise between the creator and his creatures. If however the mind sets out on its own journey through the universe, wishing to understand natural laws by relying on rational understanding, the original harmony between love and knowledge, between theology and astronomy, could easily be destroyed and dissolve into nothing:

Svelata si rimira
Quant'è possibil la Divina Sfera
Da queste intellettive agil sustanze,
E come un solo agente al tutto impera
Nello speglio ideal si vede; e ammira.
Più s'interna il più degno,
Sì che scerne l'huom Dio, ch'ogn'altro avanze.
Temerario, Ei s'oppone, osa e delira,
Se duce, ardisce, onde 'l celeste Regno
Divien Campo di Marte,
E se ben perde, indi ostinato parte.[23]

In this *Canzone* man's fall and the origin of evil are partly seen as consequences of the mind's disaffection with the picture of the perfect and ordered cosmos adopted by the Christian tradition, made up of well-established celestial orbs and sheer, inalterable quintessence. As one can immediately grasp, the occasion that gave rise to these poetic songs was the appearance in 1604 of the supernova, recalled as a *fiammante stella*, which sparked discussion and quarrels, and eventually threatened the ideal universe that was so lovingly nurtured in the minds of Marucelli and Marzimedici.[24]

Marucelli's songs tied together the theological and astronomical thinking in Marzimedici's circle. From the archbishop's intellectual perspective, which was shaped by a subtle blend of Christian and Neoplatonic ideas, the exhortation to divine love could be combined with astronomical and cosmological knowledge. Such a meaningful connection is hardly surprising, and is on full display in Marucelli's second *canzone*, where man's contem-

[23] *Ibid.*, p. 9.

[24] For a broad assessment of the cultural debates in Florence about the connection between poetry and natural philosophy in the age of Galileo one must still see ERIC COCHRANE, *Florence in the Forgotten Centuries 1527-1800. A History of Florence and the Florentines in the Age of the Grand Dukes*, Chicago and London, The University of Chicago Press, 1973, pp. 198-203.

plation of a pure and untarnished universe, devoid of any form of generation or corruption, is the note that tunes the human mind, allowing it to comprehend the Christian mysteries:

> *Ivi gli effetti ascosi*
> *Stanno conforme all'esser loro espressi*
> *Gl'Argentati d'Amor cristalli eterni*
> *Per infinita quantitade impressi*
> *Dell'invisibil deità pomposi.*
> *Ivi eterna s'infiora,*
> *Senz'oltraggio sentir d'orridi verni,*
> *(Ove godendo alma beata posi)*
> *Primavera, che mai non si scolora.*
> *Così lucido, e terso*
> *L'ornamento si sta dell'universo.*[25]

Marucelli's poem goes on to argue that the link joining the universal order of the skies and human love is at the core of the Christian mystery of Creation. Indeed, human and animal life can only be preserved and perpetuated down through the ages by means of an earth-centered universe whose skies rotate eternally about a fixed point:

> *In questo eterno Delo*
> *Nel primo già de' sette giorni santi*
> *(Al Divino voler) gemina luce*
> *Natal s'aperse di celesti amanti,*
> *La cui virtù muove intendendo il Cielo,*
> *Ov'ha ciascuna sua stella*
> *Conforme a sé, che raro effetto adduce,*
> *Altri l'amor di un infiammato zelo*
> *A custodir l'alme viventi appella,*
> *Altri le specie, e 'l seme,*
> *Che 'l mondo ha, conservar lor cura preme.*[26]

With their nuances and richness of meanings, Marucelli's poems allow us to grasp the heavy weight of the spiritual, cultural, and aesthetic ideology characteristic of the Marzimedici circle, which was not as open-minded and receptive as the milieu surrounding the Medici court. Even more, these poems help us to understand the interconnection between the astronomical and theological fields embraced by clergymen in Florence. The

---

25 Marucelli, *Canzoni* (cit. note 6), p. 12.

26 *Ibid.*, p. 13.

model of the cosmos which they adopted was the one embedded in Biblical tradition, and influenced by Renaissance Platonism and Neoplatonism. The archbishop and his theological advisers followed the lead of Biblical commentators such as Francesco Cattani da Diacceto the Younger, bishop of Fiesole between 1570 and 1595 and the author of *L'Essamerone*, a work published in Florence in 1563 that adhered closely to the Ambrosian tradition.[27] It contains recurrent references to the earth's immobility and an elaborate theory of the elements, considered to the *sine qua non* conditions for the development of any form of life. Diacceto sketches a cosmological framework in which the highest heaven is motionless, while all the stars and planets attached to their transparent spheres revolve endlessly around the earth. The latter remains absolutely still at the center of the cosmological picture, the Empyrean and the earth being regarded as the ultimate limits of a finite universe. At some point, Cattani da Diacceto argues that within our cosmos:

CANZONI
DI CARLO MARVCELLI,
ALL'ILLVSTRISSIMO, E
REVERENDISSIMO
MONSIG. ALESSANDRO
MARZIMEDICI.
ARCIVESCOVO DI FIRENZE.
IN FIRENZE,
Appresso Francesco Tosi, MDCV.
Ad istanza di Simone G.enier, & Iacopo Fabeui compagni.

Frontespice of Carlo Marucelli's *Canzoni* (Florence, 1605).

all other things are mobile, and are [situated] in the middle between the empyrean Heaven,

[27] Francesco Cattani da Diacceto, *L'Essamerone del reverendo*, Fiorenza, Torrentino, 1563. Cattani da Diacceto also published a *Discorso sopra la superstizione dell'arte magica* (Fiorenza, Panizzi and Peri, 1567) and a *Discorso dell'autorità del Papa sopra il Concilio* (Fiorenza, Giunti, 1562), which had favorable outcomes among contemporary readers. See Agnese Fantozzi, *Cattani da Diacceto Francesco*, in *Dizionario biografico degli italiani* (cit. note 21), 22, 1979, pp. 510-511.

which is the highest, immobile extremity, and the earth, the lowest and equally immobile extremity.[28]

If we take into account this cross-cultural atmosphere, our historical understanding of the archbishop's most urgent priorities in the very early years of the new century can be enhanced. He refused to admit that anyone could prove the fluidity of the heavens and the presence of new stars in the firmament. He did not approve of any form of independence in mathematical and astronomical studies, declining support for their intellectual emancipation from theology. In Marzimedici's mind, the debates regarding novelties in the heavens should be guided by competent religious authorities. Only they could reassert the validity of the Aristotelian and Ptolemaic cosmology, and of the Biblical account of Creation, drawing on principles and values originating from both the Christian tradition and from Platonic thought as Christianized during the Renaissance. Even though the veracity of some of the astronomers' observations could not be denied, their significance had to be interpreted within a traditional philosophical and scientific framework in which extraordinary phenomena were issues to be sorted out, and not useful evidence of the dissolution of the old universe.[29]

3. Marzimedici's Theologians and Ludovico delle Colombe

Marucelli's dedication to Marzimedici bears the date December 21, 1605. Exactly one week later, on December 28th Ludovico delle Colombe published his own *Discourse* concerning the origin of the new star. This book as well was dedicated to the Archbishop of Florence,[30] and contained a clear declaration that Marzimedici personally «spoke his mind and argued

---

28 Cattani da Diacceto, *L'Essamerone* (cit. note 27), p. 76.

29 Although centered on sixteenth-century Germany, very useful suggestions can be found in Charlotte Methuen, *Science and theology in the Reformation: Studies in theological interpretation and astronomical observation in sixteenth-century Germany*, London, T&T Clark, 2008 and Id., *De la sola scriptura à l'Astronomia nova: principe d'authorité, principe d'accomodation et réforme de l'astronomie dans l'oeuvre de Jean Kepler*, in Granada – Édouard Mehl (eds.), *Nouveau ciel, nouvelle terre: la révolution copernicienne dans l'Allemagne de la Réforme (1530-1630)*, Paris, Les Belles Lettres, 2009, pp. 319-338.

30 See Maria Muccillo, *Delle Colombe (Colombo) Ludovico*, in *Dizionario biografico degli italiani* (cit. note 21), 38, 1990, pp. 29-31; Bucciantini, *Reazioni alla condanna di Copernico. Nuovi documenti e nuove ipotesi di ricerca*, in *I primi Lincei e il Sant'Uffizio. Questioni di scienza e di fede, Roma 12-13 giugno 2003*, Roma, Bardi, 2005, pp. 301-319; Guerrini, *Galileo e la polemica anticopernicana* (cit. note 1), *passim*.

about the subject», namely the origin and position of the new star.[31] Delle Colombe was fully aware that Marzimedici had always taken a great interest in astronomy, as well as the cosmological implications involved. He also knew that the head of the Florentine Church had been stunned and stricken by the appearance of the supernova, and that his own testimony would be viewed as unimpeachable and highly credible. From his own words, we can assume that Marucelli personally took part in one or more meetings in which Marzimedici actually talked about the new star, seeking an orthodox explanation that would make sense of its problematic nature and uncertain position in the sky. Like Marucelli's *Canzoni*, delle Colombe's *Discourse* reflects the threads of the debate that was led by Marzimedici within the Catholic intellectual community between 1604 and 1605. The two works present in their own fashion an account or a continuation of this discussion, which was conducted for the purpose of coming to terms with a problem that was intensely puzzling not only to theologians but also to natural philosophers and mathematicians. However, this discussion guided by the archbishop ended without reaching any final conclusions.

If we read delle Colombe's *Discourse* carefully, without being unduly influenced in hindsight by Galileo's sarcastic remarks about its author and its contents, we can come to appreciate not only that this short work was written in an extraordinary rich and sophisticated vernacular, but that it also presents a broad range of divergent voices. These voices mirror the ideas of those who dealt with the issue within Marzimedici's intellectual circle. Very likely, delle Colombe cross-questioned the theologians, philosophers, poets, and even the mathematicians who surrounded the Archbishop during the entire course of the composition of his work in order to get a full and exact account of the story.

I will now focus on what delle Colombe actually writes about the opinions expressed by the archbishop's theological advisors on the new star, specifically the members of his *Collegio teologico*. Taken together with what has already been said, this could lead to a better understanding of the ways in which other related issues, such as the controversy over the hypothesized double movement of the earth, were addressed in the cultural milieu of Florence during early seventeenth century. As many may know, the solution put forward by delle Colombe in his *Discourse* fits into a broader attempt to reconcile various recent celestial phenomena with Aristotelian precepts. To the author's mind, the star that appeared in 1604, as well as all the other stellar bodies observed in the skies by astronomers

---

[31] Delle Colombe, *Discorso* (cit. note 2), p. 4.

since the Antiquity, were to be considered true phenomena. The supernova was:

a true star like those that have existed from the very beginning in the sky, not created anew, not generated, not appearing by reflection from other stars, not due to a displacement of air, nor to a meeting of many star rays embedded in the sky, nor to the condensation of any heavenly parts, and which is made bright.[32]

In the book, this judgment comes at the end of a comparative display of different opinions, all of which are carefully reported by delle Colombe. In Florence these views were held by a wide range of individuals very much interested in the astronomical debate, and all more or less personally acquainted with the author of the *Discourse*. Delle Colombe does not mention their names, but before his own assessment he reveals the judgment formulated by a congregation he calls a 'theological college'. In fact he was referring to the congregation that regularly advised the Archbishop of Florence, a small group of experts on the Bible who were engaged in exegesis and the study of canon law, and whose work was desultorily checked by Marzimedici. They met to deliberate when a matter was put forward to them and insofar as their expertise could be useful. In the dispute over the new star they adopted a particular approach, which was to evaluate the relationship between miracles and the ordinary laws of nature. They decided that the appearance of the supernova was an occurrence that ought to be deemed miraculous, as it was an extraordinary event lying outside the natural order of the world. According to these religious experts its appearance must have had a divine cause and its real nature was an enigma lying beyond the ken of human reason and as such could not be investigated by natural philosophers or mathematicians.

To these thinkers with their Christian outlook the only legitimate explanation was one involving God's direct intervention and they therefore suggested that the new star should be regarded as a wondrous event that illustrated the omnipotence of the divine, a sign of the kind that was very much present throughout the Old and the New Testaments. In the end, the perfect Christian attitude towards this event was not to set out on a scientific inquiry, pushing the powers of the human mind to its limits, but rather to believe serenely in the existence of miracles. Their judgment is accurately relayed by delle Colombe in his book:

32 *Ibid.*, p. 32.

Finally the sentence of an entire College is that such stars did not newly appear through physical generation in those incorruptible spheres, but were miraculously created by the Author of Nature. Nor therefore, they say, should guidance on this fact be drawn from human knowledge. For this reason it may come about that the cause of some prodigious events is hidden, many extraordinary effects, willed by God, happen up above in the heavenly World of which our capacity cannot fathom the ends. An example may be found in the halting of the Sun in its path in favor of Joshua; the backwards movement of the same as a sign to King Ezekiel, and the Eclipse of the same Sun upon the death of the World's Redeemer, when the Moon poised at the diametrically opposite point in the sky facing it immediately caused it to be eclipsed.[33]

This opinion, though, did not convince delle Colombe. Unlike the *Collegio*, he did not believe that the supernova was just a miraculous appearance that everyone looking up at the sky could see. Rather, he strongly maintained that natural philosophers should employ their reasoning in order to get to the bottom of the issue and discover the cause, the nature, and even the true position of the new star:

> May I dare, among these wandering minds, illustrious persons and men of letters, to intervene with my opinion on a thing that is no less precious for us to know as it is difficult to investigate, such as the search into what is the substance that resembles a star, and in what way it came to make itself visible among the last lights in the sky so resplendent and so immense to our eyes.[34]

Hence, within the circle surrounding Marzimedici there were different opinions about the origins, nature, and position of the new star, and in some way these mirrored the opinions held at the Medici court as described in the first paragraph of this paper. At least two points of view stand out. On one side, the theologians and Biblical commentators unanimously thought – and actually declared – that the star should be considered a miraculously generated event brought about by God in the celestial realm. According to them, men should not rely on their own, limited means as they seek the solution to the issues involved in its appearance. They should instead step back and satisfy their natural desire for knowledge by reviewing the many similar episodes told throughout the Bible. On the other side, there were the natural philosophers such as delle Colombe who, unlike the theologians, agreed that seeking natural explanations for this particular phenomena was entirely licit. They thought that placing reli-

---

33 *Ibid.*, p. 30.

34 *Ibid.*, pp. 31-32.

ance on a miraculous cause was utterly mistaken and that human reason could find the explanation. In the face of these quite different solutions the archbishop, far from acting as an impartial referee, came down on the side of his theologians. Nonetheless, he did not silence or shut out the natural philosophers, accepting them into his circle and allowing them to participate in the debate.

Who were the Florentine theologians mentioned by delle Colombe? They were doctors in sacred theology and canon law who taught at the theological faculty of the *Studio* in Florence or at the university in Pisa. They also exercised control over everything that was published in Florence and in some cases in other parts of the Grand Duchy of Tuscany, examining the manuscripts of philosophical, theological, and scientific works sent to the ecclesiastical authorities by authors seeking permission to have them published. The learned theologians decided whether these works should be submitted to the censors (generally the Jesuits of the Florentine College) for the final *imprimatur*. Among the members of the *Collegio* I would mention at least two prominent Dominicans. Filippo Guidi was a lecturer in theology in the capital of the Grand Duke,[35] who oversaw the publication of Ludovico delle Colombe's *Discourse*, praising the book because it complied with «the true philosophy and the principles of Aristotle» thus contributing to stem the tide of the new astronomy. The other, Domenico Gori, was a well-known preacher and a prominent figure even at the Medici court.[36] Gori reviewed Carlo Marucelli's *Canzoni* and sent the manuscript to the official censors for their approval.[37] Many other theologians frequented the palatial residence of the archbishop close to the Duomo, regularly providing exegetical and legal advice. Among them we may mention Giacinto Gucci, well known for his sacred oratory;[38] Carlo Macigni, the author of a massive work on the liturgy of the Canonical Hours;[39] and Agostino Vi-

35 See Guerrini, *Galileo e la polemica anticopernicana* (cit. note 1), pp. 29 and 39-42.

36 See Id., *Cosmologie in lotta. Le origini del processo di Galileo*, Florence, Polistampa, 2010, pp. 104-130. On Gori see also Id., *Inquisizione e università. Teologia e cosmologia in una controversia fra Pisa e Firenze nel 1616*, in Franco Tomasi – Salvatore Lo Re (eds.), *Varchi e altro Rinascimento. Studi offerti a Vanni Bramanti*, Roma, Vecchiarelli, 2013, pp. 403-421.

37 Marucelli, *Canzoni* (cit. note 6), p. 20.

38 Giacinto Gucci, *Oratio de laudibus theologiae habita in ecclesia sanctiss. Trinitatis Florentiae decimotertio Kalendas Novembris, iussu et auspiciis domini Fidelis de Corsettis Ordinis Vallisumbrosae doctoris theologi, ac tum temporis theologorum decani*, Florentiae, Sermantelli, 1614.

39 Carlo Macigni, *Trattato dell'Ore canoniche nel quale si ragione del nome, difinizione, origine, quantità, e qualità di esse. Di coloro che sono obbligati a dirle e delle pene, in che incorrono non le dicendo. Del tempo, del luogo e dell'attenzione che si dee avere nel recitarle, e in breve di tutti i quesiti e dubbi che possono accedere in cotal materia [...]*, Florence, Giunti, 1607.

giani, an expert on canon law, as well as the author of an influential treatise on ecclesiastical immunity.[40]

Some, indeed perhaps all of them formed part of the *Collegio* mentioned by Delle Colombe in his *Discourse* on the new star and all of them shared the view that its appearance was due to a transcendental intervention similar to the many described in the Old and New Testaments. As just stated, their position was that the origin and nature of this type of divine intervention lay beyond the power of human reason to comprehend. The appearance of the star was a physical reality, but should be accepted without trying to search for a rational explanation. Indeed, theologians did not deny the authenticity of such events; they simply underlined their miraculous origin, having been brought about by God in a temporary suspension of the course of natural laws. Why did such events occur? In their opinion, it was in order to fulfill a divine and providential design that could not be grasped by man and therefore should be accepted based on faith. Naturally, this did not mean that one should suspend belief, even temporarily, in the truth of the story about the birth of the cosmos contained in the book of *Genesis* or in the authenticity of the finite and earth-centered universe ordered according the Aristotelian theory of elements as well as Ptolemy's astronomy. On the contrary, divine intervention from the heavens, which implied a temporary suspension of natural laws, was not considered to be at odds with the Biblical, Aristotelian, and Ptolemaic views.

Though in appearance it might be deemed most illiberal, one consequence of this stance was a kind of tolerance, if not a genuine open-mindedness, towards different viewpoints. Unlike the Aristotelians, those who shared this stance did not feel compelled to jump down their adversaries' throats. If anything, from these kinds of theological conclusions it was possible to argue that an indirect confrontation with those who were considered *recentiores* was viable.[41] The more up-to-date astronomers could maintain their opinions as long as these fell within the purview of a mathematical hypothesis. However, for Aristotelian philosophers such as Ludovico delle Colombe it was just the other way around. Proponents both of a fluid and corruptible sky and of a sun-centered cosmos with a ro-

---

40 Agostino Vigiani, *De ecclesiastica immunitate disputationes sex. Quibus adversus huius temporis novatores clericorum exemptio a potestate civili ex iure divino, pontificio, et caesareo; necnon ex sanctis Patribus et rationibus demonstratur [...]*, Bononiae, Bellagamba, 1607.

41 See also Thomas F. Mayer, *The Roman Inquisition. Trying Galileo*, Philadelphia, University of Pennsylvania Press, 2015, pp. 27-31.

tating earth had to be ruthlessly attacked. The unbiased Aristotelian appeal to the use of human reason in the study of natural philosophy inevitably entailed a fight against opponents.

In Florentine theological circles, therefore, we face a situation in which natural philosophers and theologians, both supporters of the traditional view of the universe, adopted different attitudes towards the appearance of the new star in 1604. This would continue in the following years, when the issue came to form part of the reality of the Copernican cosmos, so that in the decade between the polemic on the new star and Tommaso Caccini's public denunciation of Galileo's heliocentrism (December 1614), contrasting tendencies would crystallize within the camp of the enemies of the new astronomy. The least we can say is that within the traditional camp there was no shared strategy against those who in 1604 were convinced of the fluidity and corruptibility of the sky (Raffaello Gualterotti) and those who, after 1610, accepted the Copernican universe (Galileo and Benedetto Castelli). Some, such as Marzimedici and his theological advisors, undertook to stem the spread of the new ideas by promoting and backing the Biblical cosmology and the Aristotelian universe through sermons, lessons and even books, but without striking a direct blow at their adversaries. In contrast, others repeatedly called for a move on the part of the ecclesiastical authorities to punish these mavericks and troublemakers who were endangering the orthodox view of the universe. Among the most unbending and least tolerant opponents of any innovation in astronomy were the Dominicans Tommaso

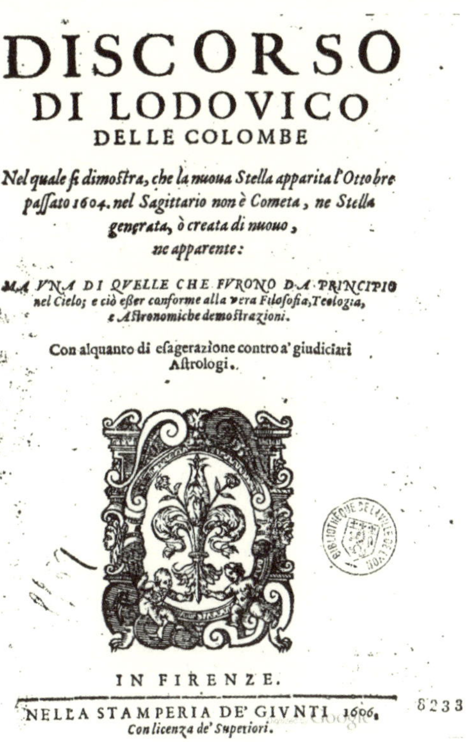
DISCORSO
DI LODOVICO
DELLE COLOMBE
Nel quale si dimostra, che la nuoua Stella apparita l'Ottobre passato 1604. nel Sagittario non è Cometa, ne Stella generata, ò creata di nuouo, ne apparente:
MA VNA DI QVELLE CHE FVRONO DA PRINCIPIO nel Cielo; e ciò esser conforme alla vera Filosofia, Teologia, e Astronomiche demostrazioni.
Con alquanto di esagerazione contro a' giudiciari Astrologi.
IN FIRENZE.
NELLA STAMPERIA DE' GIVNTI 1606.
Con licenza de' Superiori.

Frontispiece of delle Colombe's *Discourse on the new star* (Florence, 1606).

Caccini and Raffaello delle Colombe, Ludovico's brother.[42] They slowly but steadily assumed an increasingly hostile attitude towards Galileo and his pupils. And after the publication of the *Sidereus Nuncius* in 1610 and the return to Florence of its author, they decided to actively engage in a campaign against heliocentrism, eager to have Galileo condemned as a heretic.[43] They soon realized however they could not attain this goal by adopting Marzimedici's moderate cultural policy and attitude.

4. Conclusion

From what has been said so far, it appears that a deeper historical understanding of the events that occurred in Florence between 1610 and 1616 after the diffusion of *Sidereus Nuncius* and the trial in Rome of the supporters of heliocentrism (in which the Congregation decided that Copernicus' *De revolutionibus* was to be banned, 'until corrected') hinges on the events that unfolded between 1604 and 1606 with regard to the new star. On December 16, 1611 the painter Ludovico Cardi da Cigoli wrote a letter to warn Galileo about the existence of a league of opponents, nicknamed the "Pigeon league" (due to the presence of the brothers delle Colombe or 'of the Doves'),[44] which met at the residence of the Archbishop Marzimedici near the *Duomo*:

> From a friend of mine, who is a gallant Priest and much attached to you, Sir, I have learned that a certain band of persons who wish you ill and are envious of your virtue and achievements meet regularly to plot their opposition at the house of the Archbishop, and as they are angry they are seeking whether they can accuse you of anything regarding the movement of the earth or some other matter.[45]

These opponents of the new astronomy (of at least some of them) already mustered when the quarrel flared over the new star between 1604

[42] See my studies on him: Guerrini, *Raffaello delle Colombe et les origines de la polémique anti-galiléenne à Florence*, in Gian Mario Bravo – Vincenzo Ferrone (eds.), *Il processo a Galileo Galilei e la questione galileiana*, Rome, Edizioni di Storia e Letteratura, 2010, pp. 185-195; *Galileo e Raffaello delle Colombe*, in Bucciantini – Michele Camerota – Franco Giudice (eds.), *Il caso Galileo. Una rilettura storica, filosofica, teologica. Convegno internazionale di studi, Firenze 26-30 maggio 2009*, Florence, Olschki, 2011, pp. 109-126; *Echoes from the pulpit: A preacher against Galileo's astronomy, 1610-1615*, «Journal for the History of Astronomy», 43, 2012, pp. 377-389.

[43] See Pagano (ed.), *I documenti vaticani* (cit. note 3), pp. 21-41.

[44] Galileo to Cesi, 5 January 1613, OG, XI, p. 461.

[45] Lodovico Cardi da Cigoli to Galileo, 16 December 1611, OG, XI, pp. 241-242.

and 1606. However, they held different positions concerning the role of scientific inquiry and man's ability to understand God's provisions with regard to natural phenomena. In 1611 the foes of the new astronomy were more or less the same as in 1604, but Cigoli, who was in Rome and not in Florence, could not see all of the facets of the opposition. Within the group there were longstanding disagreements that, especially after the publication of Galileo's *Sidereus Nuncius* in 1610, could no longer be reconciled and eventually led to a definitive split. Everyone in the archbishop's camp agreed that the unadulterated nature of the sky, as well as the notion of the earth-centered universe, should be protected by any means possible, but there was endless contention over the limits of the use of human reason in matters of natural history, which were mirrored in the strategies used against the advocates of the new astronomy. Marzimedici and his theologians played a key role in the historical events that took place in Florence. They strongly believed in miracles and in God's direct and providential intervention in human affairs and in natural events. They never gave up defending notions of the impenetrability of the spheres and the quintessential nature of the skies, as well as the Biblical story of creation, which was melded with Aristotelian physics and Ptolemaic astronomy into a single theory, but they did not lead a public campaign against Galileo. Their opposition to heliocentrism never turned into a personal quarrel with the author of the *Sidereus Nuncius*. On the contrary, it was those who strongly believed that the senses and human reason were the most valuable guides to the study of natural history who were the most tenacious persecutors of Galileo.

Federica Favino

ALCHEMICAL IMPLICATIONS OF 1616 *AFFAIRE*
On the Parish Priest Attavanti, the Knight Ridolfi and the Cardinal Orsini

In memory of Germana Ernst.

1. Microhistory of a Deposition

In the procedural events that led to the promulgation of the Anti-Copernican decree of 1616, Giannozzo Attavanti, the priest of Castel Fiorentino parish church and a member of a notable local family, was – *malgré lui* – a key player.[1] It had been the Dominican friar Tommaso Caccini who called upon Attavanti from the very beginning when, on 20 March 1615, with the accord of Cardinal Galamini Aracoeli, he presented himself to the General Commissioner of the Holy Office, Michelangelo Seghezzi, to accuse Galileo of supporting the motion of the Earth and the immobility of the Sun, theories that, he said, «according to my conscience and understanding, are repugnant to the divine Scripture expounded by the Holy Fathers and consequently to the faith».[2] Questioned by Seghezzi about how Caccini had acquired this information on Galileo's opinions concerning

---

[1] Translations are mine, unless otherwise indicated. List of abbreviations:

ASC, AO = Archivio storico Capitolino, *Archivio Orsini*, Rome; ASF, MdP = Archivio di Stato di Firenze, *Mediceo del Principato*, Florence; DBI = *Dizionario Biografico degli Italiani*, 90 voll., Roma, Istituto dell'Enciclopedia Italiana, 1960-2017 (in the bibliographical references, the pages are not given as they refer to the on-line versions of the work).

On Attavanti, see the entry by Antonella Dolci in DBI, 4 (1962).

[2] Maurice A. Finocchiaro, *The Galileo Affair. A Documentary History*, Berkeley and Los Angeles, Univ. of California Press, 1989, p. 138. Finocchiaro translates the documents according to Favaro's edition in OG, XIX. Therefore, from here on I will report also the corresponding page in OG (here, p. 318). You can also view the documents of the trial in *I documenti del Processo di Galileo Galilei*, ed. by Sergio M. Pagano, with the collaboration of Antonio G. Luciani, Città del Vaticano, Pontificia Academia Scientiarum, 1984.

Copernican ideas, Caccini first and foremost put forth the idea of *'publica fama'*, and he did it not by chance.

The concept of 'publica fama' was established in the 13th century in the legal sphere to designate «that complex phenomenon according to which the news of a given fact spreads and takes roots in a certain social and geographic context as far as to become a matter of concern to the judge, who makes it the motor and the reason to open an investigation».[3] Caccini indicated as one of his sources the oral testimony of Monsieur Filippo de' Bardi di Vernio, then bishop of Cortona, whom he had first met when he was a priory in the monastery of Cortona and then later in Florence.[4] Above all, however, Caccini had claimed to have heard of Galileo's opinions «from a certain Florentine gentleman of the Attavanti family, – he continued – a follower of the same Galilei, who said to me that Galilei interpreted Scripture in such a way as not to conflict with his opinion» and that this exchange had taken place in the monastery of Santa Maria Novella, «in Father Ferdinando Ximenes' room [*author's note*: Ximenes was the rector of that monastery], the occasion being that Father Ximenes was telling me that I should not take too long discussing the miracle of the stopping of the Sun when he (Ximenes) was around».[5] Attavanti's name had come up at other times during Caccini's deposition, because he was, according to Caccini, one of those 'petulant minds' adhering to the Galilean 'sect' and from whom Ximenes (and therefore Caccini) had heard other theologically heretical doctrines that could clearly be traced back to his maestro and which claimed that «"God is not otherwise a substance, but an accidens"; "God is sensuous because there are in him divine senses'; and 'in truth the miracles said to have been made by the saints are not real miracles"».[6]

As is well known, on 2 April 1615 the Holy Office ordered that the official deposition containing Caccini's testimony be transmitted to the Flo-

---

[3] Antonio Stopani, *Testimoniali e la prova per «fama» in una disputa territoriale del XVI secolo*, «Quaderni Storici», 139, 2, 2012, pp. 223-247: 243.

[4] Son of Giovanni Bardi of the Counts of Vernio and of Lucrezia Salviati di Pietro, he was canon of Florence Metropolitan Church (1591), referendary of the two Signatures and a member of the private *familia* of Pope Clement VIII. Governor of Terni, Narni, Forlì (1596), he was elected bishop of Cortona in 1604. He was a member of the Florentine Academy and of the Accademia della Crusca (OG, XX, pp. 383-384). Censor from 19 February to 5 August 1592, between 1592 and 1593 he took an active part in work on the *Vocabulary*, preparing about 108 items of the letter A. On his proposal, on 8 November 1589, the verse of Petrarch's "the most beautiful flower" became the motto of the Crusca (http://www.accademicidellacrusca.org/scheda.asp?IDN=83).

[5] OG, XIX, p. 308; Finocchiaro, *The Galileo Affair* (cit. note 2), p. 137.

[6] *Ibid.*

rentine inquisitor, who was entrusted with the task of interrogating the persons mentioned in the complaint. Only in November was it possible to acquire the testimony of Father Ximenes, who substantially confirmed Caccini's account and even providing further incriminating elements. Though Ximenes did not confirm Galileo's alleged temerarious theological opinion about miracles, he did add to the charges ascribable to Galileo the dangerous philosophical thesis regarding the discontinuous nature of matter: «that neither substance of the things nor continuous quantity do exist, but that everything is a discreet quantity, made up of empty spaces».[7] Moreover, this remark hinted that Galileo subscribed to the atomistic conception of matter dating back to Democritus or Lucretius, and in an underhand way it compared Galileo to the 'sectarian' Bruno, the memory of whose pyre was still very much alive.[8]

Like Caccini, Ferdinando Ximenes named Attavanti as the main source of his information on Galileo's doctrinal beliefs, although it transpired that he did not know whether the parish priest was expressing his own opinions or relaying those of Galileo.[9] Moreover, with regard to the circumstances of his conversation with Attavanti, Ximenes was only able to state that it had taken place sometime in the previous year – 1614 – without being able to specify the exact day. We know nevertheless that it must have been subsequent to a series of lectures on the Book of Joshua delivered by Caccini at the monastery of Santa Maria Novella, where he was at the time a lecturer on the Sacred Scripture. In particular it had certainly followed his lessons on the miracle that took place during the Israelites' defeat of the two kings of the Amorites – '*stetit Sol*' (Joshua 1, 12-13) – since precisely those lessons provided the starting point for the talks between Ximenes and Attavanti.

On 14 November 1615, Attavanti's turn finally came. The priest was the only witness among those interrogated by the Florentine inquisitor who did not speak from hearsay – but «with certain and through my own hearing» [10] – and who truly did not have any bias against Galileo,[11] although he did declare that he was not a disciple of the scientist. Attavanti therefore was in the best position to clarify all the details so confusingly reported by

---

7 OG, XIX, p. 316; FINOCCHIARO, *The Galileo affair* (cit. note 2), p. 141.

8 MASSIMO BUCCIANTINI, *Contro Galileo. Alle origini dell'*affaire, Firenze, Olschki, 1995, pp. 60-68. For an updated overview on the real basis of these accusations, see now PAOLO GALLUZZI, *Tra atomi ed indivisibili: la materia ambigua di Galileo*, Firenze, Olschki, 2011.

9 OG, XIX, p. 317; FINOCCHIARO, *The Galileo affair* (cit. note 2), p. 142.

10 OG, XIX, p. 319; FINOCCHIARO, *The Galileo affair* (cit. note 2), p. 145.

11 *Ibid.*

Caccini and Ximenes. First of all, he confirmed that he had heard Galileo assert opinions in support of a heliocentric and heliostatic cosmology – consistent with what was sustained in his *Discorsi e dimostrazioni sopra le macchie solari* –, and that he had heard Galileo attempt to interpret Job's miracle in accordance with that cosmology.[12] He then stated that he had reasoned with Father Ximenes during one of the lessons on 'cases of conscience' which Ximenes was giving him at that time in his monastic cell – precisely on a day in August or September 1613. He also remembered that those talks between Ximenes and himself instigated an energetic censorship by Father Caccini, who had been listening to their conversations from the next cell and who announced his intention to admonish the heretical character of heliocentrism in a sermon from the pulpit. Lastly, he explained that, due to this precedent, Caccini had also mistakenly attributed to Galileo certain propositions – *An Deus sit substantia vel accidens... An Deus sit sensitivus*, etc... (but not those on the miracles of the saints) – which were actually some 'absolutes' by the venerable Doctor of the Church Saint Thomas Aquinas (beginning with the absoluteness of God), and which served as topics for debate – purely *per modum disputationis et addiscendi gratia* – between Attavanti and Ximenes during later lessons on which Caccini had also eavesdropped.[13]

In this way Attavanti described the time, place and circumstances of the discussions that provided Ximenes with the material for his accusations against Galileo. As for the time, he situated the conversations with Ximenes during some lessons in the summer of 1613 and not on an unspecified day in 1614, as the 'great accuser' had claimed. On this point it must be conceded that the priest's memory may be suspect, considering how long ago the events had taken place, but his testimony must have suggested to the Inquisitor and certainly suggests to the historian today a more complex chronology and concatenation of the facts. I do not think it was ever noticed by scholars that it was Attavanti's talks (which themselves point to the positive reception that Galileo's Copernicanism met with in Florentine intellectual circles) that provided Caccini with the pretext for the sermon which he delivered in the Basilica of Santa Maria Novella in December 1614. The impetus for Caccini's vehement homily can be traced back to the provocative statements made by Attavanti during the Sacred Scripture course held at the monastery of Santa Maria Novella. This allows us to add Caccini's name to the list of Dominican friars who were engaged in

[12] OG, IX, p. 318; Finocchiaro, *The Galileo affair* (cit. note 2), p. 144.

[13] *Ibid.*

defending and propagandizing Mosaic physics and the creationist doctrine based on the Book of Genesis in the Old Testament against Copernicanism in the monastic schools and the theological faculty in Florence, drawing their arguments from the work of Giovanni Maria Tolosani.[14]

The details provided by Attavanti rendered many of the already confused particulars on those facts offered by Ximenes in the course of the deposition now superfluous. In fact the rector of the convent was not only uncertain about the dates of the events in question, but was also unsure about the circumstances under which they may have taken place. When questioned by the Inquisitor regarding the controversial theological propositions that had been advanced, he seems to have declared that he heard them and disputed them with Attavanti, adding: «at which discussions a certain Mr. Ridolfi, a Florentine and a Knight of Santo Stefano, was present...Regarding the place, it was in my room, here at the convent of Santa Maria Novella...witnesses present were the above mentioned knight, sometimes, and some of our friars whom I do not remember exactly».[15]

Once the precise context was clarified by Attavanti, these confused details probably were judged by the Florentine Inquisitor to be irrelevant from a juridical standpoint. Furthermore, as noted by Massimo Bucciantini, while the minutes of the interrogations were forwarded to the Holy Office in Rome, triggering the procedure that would ultimately lead to the decree suspending *De revolutionibus*, «for its part the Florentine Inquisition did not promote any independent initiative, such as for example to summon the other person mentioned by the Dominican father to testify», that is, the Cavalier Cosimo Ridolfi, apparently the only other direct witness to the facts.[16] In short, the Florentine Inquisitor, Lelio Marzari, did not consider it worthwhile to get to the bottom of those many other discussions centered on Galilean themes which were possibly stored in the memory of Ximenes but also of Caccini, or of those 'flashes' floating in the memories of the other witnesses, which always portrayed 'the Knight Ridolfi' as

---

14 LUIGI GUERRINI, *Cosmologie in lotta: le origini del processo di Galileo*, Firenze, Edizioni Polistampa, 2010, *passim*. On the doctrinal cotinuity in the anticopernican polemic of preaching fathers throughout the period from 1546 to 1616, see also Guerrini's contibution in this volume. In fact, in the chronicles of the convent there are no documents of Caccini's preachings, probably because it was not an event (GIUSEPPE GALLI, *Card. Maculano al processo di Galileo*, «Memorie domenicane», nuova serie XLI, 1965, pp. 24-42: 29). For information on the monastic school at the convent of S. Maria Novella see *Santa Maria Novella. Un convento nella città. Studi e fonti*, «Memorie domenicane», 11, 1980. On its library, STEFANO ORLANDI, *La biblioteca di S. Maria Novella in Firenze dal secolo XIV al sec. XIX*, Firenze, Edizioni Il Rosario, 1952.

15 OG, XIX, pp. 316-317; FINOCCHIARO, *The Galileo Affair* (cit. note 2), p. 142.

16 BUCCIANTINI, *Contro Galileo* (cit. note 9), p. 70 e n.

present side by side with Attavanti (precisely because none of them were recalling a specific moment). A recent scholarship has shed a fresh light on the cosmological discussions in Santa Maria Novella,[17] while the second point has long been ignored not only by Inquisitors, but even by historians. Ridolfi's ties with Attavanti and the Dominican community of Santa Maria Novella have gone unnoticed by historians of Galileo's 'first trial'. They have tended to look at the procedural acts generated by the Galileo affair *en bloc* as a huge corpus to be examined in its entirety, without attempting to separate the different components or identify the different motives and strategies of the various protagonists. But the same has been true also for historians of the first Academy of the Lincei. Until recently even they have dismissed Cosimo Ridolfi as no more than a name mentioned a few times in correspondence between the academicians, and whose signature appears in the Academy's registers, which were faithfully maintained and bequeathed from generation to generation and still can be found in the Academy's archive today.[18] Only a completely new and unexpected document, such as the one which I will attempt to account for in these pages, could have spurred scholars to open a 'Ridolfi dossier', and in doing so shed fresh light on the debates being conducted in 1614 in the monastery of Santa Maria Novella on Copernicanism and – as we shall see – also on other issues of the day.

2. 'A certain Mr. Ridolfi, a Florentine and a Knight of Santo Stefano'

In my view, the behavior of Marzari clear demonstrates how little information and how little awareness the Florentine 'inquisitorial police' still had at the time about the spread of heliocentrism and the 'new' philosophy in general across its territory. After all, Caccini had offered Attavanti to

---

[17] See above, note 14, and also: Guerrini, *Galileo e la polemica anticopernicana a Firenze*, Firenze, Polistampa, 2009; Id., *Galileo e Raffaello delle Colombe*, in Bucciantini – Michele Camerota – Franco Giudice (eds.), *Il caso Galileo: una rilettura storica, filosofica, teologica: convegno internazionale di studi*, Firenze, Olschki, 2011, pp. 109-126.

[18] Baldassare Odescalchi, *Memorie istorico critiche dell'Accademia de' Lincei e del principe Federico Cesi secondo duca d'Acquasparta fondatore e principe della medesima...*, Roma, nella stamperia di Luigi Perego Salvioni, 1806, pp. 117, 119; Domenico Carutti, *Breve storia della Accademia dei Lincei*, Roma coi tipi del Salviucci, 1883, p. 163; OG, XI, XII, *ad indicem*; Giuseppe Gabrieli, *Contributi alla storia della Accademia dei Lincei*, 2 vols., Roma, 1969, *passim*; Id., *Il carteggio Linceo*, Roma, 1996 (già *Il carteggio Linceo della vecchia accademia di Federico Cesi (1603-1630)*, 3 vols., Roma, 1938-1942), nn. 240, 242, 252, 253, 248, 249, 266, 269, 271, 383, 435, 451, 469, 525, 532.

the Inquisition vicars as a 'small fish'. Attavanti's testimony could not only have potentially unmasked Galileo; it would also have helped the Florentine judges to dismantle the 'sect' of his 'disciples and scholars', if only they had perceived it to be a danger. And yet to exonerate himself (Ximenes having already criticized the parish priest for his weak grasp of philosophical and theological issues[19]), a simple denial on Attavanti's part was sufficient. When asked whether he knew a doctor named Galileo residing in Florence and if he, Attavanti, had been his pupil, Attavanti replied: «I have never studied under him as his pupil. I have indeed discussed scholarly topics with him, as I ordinarily do with men of letters; in particular, I have discussed philosophical matters with him».[20] As for Ridolfi, he was never even questioned, and this is even more surprising. In the summer of 1614 Ridolfi was already a member of the Academy of the Lincei and in his deposition before the judges in Rome Caccini had without prompting mentioned the Academy as another center of heresy, hinting at its ties with Protestant Germany as demonstrated (in his eyes) by the letters exchanged between Galileo and Marcus Welser on sunspots.[21]

The lack of zeal of the Florentine inquisitorial police can perhaps be explained by «the complicity» and by «a certain family likeness» shared by those who identified themselves with «that unitary and compact physiognomy of a culture in which both clerics and laymen participate without distinction», a distinguishing feature of this period in Tuscany.[22] Yet some crucial questions remain open to the historian: who was this «Mr. Ridolfi, a Florentine and a Knight of Santo Stefano» and how did he happen to be in the monastery of Santa Maria Novella in the company of the parish priest Attavanti? Did he belong to the Galilean 'sect' as well and, if so, in what way did he participate as a member of this group? And again, what might his intellectual profile reveal about the positions and re-positionings of the interested parties at the Medici court who supported or opposed the emergence of a 'new image of the world'?

Let us begin with some biographical information. Cosimo Ridolfi was born in Florence on 23 August 1570, to Pietro di Lorenzo – who belonged

---

19 OG, XIX, p. 317; Finocchiaro, *The Galileo affair* (cit. note 2), p. 142.

20 OG, XIX, p. 318; Finocchiaro, *The Galileo affair* (cit. note 2), p. 144.

21 OG, XIX, p. 310; Finocchiaro, *The Galileo affair* (cit. note 2), p. 140.

22 Adriano Prosperi, *L'inquisizione fiorentina al tempo di Galileo*, in Galluzzi (ed.), *Novità celesti e crisi del sapere*. Atti del Convegno internazionale di studi galileiani, Firenze, Giunti Barbèra, 1984, pp. 315-325, now in Prosperi, *L'Inquisizione romana: letture e ricerche*, Roma, Edizioni di storia e lettertura, 2003, pp. 183-198: 197. By the same author, see also: *L'età dell'Inquisizione romana a Santa Croce di Firenze*, in *ibid.*, pp. 199-217.

to the di Piazza' branch of the family – and his wife Maddalena Salviati.[23] His father received a death sentence in 1575 for having conspired with Orazio Pucci and other young Florentines against the Medici government, but was eventually pardoned and his sentence was commuted to life imprisonment in the Bargello, where he died in 1589. In 1594 Cosimo was appointed a knight of the Order of St. Stephen after having obtained the return of various assets and rights that his father had lost following his conviction. While still young he married Alessandra, the daughter of Alessandro Capponi, who died childless in 1595. That same year he married Laudomia, the daughter of Giuliano Ricasoli, who bore him a son, Niccolò, in 1603. Cosimo Ridolfi enrolled in the Accademia della Crusca, under the soubriquet of 'Colmo', and served as its *consolo* from 6 August 1591 to 19 February 1592 and then as president (*arciconsolo*) from 4 August 1593 to 9 February 1594. From 1593 he was engaged in the revision of *Purgatory* for an edition of Dante's *Divine Comedy* being produced by the Accademia della Crusca (Florence, Manzani, 1595).[24]

The most salient fact known about Ridolfi however is that he was a member of the prestigious Academy of the Lincei. In May 1613 his candidature was proposed jointly by Filippo Salviati and Galileo in response to the request of Federico Cesi, the founder of the Academy, for recommendations from fellow academicians for potential new members. Speaking for Galileo and himself, Salviati described Ridolfi to Prince Cesi as a «free-thinking philosopher, a knight of very noble concepts, a man of great learning, and one who in all his actions gives honor and glory to his [high-ranking] family, rather than begging it for them».[25] When he received the Prince's invitation to join the Academy, Ridolfi adhered to the *topos* of modesty in transmitting his preliminary consent, pretending to lack the intellectual qualities that were, instead, to admire in his 'companions':

And although I see – he wrote to Cesi on 15 June 1613 – that placing myself in comparison with such great companions makes my inadequacy and my demerits all the more evident, however in other respects, I estimate that the loss is less than the gain, which is to show the world that although I am not capable of the acquisition of any true and valuable knowledge on my own, nevertheless I have received from God enough wisdom to procure for myself companions who, freely philosophizing, show me where to walk in order to arrive there.[26]

---

23 Here I reproduce some of the biographical information I have exposed in the entry *Cosimo Ridolfi* in DBI, 87 (2016).

24 http://www.accademicidellacrusca.org/scheda.asp?IDN=716.

25 Gabrieli, *Il carteggio* (cit. note 19), n. 240.

26 *Ibid.*, n. 363.

The nomination of Ridolfi was formally proposed by the Prince at a meeting of the Academy on 16 July 1613, and was seconded by the Lyceum of Naples and Marcus Welser of Augsburg. His enrolment was formalized (the exact date is not known) and in September he received the membership ring from the hands of the academicians who had sponsored him.[27] On the 26th of September he wrote to Cesi to express his profound thanks and send an exemplar of his autograph signature for the registers.[28] Following this, however, there is no trace of further contacts between Ridolfi and the Prince or his associates. Indeed, it was always Galileo who acted as his mediator on official Academy matters. In August 1614, for example, Cesi was still enjoining Ridolfi through Galileo to write to his fellow academicians in order to convey his salutations as a newly inducted member, an obligation that he had not yet fulfilled (although he was not alone in this respect).[29] And it was Galileo who registered Ridolfi's vote when the academicians met to decide on the candidacy of Filippo Pandolfini that same month.[30]

There is no doubt that during this period (between 1613 and 1614, as the flurry of correspondence on the candidacy of Ridolfi was being exchanged) – when the events in Florence described by Attavanti were also unfolding (from the conversation overheard by Ximenes in August 1613 to the interrogation of Attavanti by the Inquisitor in November 1615) – Ridolfi was in frequent contact with Galileo and it is relatively easy to imagine the circumstances under which the possibility that Ridolfi might become a member of the Lincei arose. From the winter of 1611 Galileo began to spend long periods at Le Selve, the country estate of Filippo Salviati in Lastra a Signa near Florence. There he continued his astronomical studies of the Medicean planets and made the «scrupulous observations about sunspots»,[31] described in the three letters to Welser – published as *Istoria e dimostrazioni intorno alle macchie solari*, in 1613 –, that were dedicated to Salviati and sent to Welser from Le Selve. In September Galileo made a special trip to Florence in order to confer the Lincei investiture upon Ridolfi, who himself had been a frequent guest at Le Selve. A glance at the genealogical records reveals that Cosimo Ridolfi and Filippo Salviati were actually

27 Federico Cesi to Galileo Galilei, Rome, 6 September 1613, *ibid.*, p. 385 and Federico Cesi to Galileo Galilei, Rome, 6 September 1613, *ibid.*, p. 386.

28 Cosimo Ridolfi to Federico Cesi, Florence, 26 September 1613, *ibid.*, p. 391.

29 Cf. Federico Cesi to Galileo Galilei, Rome, 9 August 1614, *ibid.*, p. 451.

30 Cf. Federico Cesi to Galileo Galilei, Rome, 9 August 1614, *ibid.*, p. 450.

31 *Racconto istorico di Vincenzo Viviani*, in OG, XIX, p. 616.

related, since both belonged to a branch of the Salviati family descended from Alamanno (1459-1510), the son of Averardo.[32] As it was customary in the families of the *Ancien Régime*, the descendants of the same lineage used to be raised all together, often in the home of the common male ancestor.[33] Orphaned at an early age, Salviati was brought up by his paternal uncle Antonio (who was a banker), on a completely equal footing with his own children,[34] and the same had to happen also with the cousins in the female line. In 1609, for example, Filippo had probably recommended his cousin Vincenzo, Antonio's son, to the Grand-Duke Cosimo II, who sent him as his ambassador to the courts of Europe to convey the news of the death of his father, Ferdinando I. On that occasion, Cosimo was one of the courtiers in Vincenzo's ambassadorial retinue.[35] Ridolfi's presence at Le Selve at the same time as Galileo is documented at least in a letter that the scientist sent to the learned Benedictine friar Benedetto Castelli in Florence, in which he recommended him to observe the Medicean planets – a letter dated by Favaro *ex suppositione* (but I believe correctly) to 1612.[36]

3. 'The Magic Circle'

These family and intellectual connections may explain how Ridolfi came to be proposed as a member of the Academy of the Lincei. What is less clear is on what grounds he would have been judged to be «a man of great learning», and a «philosopher free» from the yoke of tradition, as Salviati described him to Prince Cesi. I would suggest that Galileo and Salviati sponsored Ridolfi's candidature in 1613 because they were aware of his interest in astronomy, astrology and natural philosophy, although he pursued these interests in circles that historiographers have always

---

32 In particular, Cosimo's mother, Maddalena Salviati, and Filippo's grandfather, Filippo Salviati (1515-1572), were first cousins (Pierre Hurtubise, *Une famille-temoine: Les Salviati*, Città del Vaticano, Biblioteca Apostolica Vaticana, 1985, p. 498, genealogical tree).

33 See, as an exemple, Gigliola Fragnito, *Storia di Clelia Farnese: amori, potere, violenza nella Roma della Controriforma*, Bologna, il Mulino, 2016.

34 Allì Caracciolo, *I filosofi dispersi: storia segreta di Filippo Salviati galileista negli anni della Controriforma*, Napoli, La Città del sole, 2001, p. 42.

35 *Ibid.*, pp. 81-82. On Salviati and his familiar entourage, see also Caracciolo (ed.), *Filippo Salviati filosofo libero: atti del Convegno nel 4. centenario dalla morte: 18-20 novembre 2014*, Macerata, EUM, 2016 and Id., *Salviati, Filippo*, DBI, 90 (2017).

36 Benedetto Castelli to Galileo [at Le Selve], Florence [1612?], OG, XI, p. 456: «From Mr. Cosimo Ridolfi I was ordered, on the part of Y.L., that I made the observations of the Medicean Planets at 6 and 13 hours... I kiss the hands of Y.L., and reverence to the Ill. Mr. F.[ilippo]».

presented as detached from, if not openly hostile towards, the 'new' science.[37]

The correspondence that took place in the years 1615-1620 between don Giovanni de' Medici (the son of Cosimo I and Eleonora of Albizzi) and his 'factotum' Benedetto Blanis, a merchant and amateur scholar from one of the most prominent families in the *ghetto* of Florence, in fact, proves beyond any doubt that Ridolfi belonged to the small circle of associates who shared with Don Giovanni interests in the occult sciences (practical astrology, practical alchemy, and Kabbalistic sorcery).[38] Particularly eloquent in this sense is a letter sent by Ridolfi to Don Giovanni on 25 June 1618 from Villa Meleto, the large farming estate south of Florence, near Castelfiorentino, where Cosimo and his family spent most of their time and where at that moment he was on vacation. Complaining the loss of prince's conversation, he begged him to remedy at least by sharing with him some of his 'noble studies' by letter.

> I am thinking in particular – he wrote – of certain sort of studies that you know very well, studies to which I devote much of my time and take much pleasure. They allow me so see certain kinds of things, things denied to those in the everyday world who have only a lukewarm understanding and cannot navigate beyond the Pillars of Hercules, sed deb.is habemus quia sapienti pauca. I appeal to you, great lord that you are, and ask you to favour me with news and information. Open the treasure chest of your sovereign intellect, for the sake of one who serves you with his heart and soul. This will be a great consolation to me, since I live the life of a hermit, sweet although this life may be.[39]

According to Edward L. Goldberg, who is the latest to have examine the Blanis – de' Medici *dossier*, Ridolfi was one of those «shadowy figures

---

[37] Antonio Favaro, *Galileo Galilei e don Giovanni de' Medici*, «Archivio storico italiano», 39/245, 1907, pp. 106-121; Brendan Dooley, *Narrazione e verità: don Giovanni de' Medici e Galileo*, «Bruniana & Campanelliana», 14, n. 2, 2008, pp. 389-403; Paola Volpini, DBI, 73 (2009) and the references in the following note.

[38] The epistolary exchange has already been brought to attention by Domenica Landolfi, *Don Giovanni de Medici. Principe intendentissimo in varie scienze*, «Studi Secenteschi», 29, 1988, p. 125 and has already been examined by Bucciantini, *Reazioni alla condanna di Copernico: nuovi documenti e nuove ipotesi di ricerca*, «Galilæana», 1, 2004, pp. 3-19. It is now entirely published in Edward L. Goldberg, *A Jew at the Medici Court: the letters of Benedetto Blanis hebreo (1615-1621)*, Toronto [etc.], University of Toronto Press, 2011. It also offered to Glodberg huge materials to reconstruct the relationship between Blanis and Don Giovanni: Edward L. Goldberg, *Jews and magic in Medici Florence: the secret world of Benedetto Blanis*, Toronto [etc.], University of Toronto Press, 2011.

[39] *Ibid.*, p. 172. The original letter is in ASF, MdP 5141, c. 47r.

moving in the background of Benedetto Blanis' letters – mostly Florentine devotees of astrology, alchemy, and Kabbalah... – who little or much had in common», but who «were all fellow travellers in the realm of the occult».[40] The intertwining correspondence among the prince de' Medici and these fellows 'magicians', preserved in the Mediceo del Principato Fund of the State Archives in Florence, shows that the members of this 'magic circle' came from very different backgrounds. They included house courtiers in the household of Don Giovanni, such as Cosimo Baroncelli, one of his closest familiars as well as his majordomo, and don Orazio Morandi, Abbot of the Benedictine monastery of Vallombrosa.[41] There were also freelance agents at the service of princes, nobles and cardinals, such as Blanis and the priest Ascanio Canacci. In exchange for cash rewards or favors, they procured for their patrons books and objects related to the magical-cabalistic culture. Canacci, for instance, had already offered his services to the grandson of Don Giovanni, Antonio de' Medici, who was also a passionate student of chemistry and a disciple of Paracelsus who built a workshop and laboratory known as the *Fonderia* in the Casino di San Marco, where experiments with secret formulas and recipes were conducted.[42] Already around 1605, don Antonio offered to finance the publication of the *Theatro dell'infinito* by the Paracelsian doctor and astrologer Zefiriele Tommaso Bovio from Verona, a defense of astrology against the *Disputationes* by Pico della Mirandola, characterized by the recovery of neo-Platonic, Hermetic and Cabalistic themes.[39] Canacci, who was then in Venice, had arranged the details of the publication (which, however, was never accomplished) with Cosimo Ridolfi himself, who then was in Florence.[43]

---

40 *Ibid.*, p. 172.

41 On Baroncelli, see ROBERTO CANTAGALLI, DBI, 6 (1964). On the much better known figure of Morandi, see: LUIGI FIORANI, *Astrologi, superstiziosi e devoti nella società romana del Seicento*, «Ricerche per la storia religiosa di Roma», 2, 1978, pp. 97-162; GERMANA ERNST, *Scienza, astrologia e politica nella Roma barocca. La biblioteca di don Orazio Morandi*, in EUGENIO CANONE (ed.), *Bibliothecae selectae. Da Cusano a Leopardi*, Firenze, Olschki, 1993, pp. 217-252; BRENDAN DOOLEY, *Morandi's Last Prophecy and the End of Renaissance Politics*, Princeton, NJ, Princeton University Press, 2002; STEFANO TABACCHI, DBI, 76 (2012).

42 PIER FILIPPO COVONI, *Don Antonio de' Medici al Casino di San Marco*, Firenze, Tipografia Coperativa, 1893; GALLUZZI, *Motivi paracelsiani nella Toscana di Cosimo II e di Don Antonio dei Medici: alchimia, medicina chimica e riforma del sapere*, in *Scienze, credenze occulte, livelli di cultura. Convegno internazionale di studi*, Firenze, Olschki, 1982, pp. 31-62; FILIPPO LUTI, *Don Antonio de' Medici e i suoi tempi*, Firenze, Olschki, 2006; ID., DBI, 73 (2009).

43 GOLDBERG, *Jews and magic* (cit. note 38), p. 172. Bovio's work lies still manuscript in the Laurenziana Library of Florence (Ashb. 340) and in the Marciana Library of Venice (ms Marc. It. IV, 57). See ALFONSO INGEGNO, DBI, 13 (1971). For a more updated image, MARIA PIA VANNONI, *The 'doctor of the sword': Tomaso Zefiriele Bovio*, «Bruniana & Campanelliana», 17, 1, 2011, pp. 81-96.

A recently discovered group of documents allows us now to include even Giannozzo Attavanti in this exclusive circle of practitioners of magic and alchemy, who were *habituées* of the courts of Antonio and Giovanni de' Medici, as well as at Villa Meleto. This is a handful of 18 letters (and it is possible that others will emerge as more extensive investigations of the *Fondo* are conducted) from the Archives of the Orsini family, Archives which are now conserved in the Capitoline Archives (Archivio storico Capitolino) in Rome and have only recently been inventoried.[44] They form part of the *Corrispondenza di diversi addetti all'Ecc.ma Casa Orsini* (*Correspondence of several servants assigned to the Most Excellent House of Orsini*), papers generated by exchanges between members of the noble family and those members of their staff who were concerned with strictly private matters, such as courtiers who handled food expenses (*spenditori*), suppliers, artisans, merchants, news dispatchers (*menanti*), assistants and so on.[45]

Covering the period between June 1616 and October 1618, these 18 letters concern in particular the strictly private affairs of Alessandro Orsini (1592-1626). Son of the second Duke of Bracciano, Virginio, and Flavia Damasceni Peretti (descendant of the ruling pope Sixtus V), grandson of Paolo Giordano I Orsini and Isabella de' Medici (daughter of the Grand Duke Cosimo I), Alessandro had grown up to 14 years in Florence, where his parents resided since 1594, in one of Pitti Palace apartments, participating in courtly life.[46] As a natural allied to the Medici House and an influent member of the Roman aristocracy, Alessandro has been the addressee of one of the letters of recommendation with which Cosimo II provided Galileo who was leaving to Rome in November 1615, to defend himself from

44 Elisabetta Mori, *L'Archivio Orsini: la famiglia, la storia, l'inventario*, Roma, Viella, 2016.

45 The *Corrispondenza di diversi addetti all'Ecc.ma Casa Orsini* belongs to the second series of the Archive of the Orsini family, *Administration*. This emerged during the course of its filing, by Dr. Cristina Falcucci, to whom I owe the report. I want to express my heartfelt thanks to her for her generous and discreet support in this research.

46 Mori, *L'Archivio Orsini* (cit. note 44), pp. 70-74; Stefano Boero, *Peretti Damasceni, Flavia*, DBI, 82 (2015). So far, there is no comprehensive biography of the cardinal. Some biographical notes do exist in: Lorenzo Cardella, *Memorie storiche de' cardinali della Santa Romana Chiesa*, vol. 6, Roma, Stamperia Pagliarini, 1793, pp. 184 sq.; Michele Baldelli, *Memorie intorno i dominj e governi della città di Ravenna*, Faenza, Marubini, 1822, p. 59; Konrad Eubel – Patrick Gauchat (eds.), *Hierarchia catholica Medii aevi, sive Summorum pontificum, S.R.E. cardinalium, ecclesiarum antistitum series: e documentis tabularii praesertim Vaticani collecta, digesta, edita*, vol. 4, Monasterii, sumptibus et typis Librariae regensbergianae, 1935, p. 13. On his role in the Galileo affair, OG, XX, *s.v.* As it is well known, he financed the *Rosa Ursina* by the Jesuit Christoph Scheiner (Luigi Ingaliso, *Filosofia e cosmologia in Christoph Scheiner*, Catanzaro, Rubettino, 2005, p. 141). On his activity as a collector, recently wrote Adriano Amendola, *Cardinal Alessandro Orsini, the old Masters and the sciences*, «Journal of the history of Collections Advance Access», pp. 1-13 (published 29 June 2015).

the charges launched by Niccolò Lorini and Tommaso Caccini.[47] Although it seems that Orsini still in February 1615 was on the side of Galileo's opponents in the dispute on the floating bodies, it seems that, after he had the opportunity to meet the scientist in Rome and hear him explain his theories in person, he became reconciled to the 'new science'.[48] Not only Orsini urged the scientist to put into writing his theory on the tides in support of heliocentrism (which led to the famous essay *Discorso del flusso e reflusso del mare*), which he had heard from Galileo's own voice and judged convincing.[49] He also offered to use his newly acquired authority as a Cardinal (he was elected on December 1615) to intercede on Galileo's behalf by presenting his case to Pope Paul V in the Consistory.[50]

Giannozzo Attavanti is the author of most of the newly discovered letters in the Orsini Archives, but not of all of them. They include correspondence addressed by Attavanti to Cardinal Orsini (August 1616), to Cosimo Ridolfi at the Villa Meleto, and to Abbot Orazio Morandi (who was at the time Father General of his order), intermingled with a few letters written by Morandi – on behalf of himself and Ridolfi – to Cardinal Orsini. Nevertheless, taken together these letters ideally form a homogeneous *corpus*, concerned with a hitherto unknown affair that began just after Galileo's return to Florence from Rome in June 1616, formally absolved of any charge of heresy for supporting the truth of the Copernican hypothesis.[51]

4. The 'Mine' at Castelfiorentino

As soon as he returned to Florence, Galileo made a point of sending news to Cardinal Orsini in Rome of the warm reception he had received at the Medici court, thanks in part to the reassuring message sent to the

---

47 Cosimo II to Alessandro Orsini, Florence, 28 November 1615, OG, XII, p. 205.

48 Still in February 1615, monsignor Giovanni Ciampoli considered it appropriate to speak in favor of Galileo with the Abbot Orsini, who «listened to the usual doctrines of Doctor Grazia» (Piero Dini to Galileo Galilei, Rome, 21 February 1615, OG, XII, p. 144). As early as 1612, the Abbot Orsini had hosted Francesco Bonciani, arcideacon of Florence Metropolitan Church, in his house in Rome, who openly defended the *Considerazioni* against Galileo by the 'Accademico Incognito' (Ludovico Cardi da Cigoli to Galileo Galilei, Rome, 16 October 1612, OG, XI, p. 410).

49 Galileo Galilei, *Discorso del flusso e reflusso del mare. Avvertimento*, OG, V, p. 373. See it also in the recent edition by Michele Camerota and Patrizia Ruffo in OG, Appendice III, *Testi*, Firenze, Giunti, 2017, pp. 225-254.

50 Curzio Picchena to Galileo Galilei, Rome, 4 March 1616, OG, XII, p. 242.

51 Camerota, *Galileo Galilei e la cultura scientifica nell'età della Controriforma*, Roma, Salerno, 2004, pp. 319 sq.

Grand Duke by the Cardinal just a few days before Galileo's arrival.[52] In his response to Gaileo, dated 26 June, Orsini expressed his pleasure at his friend's safe return and, just before his closing salutations, informed him that he was waiting «with much desire» for Giannozzo Attavanti, who evidently was a mutual acquaintance.[53] But what exactly was Orsini expecting from this encounter, and why was he waiting – it is not clear whether for a letter or for Attavanti in person – with such trepidation, a trepidation that he knew Galileo would understand? The Cardinal's letters provide the explanation, disclosing the unsuspected background to an affair that was running parallel to the debate over astronomical issues in Florence between the anti-Copernican edict issued in 1616 and the beginning of the 1619 dispute over the comets. Not only do the Orsini letters clarify the roles played by the various individuals in this affair; they disclose a far more complicated network of relationships than historians have hitherto suspected.

On 2 July 1616, Attavanti wrote from Florence to Cosimo Ridolfi:

> I wrote to P. Abbot about what he had asked of me in particular about the mines, which I thought was what he wanted, since I would not have dared to act as a Master in a matter so well known by Your Most Illustrious Lordship. I have greatly appreciated that the practices in this noble art have come into his hands (having always judged them suitable for a prince like him) and that in these he goes so willingly experimenting [...] I hope that with his most profound judgment, he will be more pleased with them every passing day, as I see that he had in mind to do with the mixed mine, in which – God willing – I will enjoy the favour to serve his Most Illustrious Lordship.[54]

The Abbot in question was Orazio Morandi. On that same day that he wrote to Ridolfi, Attavanti sent a letter also to Morandi in order to «confide to you what I knew you should enjoy».[55] In doing so the priest was adopting a somewhat ambiguous position for – as he told Morandi – in writing to him he was contravening Ridolfi's will. The letter dated 2 July was not

52 Cardinal Alessandro Orsini to the Grand Duke Cosimo II, 1 June 1616, OG, XII, pp. 263-264.

53 Cardinal Alessandro Orsini to Galileo Galilei, Rome, 26 June 1616, OG, XII, p. 266. According to Antonio Banfi, this excerpt proves that it was Attavanti, who had close contacts with the Medici court, who informed Galileo of what was going on within the Holy Office and persuaded him to leave for Rome (Antonio Banfi, *Vita di Galileo Galilei*, Milan, Feltrinelli, 1962, p. 148).

54 Giannozzo Attavanti [to Cosimo Ridolfi], Florence, 2 July 1616, ASC, AO, *Corrispondenza di diversi addetti all'Ecc.ma Casa Orsini*, b. 351, letter n. 464c.

55 Giannozzo Attavanti to Orazio Morandi, Florence, 2 July 1616, *ibid.*, b. 351, letter n. 464b.

the first to be exchanged between the Abbot and the parish priest regarding the activities to which Morandi had recently dedicated himself. Already on 11 June, following his return to Florence from Pieve di Castelfiorentino, Attavanti had replied to Morandi who was appealing for assistance with some difficulties arising from the «undertaken handling of the mines». On that occasion, the priest was at pains to inform the Abbot that «Mr. Galilei greets you with all his heart, after having arrived last Wednesday in excellent health, who is ready to serve you with all what is necessary regarding the Ephemerides».[56] This information adds to the *dossier* on the relationship between Galileo and Morandi, which currently consists of only 5 letters, interrupted by a long hiatus between 1613 and 1626.[57] Above all the Orsini letters allows us to establish just when Abbot Morandi gave himself over to the practice of alchemy or, perhaps more precisely, when he placed himself under the expert and attentive guidance of Cosimo Ridolfi ('the Master') and his partner Attavanti, following the transfer of Don Giovanni de' Medici to Venice and the beginning of the exclusively epistolary relationship between the Don and Morandi.[58] *Last but not least*, the letter marks the moment (June 1616) when Attavanti, thanks to a recommendation from Morandi, entered into the service of Cardinal Orsini – whom he referred to as «his Most Illustrious Lordship» in the passage quoted above – as a consultant in the setting up of the Cardinal's 'mixed mine' (in which perhaps even the Abbot was involved). And this last circumstance is probably the reason why the Cardinal was waiting so eagerly to get in touch with the priest at the end of June 1616, a reason that Galileo was well aware of.

According to the correspondence, Attavanti in fact «enjoyed the favour of serving his Most Illustrious Lordship» and held the position of consultant-alchemist certainly for the whole of 1616 and the autumn of 1618.[59]

---

56 Giannozzo Attavanti to Orazio Morandi, Florence, 11 June 1616, *ibid.*, b. 351, letter n. 464.

57 One letter dated 6 July 1613 (OG, XI, p. 530); and four dated respectively 17 April 1626 (OG, XIII, p. 319), 2 May 1626 (*ibid.*, pp. 320-321); 20 June 1626 (*ibid.*, p. 327); 27 May 1630 (OG, XIV, p. 107). On the epistolary exchange, see Ernst, *Scienza, astrologia, e politica* (cit. note 41), pp. 227-229.

58 The letters addressed from Morandi to Giovanni de' Medici, between 1615 and 1620, are in the Florence State Archives, *Mediceo del Principato*, f. 5140 (c. 55) and f. 5149, cc. 625-758. They have been partly examined by Ernst, *Scienza, astrologia, e politica* (cit. note 41), pp. 227-229 and Bucciantini, *Reazioni alla condanna* (cit. note 38), pp. 314-317.

59 Eleven letters from Attavanti to Orsini (all sent from Florence), have been preserved, specifically: 1 August 1616 (ASC, AO, *Corrispondenza di diversi addetti all'Ecc.ma Casa Orsini*, b. 351, n. 462); 13 August 1616 (*ibid.*, n. 462b); 20 August 1616 (*ibid.*, n. 461); 27 August 1616 (*ibid.*, n. 461b); 3 September 1616 (*ibid.*, n. 466); 1 October 1616 (*ibid.*, n. 466b); 9 November

In 1617, however, Attavanti was side-tracked and his time was taken up by a contentious legal affair concerning some of his private interests. For this reason he was temporarily replaced in his post by Morandi, who had risen to the position of procurator of his order and prior of the S. Prassede monastery in Rome, but who unfortunately found himself exiled to Ripoli just outside of Florence beginning on 8 April 1616 due to an event that was never recorded in his official biography and on which even this correspondence fails to shed any light.[60] Unlike Attavanti, Abbot Morandi worked for Orsini in many capacities. He satisfied alchemical curiosities but he also mediated and negotiated courtly affairs on Cardinal's behalf, especially at the Medici court; in both matters Morandi always seemed to be acting in concert with, if not in the shadow of Ridolfi. On 10 June 1617, for example, having gone to the room of Fabrizio Colloredo to request an audience with Grand Duke Cosimo II regarding a matter that was of interest to the Cardinal Orsini, he reported to the Cardinal that:

> In the meantime, I had the opportunity to be with Mr. Knight Ridolfi, whom I involved in this matter (since he is so much Your servant), and whom I partly consulted on how to deal with it. I did so because I am somehow hindered by the fact that His Highness is currently not interested in any affair, as well as by the fact that, I know for certain, even after I have dealt with him, he will not do anything without first sharing it with Serene Madam Mother. And we [Mr. Ridolfi and I] were just arguing as to whether perchance it would not be a good thing to first inform Madame about this matter, so that, when His Highness will discuss it with her, he will find Madam willing to consent.[61]

Particularly on such questions regarding the practice of alchemy, Morandi shows himself to be totally dependent on Ridolfi:

1616 [from Castelfiorentino] (*ibid.*, n. 466c); 14 September 1618 (*ibid.*, b. 352/1, n. 74); 30 September 1618 (*ibid.*, b. 385/2, n. 304); 14 October 1618 (*ibid.*, b. 351/1, n. 69).

[60] Orazio Morandi [to Alessandro Orsini], from Florence Ripoli, the first day of October 1617: «About my coming to Rome, Your Most Illustrious Lordship knows, that I was assigned a year of exile, which ends on the 8th of April, and as it is necessary to break the boundary, it takes more authority than mine. Most illustrious Lord, my only Lord, Master and Protector, I have had some troubles in my greatest happiness, but I thank the Lord, who gave me strength to bring them, and such a light of the intellect, that those things that others thought should damage me and being harmful to me, I have been able to take advantage from them in such a manner, that I have obtained the greatest usefulness and reputation for myself and for my Princes and for Your Most Illustrious Lordship, to whom particularly, both in good and in bad fortune, I will procure to be [...] so that the world may know that, since You have favored a poor and unfortunate wretch, this one has at least succeeded in integrity, sincere faith and incomparable love in spite of those who want to hurt us» (*ibid.*, b. 361, n. 41).

[61] Orazio Morandi [to Alessandro Orsini], 10 June 1617, *ibid.*, b. 361, n. 42.

Mr. Knight Ridolfi answers You with the attached letter – Morandi continued in the same epistle – and as soon as he has accomplished some of his much needed business, he will return to the Villa [Meleto], where I will also go, to take both the Azot and all the writings dealing with it, and I will inform in a timely manner Your Most Illustrious Lordship, so that You can command to whom it should be delivered, and in what way.[62]

After the 'Azot' was sent – because of Ridolfi's great desire to be of service to the Cardinal –,[63] at the beginning of December the two again appear together, now intent on examining the doubts and difficulties encountered by the Cardinal in 'manipulating' the nitrogen, with Ridolfi proposing his services in person should they prove necessary, he being the true *maestro* of the two.[64] On the contrary, the only two letters addressed directly by Attavanti to Ridolfi that have come down to us show that they treated each other entirely as equals, when comparing notes at a distance on the operative techniques to manipulate metals that were suggested by canonical texts or practiced by common acquaintances, but also when working together on experiments, probably at the forge of Villa Meleto in Castelfiorentino. On 2 July 1616 the pair discussed the «making of flowers with animated mercury and earth or with the calcinated grey powder», based on the «third question of the second part» of an unspecified text that has proven difficult to identi-

62 *Ibid.* Azoth is the term used in Rosicrucian and later alchemical literature to designate mercury as a principle which acts on matter like the soul. The *corpus* of Paracelsean treatises contains a *Liber Azoth sive De Ligno et Linea Vitae*, whose authorship is however controversial (Walter Pagel, *Paracelsus: an Introduction to Philosophical Medicine in the Era of the Renaissance*, Basel, Karger, 1982, pp. 102-103 n).

63 Orazio Morandi [to Alessandro Orsini], 1 July 1617, *ibid.*, n. 40: «Mr Knight Ridolfi can not wait to be able to serve Your Most Illustrious Lordship with the craved Azot, and I assure you that You have the true terms to enchant the gentlemen's hearts to make them eternally slaves, since all those who know you are competing with each other to be the one that goes further to serve you better».

64 «Mister Cosimo and I have made good reflection on the details considered by Your Most Illustrious Lordship around the successful manipulation of the achievement of the Azot sent, and in fact we respond as in the attached note... And if Your Most Illustrious Lordship, as soon as You arrive to fix it *ad album*, will stop there, and if You will be pleased to want the rest of the material to proceed to *rubeum*, Mr. Cosimo will keep it to Your request, and will send it to You, because this gentleman is so much in love with the rare qualities worthy of any eminent Prince that he saw in Your Most Illustrious Lordship, that he would even be disposed to destroying himself in order to find the opportunity to serve You. Work, then, Your Most Illustrious Lordship and notify us if you seem to be faced with difficulties of any kind, that we will not fail to serve you in all that you will command and will desire» (*ibid.*) It should be noted that, once back in service, Attavanti continues to use Morandi to send inert materials or treated materials to Orsini (see letter from Giannozzo Attavanti to Alessandro Orsini, Florence, 14 September 1618, *ibid.*, b. 352/1, n. 7 *r-v*).

fy.[65] The second letter, written on 1 September 1616, deals with testing «the technique of ridding gold of silver» [i.e., of extracting gold from silver], which was already being used by «a certain brother Apolinare» who was well known to them but unfortunately not to us. The letter also deals with the preparation of the ingredients for that recipe, which Attavanti urged Ridolfi to begin working on, «because I will gladly try it upon my arrival, which will be in four or five days, having experienced the delay of a month in the dispute».[66]

It is interesting to note how these letters portray the two associates framed by a city landscape apparently swarming with 'artists' of the forge, with whom Attavanti and Ridolfi appear engaged in a constant critical dialogue, such as the above-cited 'Brother Apollinare', or an 'Elia artist', who «every day... is discovering more new things in this art».[67] At least some of these artisans certainly must have had contacts with a well-known, and important center of alchemical interests, such as Don Antonio de Medici's *fonderia*.[68] In the letter quoted above – which is damaged along the right margin and therefore difficult to decipher – it would seem that Attavanti claimed primacy to the friar Apollinare for the recipe which transmuted silver into gold «which they do there [...], and where among others there is also Mr. Agnolo della Casa [...]».[69] From the studies of Paolo Galluzzi we know that Della Casa – an unknown Florentine alchemist, and the author of countless manuscripts disseminated through out the city libraries – was the one who spent the most time trying to decipher the mysterious treatise *Donum Dei* for Don Antonio de' Medici.[70] *Donum Dei* was the recipe for the transmutation of metals into gold, written in a hermetic language by the priest Antonio Neri – an alchemist of considerable fame, as well as an expert glassmaker who worked at the Casino of S. Marco –, which remained, because of the sudden death of its author, without a 'key', and therefore incomprehensible.[71] Some of the techniques by which Neri «generated some

65 Giannozzo Attavanti [to Cosimo Ridolfi], Florence, 2 July 1616, *ibid.*, b. 351, n. 464c.

66 Giannozzo Attavanti to Cosimo Ridolfi, Florence, 1 September 1616, *ibid.*, b. 351, n. 473.

67 Giannozzo Attavanti [to Cosimo Ridolfi], Florence, 2 July 1616, *ibid.*, b. 351, n. 464c.

68 See above, note 43.

69 Giannozzo Attavanti [to Cosimo Ridolfi], Florence, 1 September 1616, ASC, AO, *Corrispondenza di diversi addetti all'Ecc.ma Casa Orsini*, b. 351, n. 473.

70 Galluzzi, *Motivi paracelsiani* (cit. note 42), p. 52 e n. According to Galluzzi, Della Casa lived between 1574 and a date before 1621. On Della Casa, see also Paul Engle, Conciatore (blog): http://www.conciatore.org/2016/05/agnolo-della-casa.html (accessed 11 January 2018).

71 On the priest Antonio Neri, author of a successful *De Arte Vetraria* (On glassmaking,

gold in his silver» were also known to Attavanti and Ridolfi.[72] Orazio Morandi is an even more explicit source regarding the contacts between the two laboratories. As he wrote to Orsini on 1 December 1617:

> Mr. Giannozzo Attavanti sometimes dispute in courts, which is his main occupation at the moment, sometimes looks after pulling beautiful and unique things out of a German, a most talented man, who currently works with Mr. Don Antonio [de' Medici], and among other things he seeks to extract an exquisite core of gold, and if anything good comes of this, His Most Illustrious Lordship will be immediately served...».[73]

The identity of this particular 'German' in the service of the Medici is not known, but – as Galluzzi has already noted – the 'German' is «an often nameless figure who crops up frequently in these kind of accounts and probably alludes to Germanic matrix of the spagyric art».[74] Moreover, another unpublished document also conserved among the letters of the agents of Casa Orsini testifies to direct contacts between Antonio de' Medici and a courtier in Orsini's retinue concerning 'Priest Neri'.[75] This is a letter written on 30 October 1612 by the prince Antonio de' Medici from his estate in Calapiano and addressed to an unknown recipient. With it the prince enclosed the copy of a letter sent to him from Brussels on 10 August 1612 probably by Emanuele Ximenes, a Portuguese knight who had hosted Neri during his extended sojourn in Holland (1609-1611) and who had remained in his correspondence for many years.[76] The document reveals that

---

1612), active alchemist and advocate of a philosophical engaged paracelsism, see: Antonio Neri, *De Arte Vetraria*, introduced and edited by Ferdinando Abbri, Firenze, Giunti, 2001; Ferdinando Abbri, *Antonio Neri*, DBI, 78 (2013); Antonio Neri, *Discorso sopra la Chimica: the Paracelsian philosophy of Antonio Neri*, ed. by Maria Grazia Grazzini, «Nuncius», 27, 2, 2012, pp. 411-467; Paul Engle, *Depicting Alchemy: Illustrations from Antonio Neri's 1599 Manuscript*, in Dedo von Kerssenbrock-Krosigk (ed.), *Glass of the Alchemists*, Corning, NY, Corning Museum of Glass, 2008, pp. 49-62; Id., *Conciatore: The Life and Times of 17th Century Glassmaker Antonio Neri*, Hubbardston, MA, Heiden & Engle Publications, 2014; Id., *Conciatore* (blog) (cit. note 70); Marco Beretta, *Glassmaking Goes Public: The Cultural Background to Antonio Neri's* L'Arte Vetraria, «Technology and Culture», 58, 4, 2017, pp. 1046-1070.

72 Giannozzo Attavanti [to Cosimo Ridolfi], Florence, 1 September 1616, ASC, AO, *Corrispondenza di diversi addetti all'Ecc.ma Casa Orsini*, b. 351, n. 473r.

73 *Ibid.*, b. 361, c. 41.

74 Galluzzi, *Motivi paracelsiani* (cit. note 42), p. 53.

75 ASC, AO, *Corrispondenza di diversi addetti all'Ecc.ma Casa Orsini*, b. 351, n. 83. It is reasonable to suppose that the unnamed addressee was the Duke Virginio Orsini (1572-1615), Cardinal Alessandro's father. Virginio had grown up with Don Antonio at the Medici court and had served with him in the expedition in support of Emperor Rudolf II against the Turks (1594) (Mori, *L'Archivio Orsini*, cit. note 44, p. 70).

76 Abbri, *Neri, Antonio* (cit. note 71). The letters that Ximenes wrote to Neri from An-

Antonio engaged Ximenes as well as Della Casa – in vain in both cases – to search for the key «to the recipes that he [i.e., Neri] had left to Your Excellence. *In verbo veritatis. Amirabile, et Donum*». With just as much evidence this document shows that Don Antonio's investigations were also followed with interest at Orsini's court.[77]

5. Which alchemy? Histories of books

Most of the letters contain recipes, written in the allegorical language of the alchemical tradition. It is important to note that, although he had hired Attavanti himself, Orsini could not overcome the suspicion that his court alchemist might fail to share with him news of his discoveries, or at least put off doing so. In order to excuse himself Orsini often felt obliged to shower the priest with florid promises and declarations of confidence.[78] Given the arcane nature of the alchemical terminology and above all the poor state of conservation of the letters, it is not easy to identify the sources which Attavanti and Ridolfi draw their knowledge from, nor the speculative dimension that underlay their frenetic transmutatory activity. The long letter addressed by Giannozzo to Cardinal Orsini on 1 August 1616 offers some food for thought.[79] This reads almost like an introduction

---

twerp – 28 – are conserved at the National Library in Florence, Fondo Nazionale, II, I, 391 (Galluzzi, *Motivi paracelsiani*, cit. note 42, p. 50n). I suggest identifying the 'Friend from Antwerp' – as Antonio de' Medici calls him in the Orsini letter – who was the author of the 10 August 1612 letter, with Ximenes because the 'Friend from Antwerp' alludes to a stay of Neri 'here' as well as to some «little secrets that he learnt with me, and at my expenses and thanks to my excellent food».

[77] The letter by Ximenes to Antonio de' Medici calls into question Neri's date of death, set by his biographers at 1614 (see above, note 58). Dated, as we have seen, 10 August 1612, Ximenes speaks of Neri as having already passed away to a better life («As, then, to the said Priest, to whom God give peace [...]»). The date of 1612, in fact, could explain why the letters of Ximenes to Neri were interrupted in 1611.

[78] In this regard, the letter that Giannozzo Attavanti sent to Alessandro Orsini from Florence on 10 September 1616 is particularly eloquent: «Lord Most Illustrious. Everything I have written about the Mine – made according to the abbreviated, or the long procedure – you can understand from my letters that it was done because you commanded me to do it, and not because I went further. And if the matter were in another way, as I have done so far and still do, I would confer it on Your Lordship. And do not doubt this at all, nor do you think I want to hide anything from You, because it would seem to me to be too neglectful of the obligation I have made to serve You in this» (*ibid.*, 351, n. 473). On the subject, more in general, Pamela O. Long, *Openness, Secrecy, Authorship. Technical Arts and Culture from the Renaissance*, Baltimore, London, Johns Hopkins University Press, 2001.

[79] *Ibid.*, b. 351, n. 462. The following quotes are taken from here (un-numbered pages).

to the 'major art', in which the parish priest traces a parallel between the procedures that nature follows in transforming matter on one hand (the 'mine'), and on the other «the means and ways that must be followed in the practice of the major art». As much as nature's procedures, the means used by the major art are «innumerable and, depending on the skill of the craftsmen, better or worse, long or even short». Unlike the 'mine', where nature itself provides the means to break minerals down into their elementary components (first and foremost the quicksilver), the major art has to procure them through ingenious processes. In this parallel vision, in which the two poles nevertheless remain independent, 'sulphurous mercury' – which is «the principle, medium and end goal of the Mine» – corresponds in the major art to «the [ferment] of philosophers, the source of each transmutation and the principle of all wonder». More specifically, just as in nature «the medium that makes living silver turn into earth and [which] is fixed in metal, is a certain sulphurity, or metallic spirit as you may wish to call it, found in the center of the body», so «in the art, it will be the metals of the philosophers. And this is precisely the reason why they [i.e., the philosophers] say that their gold and silver are not vulgar gold and silver, because the latter are dead while their metals are alive».

According to the letters presented here, the alchemy practiced at the Villa Meleto seems to be particularly the alchemy focused on the transmutation of elements, especially metals. On the one hand, this practice rests on the branch of alchemical research represented in the Latin tradition mainly by pseudo-Geber's *Summa perfectionis magisterii* (especially with regard to the relationship between nature and art and the theme of 'mercury alone', i.e. the composition of metals by sulfur and mercury). On the other hand, according to the syncretic tendency which was common still in late Renaissance, it also was in line with the alchemical texts of the Lullian tradition (especially with regard to the noble metals conceived as principles, the ferment, and the overall definition of a *circular opus* which relied on the schema of *extrema* and *media naturae*).[80]

---

[80] In fact, Attavanti continued: «With these foundations, if one goes on to consider the matter more in particular, one will see that just as the live metal which is produced by means of art in the beginning was the water which is transformed into earth and that this water, before it becomes earth and live metal, gradually had to pass through many intermediate states and for many dispositions, in the same way this earth and live metal still had to ascend by the same degrees, if you want it to return to water, since nature never passes by extreme to extreme [...]» (ASC, AO, *Corrispondenza di diversi addetti all'Ecc.ma Casa Orsini*, b. 351, n. 473r). This brief scrutiny of the letters' content, which deserve further investigation, is based on: Michela Pereira, *Arcana sapienza: l'alchimia dalle origini a Jung*, Roma, Carocci 2002; Id. (ed.), *Alchimia: i testi della tradizione occidentale*, Milano, Mondadori, 2012; Chiara Crisciani – Michela Pereira – Wolf-

In fact, among the handful of cited texts that can be identified (others are alluded to only by a paragraph or chapter number) we find the *Summa perfectionis* of Pseudo-Geber/Paul of Taranto,[81] as well as works by the little known fourteenth-century alchemist Isaac Hollandus[82] and by Ramon Llull.[83] Paracelsus is mentioned only once – and by Orazio Morandi –, as an *auctoritas* on the manipulation of Azoth.[84] This scant list certainly does not exhaust the catalogue of alchemical interests of the protagonists of our story, who all possessed vast and reputable libraries. It is well known that one of Orazio Morandi's most prized possessions was his «*locupletissima ac selectissima bibliotheca*», the fruit of «his study, *expensa et labore*» and assembled over a period of many years.[85] It was properly this library which played a crucial role in the trial for necromancy and the possession of forbidden books to which Morandi was subjected in 1630.[86] As Germana Ernst has made clear, an important nucleus of this collection reflects the interests of the Abbot in the field of medicine, «and the most prominent element is the extremely close attention paid to the Paracelsian vein along with innovations in chemical medicine, which provided a background to the alchemical texts».[87] As for Antonio de' Medici, he possessed a 'considerable' library, with a 'chemistry' section of truly remarkable proportions,

---

Dieter Müller-Jahncke, *Il Rinascimento. L'alchimia*, in *Storia della Scienza*, Roma, Istituto dell'Enciclopedia Italiana, 2001; Antonio Clericuzio, *Alchimia, iatrochimica e arti del fuoco*, in Clericuzio – Saverio Ricci, *Il Contributo italiano alla storia del Pensiero. Scienze*, Roma, Istituto dell'Enciclopedia Italiana, 2013; Id., *Mechanism and Chemical Medicine in 17th-century England: Boyle's investigation of ferments and fermentation*, in Peter Distelzweig – Benjamin Goldberg – Evan Ragland (eds.), *Early Modern Medicine and Natural Philosophy*, Dordrecht-London, Springer, 2015 (eBook) and 2016 (printed), pp. 271-294.

81 Giannozzo Attavanti (to Orazio Morandi), 11 June 1616, ASC, OC, *Corrispondenza di diversi addetti all'Ecc.ma Casa Orsini*, b. 351, n. 464 («[...] Summa vulatilis, summa fissi superat». See *The Summa perfectionis of Pseudo-Geber: a critical edition*, translation and study by William R. Newman, Leiden, Brill, 1991, p. 306.

82 Giannozzo Attavanti (to Cosimo Ridolfi), 2 July 1616, ASC, AO, *Corrispondenza di diversi addetti all'Ecc.ma Casa Orsini*, b. 351, n. 464c: «I let Your Most Illustrious Lordship know how I acted with Saturn's salt and the salt of the Moon, and since I complied with Isaac's opinion [...]». On Isaac Hollandus, see Annelies van Gijsen, *Isaac Hollandus Revisited*, in Miguel López Pérez – Didier Kahn – Mar Rey Bueno (eds.), *Chymia: Science and Nature in Medieval and Early Modern Europe*, Newcastle upon Tyne, Cambridge Publishers, 2010, pp. 310-330.

83 Giannozzo Attavanti to Alessandro Orsini, 1 August and 27 August 1616, ASC, AO, *Corrispondenza di diversi addetti all'Ecc.ma Casa Orsini*, b. 351, 461b (on *nigredo*) and 462 (on vitriol).

84 Orazio Morandi to Alessandro Orsini, 1 December 1617, *ibid.*, b. 361, n. 41. See above, note 63.

85 Ernst, *Scienza, astrologia e politica* (cit. note 41), p. 230.

86 *Ibid.*, *passim*.

87 *Ibid.*, pp. 241-242.

«in which not only many of Paracelsus's works were present, but virtually all of the most important authors from the second half of the 1500s and the early 1600s who concerned themselves with the Paracelsian tradition, either embracing it and promoting its dissemination, or contesting it, refuting it, or trying to bring it into line with traditional doctrines».[88] We have no evidence from Giannozzo Attavanti's bequest. A thorough examination of the judicial and notary sources would help to shed light on this figure and hopefully also on his alleged library. In contrast, we know that Cosimo Ridolfi owned a notable collection of books on 'occult' matters and that for a period of some months his library, as well as the secrets of his *fonderia*, were the object of contention among the highly secretive agents of various alchemist-princes, who were always on the lookout for books on rare and forbidden curiosities, like the books on alchemy.

Ridolfi died on the night of 19 January 1619 at the Villa Meleto, from suffocation due to a 'congestion of phlegm', an event that afterwards Morandi claimed he had foreseen.[89] Great was the feeling of bewilderment, beginning with Galileo and the Lincei academicians and extending to the members of the 'magic circle', who could only console themselves that he had had time to confess and receive the extreme unction, and therefore died as a good Catholic.[90] After a few weeks, however, their concern turned to the fate of his library, a collection lovingly assembled by Ridolfi that included texts on «distillation and on astrology»,[91] and considering «the other principal books, it does not lack anything».[92] The books had remained in the – anything but safe – hands of Ridolfi's widow, Laudomia, whom they knew «to have harbored ill will against the husband who had attended to those studies» and who had decided to keep their son well away from

---

88 Galluzzi, *Motivi paracelsiani* (cit. note 42), pp. 36-43: 37-38. On the 'enormous and precious' library of Don Giovanni de' Medici, which was scattered through out Florence and the other places where he lived, see Landolfi, *Don Giovanni de' Medici* (cit. note 38), p. 127.

89 Don Giovanni de Medici to Cosimo Baroncelli, 26 January 1619, in ASF, Alessandri 2, f. 372, quoted in Goldberg, *Jews and magic* (cit. note 38), p. 171.

90 See Benedetto Blanis to Don Giovanni de Medici, 19 January 1619, ASF, MdP, 5150, c. 145; Cosimo Baroncelli to Don Giovanni de Medici, 19 January 1619, *ibid.*, 5147, c. 266. On Lincei's reaction, see Federico Cesi to Galileo Galilei, Acquasparta, 16 February 1619, OG, XII, p. 439.

91 ASF, MdP, 5150, f. 293r.

92 *Ibid.*, c. 420v. Negotiations for the purchase of Ridolfi's library by Giovanni de' Medici can be reconstructed through his correspondence with Orazio Morandi, already reported by Landolfi (*Don Giovanni de' Medici*, cit. note 38, p. 128n), read against the contemporary exchange of the same with Benedetto Blanis, reported and edited in its entirety in Goldberg, *A Jew at the Medici Court* (cit. *ibid.*), docs. nn. 128, 130, 131, 137, 138, 139, 147, 163, 164, 169, 170, 172, 173, 174, 181, 182, 183, 184, 187, 188, 189, 193, 194, 195, 199.

these dangerous books.[93] Already at the end of March 1619, Blanis entered into negotiations with the widow to acquire the books on behalf of Don Giovanni de' Medici. Laudomia managed to defy Blanis' deceptions: his insistent and pretentious visits, his stratagems, gifts, and favors. She bought time by pleading the difficulties of having the books transported to the city, made and broke promises, and set Blanis and Orazio Morandi against each other by, for example, giving Blanis the impression that the books were destined for the 'friar'. In the context of this discussion, it is interesting to note something that Blanis learned when he was seeking the identity of the person to whom the widow said she had given «what books I do not know».[94] In December 1619, as Blanis afterwards recounted to Giovanni de Medici, a solicitous informant told him «that – this – was Father Don Oratio and that he had sent them to Orsino». To Blanis' inquiry as to whether this 'Orsino' was the 'cardinal friend', the informant confirmed «that he was very intimate and that they were blowing together [i.e., they were conducting alchemical experiments together], and that he has had the best of the writings and the books of Mr. Ridolfi» and that in any case he would have sought «at his own convenience, to try to understand what the Ottavanti (i.e., Attavanti) says», since the informer knew that he was intimate with the knight.[95] In fact, even Morandi had started negotiations with the widow at this time on behalf of Don Giovanni, and it is from his informative report to the Medici prince that we know that among these books «there are works by Paracelsus, there is Giovanni Reuclino [Johannes Reuchlin] and others on cabala... among the books of philosophy there is Telesio».[96] Blanis, for his part, mentioned to Don Giovanni that he had noted among Ridolfi's books «all the works by Paracelsus and Cardanus», as well as the *Albero della Kaballah* transcribed by a certain Lippiano.[97] The negotiations of both agents came to no naught, however. According to the latest evidence, which dates to June 1620, the bulk of the library came into pos-

---

93 Benedetto Blanis to Don Giovanni de Medici, 24 March 1618 (ASF, MdP, 5150, c. 251r), *ibid.*, n. 130.

94 Benedetto Blanis to Don Giovanni de Medici, 22 December 1619 (ASF, MdP, 5150, c. 270v), *ibid.*, n. 168.

95 *Ibid.*

96 Orazio Morandi to Don Giovanni de Medici, May 1620, ASF, MdP, f. 5149, c. 716r. It is likely that Reuchlin's text was *De arte cabalistica libri tres, iam denua adcurate revisi*, published for the first time on 1530). See, now, JOHANNES REUCHLIN, *L'arte cabbalistica (De arte cabalistica)*, ed. by Giulio Busi and Saverio Campanini, Firenze, Opus Libri, 1995.

97 Benedetto Blanis to Don Giovanni de Medici, 28 June 1620 (ASF, MdP, 5150, c. 421r-v), in GOLDBERG, *A Jew at the Medici Court* (cit. note 38), n. 199.

session of Giovanni Ronconi, a physician who attended members of the Grand Duke's court and later the Galilei family,[98] through various channels that at the time were kept deliberately hidden. The fate of the books after Ronconi falls into mystery.[99]

We have no way of knowing – not even from the Orsini papers – whether Laudomia really did hand over to Morandi some of her husband's books to give to Alessandro Orsini. Concerning the contents of Cardinal Alessandro's library, there is unfortunately no record in the family archive. That he was interested in forbidden books is not in doubt. Among the parchments in the Orsini fund there is the brief written by Urban VIII and dated 28 April 1625 in which the Pope granted the Cardinal and his librarian «the permission to read and keep forbidden books», although Orsini would probably not have kept them together with the Church-authorized texts in his library.[100] The only books in the family's possession that we know of are those listed in an inventory compiled on the death of Alessandro's nephew Lelio Orsini (1622-96). Among these, however, some volumes that must certainly have been purchased by Lelio's uncle can be identified.[101] The works on chemistry published when the Cardinal was live and engaged in scientific pursuits can not really be said relevant – 9 texts out of a total of about 1,400, and not all of these can be identified today. It constitutes a minimal collection, but reveals that its owner had some knowledge

---

[98] See biographical notes on Ronconi and the references to his epistolary exchanges with Galilei (1629-30) in OG, XX, *s.v.*

[99] Only one of the books owned by Laudomia, *Prediche di Frate Hieronymo da Ferrara*, 1496, resurfaced in the personal library of Prince Franz Joseph II of Liechtenstein (1906-1989). It was subsequently purchased at the antiquarian market by the Library of the University of Missouri (http://library.missouri.edu/news/special-collections/qo-book-and-dela-signora-laudomia-ricasoli-ridolfi).

[100] Archivio Orsini, s. II.A.30,003 (http://old.archiviocapitolinorisorsedigitali.it/scheda_doc.php?IDA=71&IDF=1997&PS=1&PR=25&PB=1&SF=Regesto_Pressutti&SV=Urbano%20VIII&OB=Segnatura&OM=). The books «on the art of the Cabala, and on Magic» were held by Don Giovanni de' Medici within a coffer which caused great concern to the Medicean ambassador in Venice (where De' Medici died) «for the reputation of His Excellence's memory» (letter from Niccolò Sacchetti to the Grand Duke, 18 September 1621, quoted in BUCCIANTINI, *Reazioni alla condanna*, cit. note 38, pp. 308-309).

[101] In particular: *Alexandri ab Alexandro*, by Alessandro Alessandri; *Catalogus Gloriae Mundi* by Barthelemy de Chasseneuz; *Uccelleria* by Giovanni Pietro Olina, purchased by Alessandro on 25 February 1625 (ASC, AO, II s., 1311, c. 80r, reported by AMENDOLA, *Cardinal Alessandro Orsini*, cit. note 46, p. 103n). The inventory of the books owned by Lelio Orsini – which is preserved at Univerity of California Los Angeles, Library Department of Special Collections, Orsini Family Papers, box. 44, f.1, old signature I.B.LXXI.5 – is entirely edited in AMENDOLA, *La collezione del principe Lelio Orsini nel palazzo di piazza Navona a Roma*, Roma, Campisano, 2013, pp. 146-157, which I refer to for biographical notes. The above mentioned volumes owned by Alessandro Orsini, are listed *ibid.*, p. 146.

of the basic texts of medieval and Renaissance alchemy (it included the five-volume work *Theatrum chemicum* published by Lazarus Zetner,[102] *De magno lapide antiquorum sapientum* by Basilio Valentinus,[103] *Introductio in artem chimicam* by Pietro Bono from Ferrara,[104] and Roger Bacon's *Thesaurus Chemicus* [105]), and was also aware of the controversial nature of that art (he owned a copy of *De interitu alchimiae metallorum transmutatoriae* by Nicolas Guibert).[106] In addition, the inventory lists Oswald Croll's *Basilica chymica* (Frankfurt 1609), which testifies to Orsini's interest in Paracelsism.

6. Concluding Considerations

The documents found recently during the reorganization of the Orsini Archives confirm certain historiographical conclusions that have been taken for granted, but they also raise many questions. The presence of a laboratory for the transmutation of metals on the Orsini estate adds yet another detail to the map of 'chemistry sites in Rome' that has been compiled over the past twenty years by historians in research that is still ongoing. This map shows the location of the sites where some members of the Academy of the Lincei experimented with the remedies of chemical medicine,[107] and the residences of nobles and prelates where distillation was practiced.[108] That the theories of Paracelsus took root in Tuscany, par-

---

102 Apparently in the four-part edition (*ibid.*, pp. 150 and 155). The edition in 4 vols. was published in Strasbourg in 1613. However, the double recurrence of the work could also allude to two copies of the first edition in three parts (Ursell, 1602).

103 *Ibid.*, p. 146.

104 *ibid.*, p. 153.

105 *ibid.*, p. 152.

106 *Ibid.*, p. 150. See Nicolas Guibert, *De interitu alchimiae metallorum transmutatoriae*, Toul, 1614. On the controversy between Guibert and Andreas Libavius over the value of alchemy, see Bruce T. Moran, *Andreas Libavius and the Transformation of Alchemy: Separating Chemical Cultures with Polemical Fire*, Sagamore Beach, Mass., Science History Publications/USA, 2007, *ad indicem*.

107 Silvia De Renzi, *Un linceo alla Sapienza: la natura del fuoco e dei metalli in un'orazione di Johannes Faber*, in Andrea Battistini – Gilberto De Angelis – Giuseppe Olmi (eds.), *All'origine della scienza moderna: Federico Cesi e l'Accademia dei Lincei*, Bologna, il Mulino, 2007, pp. 271-316; Clericuzio, *Chemical Medicines in Rome: Pietro Castelli and the Vitriol Debate (1616-1626)*, in Maria Pia Donato – Gill Kraye (eds.), *Conflicting Duties: Science, Medicine and Religion in Rome, 1550-1750*, London, Turin, 2008, pp. 281-302 (with former bibliography).

108 Like Virginio Cesarini and Cassiano dal Pozzo (Eraldo Bellini, *Umanisti e Lincei. Letteratura e scienza a Roma nell'età di Galileo*, Padova, Antenore, 1998, pp. 261-264; Oreste Trabucco, *Scienza e comunicazione epistolare: il carteggio fra Marco Aurelio Severino e Cassiano Dal Pozzo.*

ticularly in court circles, has been thoroughly documented. Alongside the names of Don Giovanni de' Medici, Raffaello Gualterotti and Benedetto Blanis, personages such as Don Antonio de' Medici and Antonio Neri, and still largely unknown figures such as Agnolo della Casa, Benedetto Punta, and Jacopo and Giovanni Macolo show how widely Paracelsian science and the esoteric arts of Llull circulated in Siena, Pisa and Florence between the end of the 1500s and the beginning of the 1600s.[109] Moreover, recent scholarship has familiarized us with the variegated and contradictory worlds that lay behind *milieux*, circles, and institutions whose members historians once assumed were all of one mind – settings such as the Medici court and institutions such as the Academy of the Lincei,[110] but even the Congregations of the Holy Roman Church.[111] Today we see each of these settings and institutions more clearly as a juxtaposition of poles, factions, and parties. The assumption that these poles, factions, and parties were completely mutually unrelated, however, is not in question, because we suppose that they were divided by diversity of formations, methods, forms of knowledge and images of the world. Given this frame of reference, Orsini's correspondence is surprising and of great interest because it reveals the ease with which individuals from very different *milieux* were able to cultivate relations with different groups. This discloses a much more nuanced picture than we have hitherto been aware of, one in which it is difficult to map clear-cut alignments, even in the events that surrounded and in part determined the outcome of the Galilean affair of 1616.

We need only consider Orazio Morandi, friend and agent of Cardinal Orsini, that is of a staunch protector of Galileo during the *affaire* and the friend to whom the scientist addressed his *Discorso del flusso e reflusso*. Even if he served to a friend of Galileo, Morandi immediately sent to Don Giovanni de Medici the *Discorso* together with the physical-astrological writing on the same phenomenon by Galileo's opponent Alessandro Padovani.[112] We may

Con un'appendice di nuovi documenti, «Giornale critico della filosofia italiana», s. 6, 76 (78), 1997, 2, pp. 217-219, 229-230), or Cardinal Virginio Spada (Giuseppe Finocchiaro, *Il museo di curiosità di Virgilio Spada*, Roma, Fratelli Palombi, 1999, especially pp. 19-23, 31-38).

109 Bucciantini, *Reazioni alla condanna* (cit. note 38), p. 313.

110 On the different souls which lived together in the Academy of the Lincei, see Galluzzi, *Libertà di filosofare in naturalibus: i mondi paralleli di Cesi e Galileo*, Roma, Scienze e Lettere, 2014.

111 See, for instance, Rivka Feldhay, *Galileo and the Church: political inquisition or critical dialogue?*, Cambridge, Cambridge Univerity Press, 1995; Maria Pia Donato, *Scienza e teologia nelle congregazioni romane. La questione atomista, 1626-1727*, in Antonella Romano (ed.), *Rome et la science moderne: entre Renaissance et Lumières*, Rome, École française de Rome, 2008, pp. 595-634.

112 Bucciantini, *Reazioni alla condanna* (cit. note 38), pp. 314-315.

also think back to Cosimo Ridolfi. The sources examined leave no doubt about his adherence to the *entourage* (or should we say the party?) of Don Giovanni de' Medici and their vision of the world including a place for alchemy and cabalism. What is more, one of the few references to Ridolfi in Galileo's correspondence describes him at lunch with Girolamo da Sommaia, provost of the University of Pisa, Piero Girolami, a lecturer on canon law, the humanist Giulio Bulingiero, Benedetto Castelli, and a certain 'Giovanni alchemist'. On that occasion the conversation turned to the *Mundus Iovialis* by Simon Mayr and Castelli fought hard to defend Galileo from attacks by 'the alchemist'.[113]

Unless one admits that hostility on just one side is possible in a war between good and evil, why would the embattled Galileo have brought Attavanti and his alchemical skills into play in an exchange of favors for protection during the months of his stay in Rome in 1615, if the priest's vision of the world and nature had been one of hopelessly outdated notions and insurmountable prejudice? An even more significant question is why it was Galileo – champion of the new science – who proposed the election of Ridolfi – an alchemist and Kabbalist, but also an aristocrat and a relative of his closest associate, Filippo Salviati – as an academician of the Lincei? And why then did he at the same time (May 1613) reject the candidature of the Sienese spagyric doctor Benedetto Punta, an exponent of a vision of the world that was probably not very different from that of Ridolfi, but who above all was sponsored by the only figure – Giovan Battista della Porta – who could possibly outshine the astronomer in the Lincean assembly?[114] The question remains open, but once again the sources invite us not to neglect the context and the 'micro-politics' of individual power strategies.[115]

113 Benedetto Castelli to Galileo Galilei, Pisa, 2 January 1615, OG, XII, p. 134.

114 On Benedetto Punta and his missed appointment to the Accademia dei Lincei, see GALLUZZI, *Motivi paracelsiani* (cit. note 42), pp. 58-60.

115 For methodical considerations on this subject, I would like to refer to FEDERICA FAVINO, *Le ragioni del patronage: i Farnese di Roma e Galileo*, in BUCCIANTINI – CAMEROTA – GIUDICE (eds.), *Il caso Galileo* (cit. note 17), pp. 163-185.

Giovanni Pizzorusso

FRANCESCO INGOLI: KNOWLEDGE AND CURIAL SERVICE IN 17th-CENTURY ROME

1. Francesco Ingoli: A Broad Cultural Background [1]

On March 5, 1616, the Congregation of the Index issued a decree imposing the correction of Copernicus's text *De revolutionibus orbium coelestium* after more than half a century in which it had freely circulated. The decision was based on the deliberation taken during the March 1st session, which was attended by Cardinal Bonifazio Caetani, who had recently become a member of the Congregation. The exact reconstruction by Pierre-Noël Mayaud of the sequence of events that led to the decision shows how the matter was dealt with in an inquisitorial context and which cardinals were involved. Mayaud also points out that since this was the first time Caetani had attended a meeting of the Index, his co-option could have been *ad hoc* for this purpose.[2] The correction of the text was then entrusted to

---

1 I would like to thank Caterina Fiorani (director of the Archivio della Fondazione Caetani, Roma) for her assistance during my research on Ingoli's letters, and Manuela Bragagnolo (Max-Planck Institut fur Europaïsche Rechtgeschichte, Frankfurt am Main) for some useful information on Giovan Battista Della Porta, as well as the two curators of this volume Natacha Fabbri and, above all, Federica Favino for her invaluable help offered in helping me understand the astronomical arguments in Ingoli's texts. The responsibility for any mistakes is, of course, my own. The following abbreviations are used in the paper: ACDF = Archivio della Congregazione per la Dottrina della Fede (Index = Archivio Congregazione dell'Indice); AFC = Archivio Fondazione Caetani Roma (FG = Fondo Generale); APF = Archivio storico della Congregazione de Propaganda Fide (SOCG = Scritture Originali riferite nelle Congregazioni Generali); BAV = Biblioteca Apostolica Vaticana; *DBI* = *Dizionario Biografico degli Italiani*, Roma, Istituto dell'Enciclopedia Italiana (in the bibliographical references, the pages are not given as they refer to the on-line versions of the work).

2 Pierre-Noël Mayaud, *La condemnation des livres coperniciens et sa revocation à la lumière de documents des Congrégations de l'Index et l'Inquisition*, Rome, Editrice Pontificia Università Gregoriana, 1997, pp. 25-35. The decree text is also published in Galileo Galilei, *Scienza e religione. Scritti copernicani*, ed. by Massimo Bucciantini and Michele Camerota, Rome, Donzelli, 2009, pp. 259-262.

Caetani, as Galileo Galilei was quite aware, since he was following these events with interest, being himself subject to the attentions of Cardinal Roberto Bellarmino at that time.[3]

A few months earlier an academic meeting was held in the palace of Cardinal Lorenzo Magalotti where Galileo, who had recently arrived in Rome, defended the Copernican system against its opponents. The spokesman for the Church on that occasion was Francesco Ingoli, Cardinal Caetani's secretary, who supported the positions of the "old mathematicians". As Ingoli himself tells us, when he returned home after that meeting, he decided to put the anti-Copernican arguments in writing, so as to compile a «vera e propria summa»,[4] the *Disputatio de situ et quiete terrae*. In this way he became fully involved in the Galilean affair and it is not surprising that at the next meeting of the Index (May 10, 1616), despite the absence of his patron Cardinal Caetani, Ingoli was appointed consultant, having been proposed by Cardinal Orazio Lancellotti, and as such became officially involved in the matter: «[The cardinal] Lancellotti also suggested that Francesco Ingoli, who is now serving the Cardinal Caetani, should be appointed as a consultant».[5]

Up until now, although historians of science have come across Ingoli's name as Galileo's interlocutor, they have never seen fit to explore who the man was, let alone investigate sufficiently to provide a full-length portrait. The first historian to provide important elements, painstakingly sifted from a sparse number of largely unpublished writings and sources concerning Ingoli, was undoubtedly Massimo Bucciantini. It is thanks to his work and subsequent studies by other authors that we have begun to catch a glimpse of the profile of Ingoli and cease to consider him as just a passing figure within the complex and many-faceted story of the Galilean controversy.[6] However, there has not yet been any research centering on Ingoli, especially for the period around the dispute.

---

3 Mayaud, *La condemnation* (cit. note 2), p. 55. May 26, Bellarmino had officially informed Galileo of Pope Paul V's statement against his heliocentric theory made at the meeting of the Holy Office on February 25, Michel-Pierre Lerner, *Copernic suspendu et corrigé: les deux décrets de la Congrégation romaine de l'Index (1616-1620)*, in *I primi Lincei e il Sant'Uffizio: questioni di scienza e fede*, Roma, Accademia dei Lincei, 2005, p. 321.

4 OG, V, pp. 403-404; Galilei, *Scienza e religione* (cit. note 2), pp. 163-166; quote from Bucciantini, *Contro Galileo. Alle origini dell'affaire*, Florence, Olschki, 1995, p. 84.

5 ACDF, Index, *Diarii*, I, 2, 93r: «Ill[ustrissi]mus Lancillottus proposuit annumerandum etiam de novo pro consultore Dominum Franciscum Ingolum ad servitia Ill[ustrissi]mi D[omi]ni Card[ina]lis Caietani demorantem».

6 See Bucciantini, *Contro Galileo* (cit. note 4), to which reference was made to the aforementioned essays by Mayaud and Lerner. See also Maurice A. Finocchiaro, *Defending Coperni-*

In fact, in 1616, when he was formally accepted into the Curia, just who was Francesco Ingoli? From the currently available sources it is not easy to assemble a sufficiently broad biographical and intellectual profile, although from careful study of the documentation and the young astronomer's manuscripts, we can draw a fairly clear picture of his competencies, his scientific status, and his intellectual position within the controversy, as well as his relationships within the Curia. From these writings a distinct intellectual personality emerges. We discern Ingoli's early (just after his departure from the University of Padua) and quite autonomous ability to re-elaborate the principles of an anti-Copernican astronomy and masterfully discuss the merits of Tycho Brahe's thesis.

Let us look back then at the stages of his biography. In 1616 Ingoli was 38 years old and for some time had been living in Rome after transferring from Ravenna, where he was born on the 21st of November 1578 into a family of the local nobility.[7]

We know nothing of his youth apart from his early fascination with languages. At 16 Ingoli transcribed some Arabic Penitential Psalms, which are still preserved in the Biblioteca Classense in Ravenna along with an annotated Arabic lexicon, testifying to his early study of Arabic, a fact which Ingoli himself confirmed in a much later document.[8] He was sent to the University

---

cus and Galileo. Critical Reasoning in the Two Affairs, Dordrecht, Springer, 2010, pp. 72-75 (which has deepened the «sophisticated and strong statement of the scriptural objection to Copernicanism», *ibid.*, p. 94), and Christopher Graney, *Setting aside all Authority, Giovanni Battista Riccioli and the Science against Copernicus in the Age of Galileo*, Notre Dame, University of Notre Dame, 2015, Appendix A (which translates into English the *Disputatio* by Ingoli and supports the prevalence of physical-mathematical motivations with respect to the theological ones that Ingoli, taking from Tycho Brahe, proposed to Galileo in the condemnation of Copernicanism).

7 Information on Ingoli is found in Girolamo Fabri, *Le sagre memorie di Ravenna antica*, Venice, per Francesco Valuasense, 1664, Part I, pp. 69-70: 163; Serafino Pasolini, *Lustri Ravennati*, part I, vol. III, p. 82; Part V, vol. XIV, per Carl'Antonio Zampa, pp. 103, 150-151, Forlì, 1684; Pietro Paolo Ginanni, *Memorie storico-critiche degli scrittori ravennati*, tome I, Faenza, presso Gioseffantonio Archi, 1769, pp. 437-442; see Josef Metzler, *Francesco Ingoli, primo segretario della Congregazione (1578-1649)*, in Francesco Ingoli, *Relazione delle quattro parti del mondo*, ed. by Fabio Tosi, Rome, Urbaniana University Press, 1999, pp. 293-332 (original German edition in *Sacrae Congregationis de Propaganda Fide Memoria Rerum*, ed. by Josef Metzler, vol. I/1, Rome-Freiburg-Wien, Herder, 1971, pp. 197-243). See also the entries by Jean Pirotte, *Dictionnaire d'histoire et géographie ecclésiastiques*, vol. 25, Paris, Létouzey et Ané, 1995, cc. 1150-1157 and by Giovanni Pizzorusso, *DBI*, vol. 62 (2004).

8 Ravenna, Biblioteca Classense, ms. 17 and 282, quoted by Msgr. Giovanni Montanari, *Francesco Ingoli e la sua città natale di Ravenna*, unpublished communication read at the symposium *Mons. Francesco Ingoli primo segretario di Propaganda Fide: memoria e attualità* (Rome, Pontifical University Urbaniana, 3 December 1999). I sincerely thank the author for making the manuscript available to me. See also APF, SOCG, vol. 388, ff. 193r-194v (letter from Ingoli to Cardinal Bandini on June 1, 1628, referring to maronite Vittorio Accurensis, stating: «I know

of Padua by his family to study canon law and at the age of twenty-three was awarded a degree *in utroque iure* by unanimous consensus on the 17th of May 1601.[9] We do not know the exact date, but it turns out that he also was awarded the title of *magister theologiae*.[10] Biographical notes suggest that he showed an interest as a young student in the sciences, especially astronomy. Even if there is no direct evidence, this seems very likely since the young Ingoli was soon writing on the subject. It would have been difficult for him to avoid being influenced by the intellectual environment in Padua, where Galileo was disputing with the local Aristotelians on science and the stars and exchanging ideas with Paolo Sarpi on the theories of Copernicus.[11] Certainly there is a dearth of potentially intriguing information; we don't know whether the student from Ravenna attended Galileo's lectures or frequented the circle of Giovan Vincenzo Pinelli, whose vast library contained volumes that fed the lively debates on Tycho's astronomical theories.[12] After his brilliant achievements as a student, however, Ingoli did not remain in Padua but returned to Ravenna, where he probably took minor orders, although there is no documentary evidence on this point. He entered the College of Doctors on the 24th of October 1601 and began to teach law.[13]

The turning point for Ingoli from a local career to one of more promising prospects in Rome arrived when Cardinal Bonifazio Caetani came to Ravenna at the end of 1606 as a legate-president of Romagna. It seems that two years later, in 1608, the cardinal took Ingoli into his service as his private secretary. Thus, the young man was granted the opportunity to work closely with an influential prelate who could help him gain access to the Roman Curia. At this stage, if due consideration is taken of Ingoli's interest in astronomy, it may be said that there was some continuity between his post-university and his later Roman period linked by a stay in Bologna, where Ingoli happened to be in 1604 when he probably made his observations regarding the *nova* star.[14]

---

so much of arabic that they will not be able to tell me one thing and mean another» [«so tanto di arabico che non mi potrà dar ad intendere una cosa per un'altra»]).

9 *Acta Graduum Academicorum Gymnasii Patavini ab Anno 1601 ad Annum 1605*, ed. by Francesca Zen Benetti, Padova, Antenore, 1987, pp. 38-39.

10 We find him so named on the opening page of *De Cometa 1618 Tractatus*, Rome, Bibl. Vallicelliana, R 56 (A), doc. 61, f. 776r.

11 For a reconstruction of the Padua atmosphere in respect to the observation of the *nova* and to the connection with the Copernican question, BUCCIANTINI, *Galileo e Keplero. Filosofia, cosmologia e teologia nell'Età della Controriforma*, Turin, Einaudi, 2003, pp. 23-48.

12 On Pinelli, see the very recent article by MARCO CALLEGARI, *DBI*, vol. 83 (2015).

13 MONTANARI, *Francesco Ingoli* (cit. note 8) and METZLER, *Francesco Ingoli* (cit. note 7), p. 296.

14 This important biographical record is derived from a reference to a subsequent letter

2. An Apprenticeship in Tycho's Astronomy

To understand this period in Ingoli's life, despite the lack of documented facts, it is important to sketch an outline of the scientific culture to which he belonged, and what better place to begin than with two of his short astronomical works, which were written at the time of two events that triggered an animated debate throughout the scientific community – the *nova* or new star which appeared in 1604 and the comet of 1607. These texts remained in manuscript form so it is difficult to assess their diffusion.[15] From an examination of the handwriting it seems certain that they are not in Ingoli's hand, but were copies prepared for circulation among colleagues and in this sense 'published', albeit not in printed form. In fact, they were for all intents and purposes scientific treatises – succinct but at the same time clearly structured and accompanied by figures, diagrams and tables, as well as citations of the sources used.

The *De stella Anni 1604 Tractatus*[16] is divided into two *partes*. The first part concerns «physica, historica et Astronomica istius stellae»; the second addresses astrological themes and predictions of the future associated with the new star that had appeared in the constellation of Sagittarius. The text opens with a *proemium* and closes with a *peroratio*. At the very end, the author has added two figures depicting the celestial sphere with the motion of the stars in relation to the sun and the planets, and an image of the *Figurae et themata coelestium domorum*.

In the preamble we find some significant indications. Ingoli writes that he too was caught up in the general wonderment at the appearance of this star, but not to a degree that prevented him from writing about it based on his knowledge of astronomy or searching for a scientific explanation for its origins. He states that he wrote this treatise at the pressing request of friends and acknowledges that it might appear presumptuous for someone so young as he to do so without even waiting for the star to expire, or for

---

to Caetani: «having stayed in student housing and then immediately going to off[ice]in Bologna and in Bologna entering the court of V.S. Ill. ma» («essendo allevato fuori a dozina, et in studio, et poi subito andato in off[ici]o a Bologna e da Bologna entrato in corte di V.S. Ill.ma»), AFC, FG, 140661 (September 27, 1613). This is also confirmed by the *De Cometa 1618 Tractatus*, quoted above, in which Ingoli claims to have met in Bologna Jacques Davy du Perron, the famous scholarly cardinal in 1604 who had also observed the *nova* in France. Du Perron was on his way to Rome to receive Cardinal insignia, Rome, Bibl. Vallicelliana, R 56 (A), doc. 61, f. 778v.

15 They are quoted in Bucciantini, *Contro Galileo* (cit. note 4), p. 144, but, as far as is known, they have never been analyzed in detail in any specific study.

16 BAV, Ottoboniano Latino 2484, Part I, ff. 139r-146r.

the more learned to pronounce their views on the phenomenon. He limited himself to discussing a few essential points, in conscious agreement with Aristotle that: «It is worth more to know or describe the minimum details of the higher issues rather than [knowing or describing] everything about all else».[17]

All the same, the decision to provide only a 'minimal' amount of detail would certainly have been his own deliberate choice. In his first venture into science, at the age of twenty-six and just shortly after graduating from university, Ingoli demonstrated a sense of modesty, pride and self-assurance all at the same time.

By citing the «demanding and insistent» («expostulantibus flagitantibusque») request from his friends he seems to be professing due prudence with respect to his ambition in addressing this issue, but he was probably referring to a real group of persons whom he frequented during these post-university years. Moreover, the hint of his adherence to objective reality was likely to have been a veiled criticism of those who wrote about the phenomenon in terms suggesting that it was a miraculous, perhaps even divine event.

In his text Ingoli gives precise dates for his observations of the *nova*, which he began studying on the 16th October 1604. Regarding its location in the heavens, Ingoli affirmed that the star was placed in the sky like any other star, quoting Girolamo Cardano (*De subtilitate rerum*, Nuremberg 1550 and *De rerum varietate*, Basel 1559), Tycho Brahe (*De cometa 1577*, written in 1578) and Marcello Squarcialupo (*De cometis dissertationes novae clarissimae*, Basel 1580),[18] and thus opposing Aristotle, who had sustained that such a hypothesis could not have been based on the incorruptibility of the sky and that the star therefore must be located below the sphere of the moon. On the other hand, Ingoli adhered without wavering to the hypothesis of the celestial nature of the *nova*, providing an explanation for Aristotle's 'error' in the inadequacy of the scientific instruments available at the time. Ingoli came to this conclusion based on various factors (such as the circular motion of the star, which would be impossible in the sublunar space according to Aristotle), but mainly using the parallax method for which he provided a long explanation, referring to Euclid and illustrating his argument with figures,

[17] *Ibid.*, f. 139r: «de altissimis minima praestat scire vel dicere, quam reliquorum omnia».

[18] Brahe had written *De stella nova*, published in Denmark in 1572. The quotes of these authors are *ibid.*, f. 139v.

from which it is necessary to conclude that this new star is not in the higher region of the sky, where the traces of the parallax are out of focus and larger than the moon, but lies above the moon itself.[19]

In addition, Ingoli relied on the observations made in other parts of the world by various mathematicians, such as Brahe who published his findings on the 1572 *nova* that appeared in the constellation of Cassiopea the following year. Following this, Ingoli demonstrated his full acceptance of Tycho's cosmological model, which he «liked more to follow», in contrast to that of Copernicus, «for the absurd things that he said on the triple motion of the earth», to that of Georg von Peuerbach, to that of the «excellent» Giovanni Antonio Magini; to that of the «other followers of Ptolemy, based on the fact that Ptolemy's theories were not consistent with the principles of mathematics».[20]

Instead, Tycho's theory according to Ingoli solved the great difficulties of ancient mathematics relating to the inequality and the apparent motion of the planets, which Ingoli examined in detail together with other aspects of astronomy. The astrological interpretation of events is clearly separated from the scientific discussion, being confined to the second part of the text in which Ingoli, in the light of various texts (including works by Cardano and Ptolemy's *Tetrabyblos*), lists various calamitous phenomena down through the ages that were attributed to unusual astronomical events.

De cometa anni 1607 Tractatus Francisci Ingoli Ravennatis[21] was just as carefully composed as his treatise on the 1604 *nova*, and like it is divided into short chapters. We must recall that it dates to a different period in Ingoli's life, when he was in the service of Caetani at the Apostolic legation of Romagna. In the Preamble we find a rather solemn incipit which distinguishes, in man's capacity to understand the designs of the Divine, between «The criterion of the order of all things», which is destined to remain an inscrutable secret, and the «execution of the order» which is accessible to men. He then presents a brief explanation of the structure (*methodus et divisio*) of his work, which like *De stella Anni 1604 Tractatus* was prepared at the request of his friends, who pressed him both in person and by letter, to make available his observations of the «Star born in the

19 *Ibid.*, f. 140v: «ex quo necessario concluditur astrum hoc novum non solum in suprema regione aeris non esse, ubi maior ignitarum impressionum parallaxis visit quam luna, sed supra ipsam lunam reperiri».

20 *Ibid.*, f. 143r, «magis placuit sequi»; «ob absurda quae de motu terrae triplici dixit», «reliquorumque Ptolomei asseclarum ob repugnantiam, quae habet, cum Mathematicis principijs».

21 BAV, Ottoboniano Latino 2484, Part II, ff. 176r-184v.

constellation of Sagittarius». There is no signature, but a brief closing paragraph in which the author states that he will say nothing more because the intelligence of his friends can fill the gaps in his observations. In this case as well, the text contains notes in the margin with references to other authors and to his earlier treatise; for example, he cites Tycho's method of calculating the parallax which he already explained in his 1604 treatise. Regarding the comet's location in space, Ingoli placed it beyond the moon, since it has a heavenly nature. Antonio Caetani, the brother of Cardinal Caetani and apostolic nuncio to the emperor, brought Ingoli up to date on the opinion of German mathematicians, who believed that this phenomenon was caused by a conjunction of the Sun and Mars in the constellation of Leo. Ingoli found this hypothesis to be of interest, but based on his observations of the pale color of the comet, he concluded, «So I think it is more true that comet came from Saturn».[22] As regards the material substance of the comet, it was certainly a heavenly body and not the earthly vapor Aristotle had posited, since otherwise it would not have been able to rise so high and last so long. Moreover, if it were an atmospheric phenomenon, its motion would not have had the circular trajectory that was so clearly observable. Its shape was determined by the condensation and illumination of the celestial matter. Ingoli understood quite well that the form of the tail could be attributed to the effect of the sun's rays: «It is true that the most reliable reason for the tail must be sought in the shining rays of the sun that pass through a less dense body» and not in Aristotle's hypothesis of an «outflow of matter». Ingoli cited a list of authors from Pietro Apiano (the Saxon humanist and man of science Peter Bienewitz) who had observed the 1531 comet, to Girolamo Cardano who had noted in more than sixteen comets that the tail was on the opposite side in relation to the sun.[23]

These two works plainly demonstrate Ingoli's knowledge in the areas of astronomy and astrology while highlighting his critical and analytical abilities with respect to the *auctores* which allowed him to indicate the characteristics of the *nova* and the comets by drawing on the conclusions made by Tycho Brahe between 1572 and 1578. By applying Tycho's conclusions to the two phenomena he was studying, Ingoli was able to establish that they were celestial bodies situated beyond the moon and, as evidenced by their circular motion, were formed of celestial matter into some constel-

---

22 *Ibid.*, f. 178r: «ratio ordinis rerum omnium»; «executio ordinis»; «stella in sagittario orta»; «verius itaque arbitror Saturnum fuisse cometae authorem».

23 *Ibid.*, ff. 178v-179r: «verum securiorem causam esse causae expetendam a solis splendentibus readijs per cometae corpus minus condensatum»; «defluxus materiae».

lation. More generally, we can observe that at the age of just thirty Ingoli had already developed positions that were quite independent of the Aristotelian tradition, which he would often refute on the basis of knowledge gained in recent times. In opposition to Aristotle and Ptolemy, Ingoli not only argued for the fluidity of the sky and adhered in various ways to the theories of Tycho, but also demonstrated that he was able to carry out a critical discussion of the various theories and, even while limiting himself to following in the footsteps of the Danish astronomer for his own demonstrations, was able to address scientists using their own language.

This observation of stars and comets, and his written accounts of these observations prompts us to ask in what cultural and scientific context Ingoli was moving in those post-graduate years. There are also questions about who these «friends» mentioned by him were, as it seems this was not just an attempt to respect the forms by showing due modesty. To these questions we can only offer hypotheses since too little is known about the period when he was living for extended periods in Padua, Bologna and Rome.

It seems likely that it was during this time that Ingoli met the Bolognese *Monsignor* Giovanni Battista Agucchi (1570-1632), who would remain an important figure in his life, as between 1604 and 1605 Agucchi held the position in Ravenna of secretary to Archbishop Pietro Aldobrandini, the nephew of Clement VIII. Agucchi, who came from a Bolognese family that counted various cardinals among its number (his brother Girolamo, 1555-1605; his uncle Filippo Sega, ca. 1537-1596), was a multifaceted intellectual. A genuine polymath with an interest in geography and science, he is now certainly best known as a theoretician of Classicism in the field of art and an expert in political matters who served as private secretary to Gregory XV and then as apostolic nuncio in Venice for the pope's successor Urban VIII. It can safely be assumed, based on the information currently available that it was Agucchi who introduced Ingoli to the scientific community in Bologna, particularly to Giovanni Antonio Magini, geographer, cartographer, astronomer and astrologer *sui generis* – 'mathematician' but not 'observer'. It was Magini whom Ingoli linked to Ptolemy in his 1604 report, but who in fact played a key role in the diffusion of knowledge of Tycho's theories in Italy.[24]

---

[24] On Magini see Ugo Baldini, *DBI*, vol. 67 (2006). On Agucchi Ilaria Toesca – Roberto Zapperi, *DBI*, vol. 1 (1960) and, on the side of art historian, Silvia Ginzburg, *Giovanni Battista Agucchi e la sua cerchia*, in Olivier Bonfait – Christoph L. Frommel – Michel Hochmann – Sebastian Schütze (eds.), *Poussin et Rome*, Paris, Réunion des Musées Nationaux, 1996, pp. 273-291.

The fact that the aforementioned works on the *nova* and comet were found in a collection of «Astronomical» documents that most likely belonged to Agucchi could shed light on the connection between the Bolognese intellectual and Ingoli, which was firmly based on their shared interest in astronomy and adherence to the ideas of Tycho Brahe. In this collection is a manuscript that has been attributed by Massimo Bucciantini (who published it for the first time) to Agucchi, entitled *Della penetrazione e della corruttibilità del cielo* and dated June 1611 (with a short note written between late 1612 and early 1613). The author sent this manuscript to Ingoli («Al S[igno]r Ingoli»), asking him to read and critique it. Though Agucchi viewed with great interest and esteem Galileo's work, which he had been introduced to in Rome by his friend Luca Valerio, according to Bucciantini, his reading of *Sidereus Nuncius* did not convince him to accept Copernicanism; indeed it had quite the opposite effect.[25]

Returning to Ingoli, it can be argued with some probability, although in the absence of any documentation, that the Bolognese scientific community had influenced the orientation of his astronomical theories and his general attitude towards science. While Magini was a figure of immense prestige, Agucchi – although he certainly could not be called a scientist – took a genuine interest in Ingoli, and shared his opinions on certain subjects, for example on the legacy of Aristotle.

3. With Bonifazio Caetani: Astrology and Career

Continuing with Ingoli's biography, we find him in Caetani's legation to Ravenna in 1608, after taking office on the 25th September 1606. He would remain there until 1611. Of this activity we have only general and laudatory descriptions that indicate a perfect agreement between Ingoli and Caetani. The latter was sometimes described as a sober, disciplined administrator who was surrounded by admirably blameless people.[26] Other sources hint that Caetani was a lively and controversial spirit, especially in his life at the Curia. Many examples point to Caetani's satirical nature, with the Archbishop of Ravenna Aldobrandini being a favorite

[25] Bucciantini, *Teologia e nuova filosofia. Galileo, Federico Cesi, Giovanbattista Agucchi e la discussione sulla fluidità del cielo*, in Catherine Brice – Antonella Romano (eds.), *Sciences et religions de Copernic à Galilée*, Rome, École française de Rome, 1999, pp. 411-442 (Agucchi's text is published on pp. 426-442). Agucchi was also Galileo's correspondent, albeit for a short time.

[26] Metzler, *Francesco Ingoli* (cit. note 7), p. 296.

target.[27] Certainly, Caetani had an established reputation as a man of immense culture and learning, and quite knowledgeable about the scientific arguments of his time. This transpires clearly from his correspondence with Francesco Ingoli. The twenty surviving letters that Ingoli sent to the cardinal between the 13th February 1613 and the 12th September 1614 were written during periods of travel that temporarily separated the two: Ingoli's journey to Naples and Salerno as a companion of «sig. Abbate», probably a member of the Caetani family whom he was tutoring in civil and canon law, and the cardinal's sojourn in his archbishopric seat of Taranto for the celebration of the provincial synod. It is due to this physical separation and the need to communicate by letter that we have these written testimonies about Ingoli's life and activities. They were in fact used by Massimo Bucciantini to illustrate, along with other documents, how the «cosmological debate that develops [...] after the publication of *Sidereus Nuncius* finds one of its most intense moments in the Roman milieu».[28] For our purposes, these letters contribute to our knowledge of Ingoli, of the man at the Curia, of the intrigues at the court of Rome and of the difficulties he experienced in his career, but they also provide a picture of the arguments he dealt with in association with a man of the Church – Cardinal Bonifazio Caetani – who was famed for his interest in science. From Naples, where Ingoli wrote on the 12th March 1613, he informed Caetani that he had met the eminent natural philosopher Giovanni Battista Della Porta while introducing his student to the refined intellectual circles of the city. Della Porta at this time had made a prediction connecting astrology and physiognomy, of which Ingoli seemed to be quite skeptical.[29] On March 23, however, he wrote from Salerno complaining about his health (he was 'melancholic' and a hypochondriac, and seems to have suffered from a number of discomforts and illnesses). Such personal information was very common in his correspondence and was often accompanied by astrological predictions with complex zodiacal calculations on the effect of the stars on his health.

---

27 On Caetani see GASPARE DE CARO, *DBI*, vol. 16 (1973).

28 BUCCIANTINI, *Teologia e nuova filosofia* (cit. note 26), p. 411: «il dibattito cosmologico che si sviluppa [...] all'indomani della pubblicazione del *Sidereus Nuncius* trovi proprio nell'ambiente romano uno dei suoi momenti più intensi».

29 On Della Porta's positions on astrology see ALFONSO PAOLELLA, *Il cielo stellato in Copernico e nella Coelestis physiognomica in Giovanni Battista Della Porta*, in *La diffusione del copernicanesimo in Italia, 1543-1610*, ed. by Bucciantini and Maurizio Torrini, Firenze, Olschki, 1997, pp. 189-202; see also MANUELA BRAGAGNOLO, *Fisiognomica, astrologia e medicina al tempo di Della Porta. La Fisionomia naturale di Giovanni Ingegneri (1606)*, «Bruniana & Campanelliana», 22, 1, 2016, pp. 97-104.

Certainly, as Bucciantini points out, the theme of astrology and horoscopes figured prominently in the discussions between Ingoli and Caetani; indeed, on the basis of these letters it can be said that to draw up horoscopes and make predictions was a form of daily therapy for Ingoli.[30]

We learn from one of his letters that in 1614 Ingoli penned «at the instance of many friends» (the same formula adopted for the texts on the new star of 1604 and the comet of 1607) a «writing» in defense «of Astrology which I find reasonable and well-founded». In it a summary of all the favourable and contrary opinions were passed in review, distinguishing the 'reasonable' aspects of astrology from the purely 'superstitious'. The treatise was very positively received by the Secretary of the Congregation of the Index, Monsignor Tommaso Pallavicini (who also participated in the meetings of the Holy Office) without his knowing who the author was: «He especially liked it and tried diligently to find out who the author was».[31] After having learned his name from the secretary of Metello Bichi, the cardinal of Soana, Pallavicini «wanted to invite me to this office». This phrase could be interpreted as the prelude to an offer of employment at Congregation of the Index, for which Ingoli had demonstrated that he was in possession of all the right qualities. That he was attracted to this possibility is apparent: «It seems to me that this [work] would be in just proportion to the variety of studies which I have undertaken, and that I should honor V.S. and in this myself the same».[32] Quite rightly Ingoli recognized in the multiplicity of his knowledge the ideal conditions for working on the Index, and his ability to deal with various types of format and subject matter.

But this was not to be. During these months, when Caetani was in his diocese at Taranto, Ingoli had many other duties to carry out for his patron. One of these was the translation of Ptolemy's *Tetrabiblos* (better known as *The Astrological Predictions*) with which Caetani sought to further con-

---

30 AFC, FG 140654 (July 26, 1613), 71998 (December 7, 1613), 140222 (December 20, 1613). The importance of astrology lies in the studies of various scientists of the time; for a picture on Rome, LUIGI FIORANI, *Astrologi, superstiziosi e devoti nella società romana del Seicento*, «Ricerche di Storia religiosa di Roma», 2, 1978, pp. 97-162; GERMANA ERNST, *Scienza, astrologia e politica nella Roma barocca. La biblioteca di don Orazio Morandi*, in EUGENIO CANONE (ed.), *Bibliothecae selectae. Da Cusano a Leopardi*, Florence, Olschki, 1993, pp. 217-252; PIERLUIGI PIZZAMIGLIO, *L'astrologia in Italia all'epoca di Galileo Galilei*, Milan, Vita e Pensiero, 2004, pp. 175-176.

31 AFC, FG 63591 (June 29, 1614), and 63609 (March 15, 1613): «ad istanza di molti amici»; «di quella parte d'Astrologia che mi pare ragionevole e fondata»; «soprammodo gl'è piaciuta et ha cercato con diligenza di saperne il compositore».

32 AFC, FG 63590 (June 13, 1614): «ha voluto invitarmi a quest'ufficio»; «mi pare che per la varietà de' studij da me fatti sarebbe cosa molto proporzionata, e che potrei far honor a V.S. ill.ma et a me medesimo».

solidate his knowledge of astrology, with Ingoli's crucial help in translating the Greek to Latin. Above all Ingoli was concerned with checking if other translations were available in Rome and whether there were scholars able to verify the translation. With this intention, he informed the cardinal on his progress with the project on various occasions between 1613 and 1614. In Venice the first book met with great success, so much so that two further books were requested, but Ingoli refused to release them because they lacked the permission of the Inquisitor. Meanwhile, he advised checking the translation with his friend Monsignor Agucchi «for the sake of the language». However, he reassured Caetani as to the quality of the translation, which had been appreciated by everyone who read it, including Magini. Finally, he stated that it would be a simple matter to publish them in Rome where «they easily allowed the translation of the Melantone *deleto nomine*» and where an incomplete version of the *Tetrabiblos* was already in circulation (indeed, it was known that Cardinal Bandini possessed a copy).[33] Another piece of encouraging news announced by Ingoli to Caetani on the 15th March 1614 was the supportive attitude of Bellarmin towards astrology expressed after reading a work by the Sienese doctor Giulio Mancini, who was then active in Rome.[34] Astrology remained a constant thread in their correspondence: for example, on July 26 and August 30, 1613, Ingoli interjected comments on the *idengrafia* (graphology) experiment conducted by Prospero Aldorisio, expressing his complete distrust of the use of the handwriting to make predictions.[35]

During the summer of 1613 Ingoli also frequently visited the palace of Prince Cesi to dispute «questions of mathematics, philosophy and theology» with «Peripatetics, Paracelsians and Telesians», as he wrote to Caetani on August 9 and 30.[36] The discussions were heated and he described them in

33 AFC, FG 140297 (November 15 1613): «per cagion della lingua» and 140222 (December 20 1613): «lasciano dar facilmente traduzione del Melantone deleto nomine».

34 AFC, FG 63609 (March 15, 1614). On Mancini see Silvia De Renzi – Donatella L. Sparti, *DBI*, 68 (2007).

35 AFC, FG 140654 (July 26, 1613) and 140754 (August 30, 1613). See Salvatore Ruzza, *Agli inizi della grafologia: Prospero Aldorisio e l'idengrafia*, «Scrittura», XIV, 50, 2 (April-June 1984), pp. 67-72. Confirming his closeness to the cultural environment of Bologna, Ingoli names «Baldo di Bologna», that is, Camillo Baldi, a professor of logic and philosophy at the University, who merely draws from the handwriting elements to define the costumes and nuisances of the writer; see Camillo Baldi, *Trattato Come Da Una Lettera Missiva Si Conoscano La Natura E Qualità Dello Scrittore*, Carpi, Girolamo Vaschieri, 1622. On Baldi see Mario Tronti, *DBI*, vol. 5 (1963), where Aldorisio's technique is called «gelotoscopy».

36 AFC, FG, 140739 (August 9, 1613) and 140754 (August 30, 1613): «questioni di mathematica, filosofia e teologia»; «peripatetici, paracelsisti, e telesiani»; Bucciantini, *Teologia e nuova filosofia* (cit. note 26), pp. 411-412.

detail to Caetani: «If one put forth the argument that the heavens were animated, he was attacked in a bad way». Ingoli countered by referring to the authority of the Book of Genesis, as well as the opinion of the philosophers of the Sorbonne. On another occasion Ingoli opposed the Jesuit Francesco Diotallevi in a theological discussion concerning «dato vacuo motus esset instantaneus». Here too his refutation was couched in theological terms, while in the same discussion Luca Valerio expressed his criticism in mathematical terms. At the beginning of September, Ingoli confided in Caetani his appreciation for the position taken by Diotallevi in his criticism of Giulio Cesare Lagalla's argument concerning the animated sky, which was in agreement with the Aristotelian Cesare Cremonini, an argument that the cardinal took a great interest in despite his complete opposition to the hypothesis of planetary souls. It may be noted Ingoli always expressed his concern with the orthodoxy of positions which «if not heretical were at least wrong».[37]

Along with this information on Ingoli's scientific-philosophical interests, which shed light on his relations in the Roman and Italian cultural community, his letters also allow us to retrace his career within the Curia. These biographical details run parallel to his 'scientific' life, but to draw a complete portrait of Ingoli they must be taken into account since the activities and interests of Caetani, Agucchi and others created a bridge between science and the Curia, a bridge that Ingoli was constantly traversing in both directions as he pursued a career in Rome.

Unfortunately, the details that emerge from the letters are often cryptic, referring to people in an oblique manner that does not allow us to identify them. It is certain that there was the prospect of Ingoli being transferred, on Caetani's recommendation, to the service of Cardinal Borghese, a member of the pope's family. However, this arrangement fell through because of the ineffectual mediation of a certain Friar Giuseppe, who at a certain point could no longer intervene because of disagreements with the superiors of his order. This theme recurs in their correspondence, which demonstrates how Ingoli, while wishing to show his loyalty to Caetani, at the same time was seeking a more and stable position at the Roman Court.[38]

---

[37] AFC, FG, 140664 (September 13, 1613): «uno che sosteneva li cieli essere animati, se li fu addosso di mala maniera»; «se non sono heretiche almeno sono erronee». On all this more thoroughly, Bucciantini, *Contro Galileo* (cit. note 4), p. 86. On Diotallevi see Matteo Sanfilippo, *DBI*, vol. 40 (1991). On Lagalla see Cesare Preti, *DBI*, vol. 63 (2004). On Valerio, see Baldini – Pier Daniele Napolitani, *Per una biografia di Luca Valerio. Fonti edite e inedite per una ricostruzione della sua carriera scientifica*, «Bollettino di storia delle scienze matematiche», 11, 1991, 1, pp. 3-157.

[38] Requesting his patron to contact Felice Centini, Cardinal of Ascoli, AFC, FG 140661 (September 27, 1613).

In this period Ingoli took holy orders, with demissorial letters issued by archbishop Pietro Aldobrandini, as Ingoli wrote on the 15th March 1614, hoping that with the support of Caetani he could take over the role of presbyterate on the day of Pentecost from Mons. Vulpio (Ulpiano Volpi, archbishop of Chieti, but in the service of the Curia).[39] However, from the few available letters that refer to this expectation, he appears to have run into some difficulties in 1614. In a letter dated 13 June (to which he attached his calculation of the *Longitudo tunc calculo tychonico computata ad annum et horam natalitiam*, citing the mathematician Giovanni di Strassoldo, a document that has since been lost), he showed his concern about some comments Caetani had made about him. In fact, Ingoli seemed to fear Caetani's reaction to some moves which he had made within the Curia. On the advice of Ludovico Ridolfi, whose home he frequented,[40] Ingoli had gone to visit the Archbishop of Ravenna, Pietro Aldobrandini, with whom Caetani was not in good standing. He even spoke to the cardinal about his concerns, assuring him that he had always sought to behave correctly: «[You can] be sure that when frequenting cardinals, prelates and lords I always proceed sincerely». Nevertheless, he incurred a suspension from Caetani's service for reasons that were never specified. Ingoli stated that he had not said anything to this patron which would have been contrary to his beliefs. Consequently, it appears that Ingoli found himself in the embarrassing position of having to act either against his conscience (which would «always disturb all that pertains to God's worship») or against the cardinal's opinion. Deciding at the outset to avoid an open confrontation with Caetani on the matter, Ingoli shielded himself behind an 'honest dissimulation', trusting in his reputation with the cardinals and especially with his patron. However, having no more to go on than a single, more or less cryptic, document we cannot provide a reliable interpretation of this delicate affair, in which Ingoli, despite everything, expressed considerable assurance that he had acted justly and did not fear to lose the confidence of his patron, to whom he «must be infinitely obliged for the help and authorization he offers my person in the honorable city of Rome». We can emphasize that even in this case astrology, a constant theme running through the communications

---

39 In July 1616 he was appointed secretary of the Congregation of Bishops and Regulars and then in 1621 Datary, BAV, Urb Lat. 1090, 76v.

40 In July 1613, returning from Salerno suffering from one of his many disturbances, Ingoli underwent a beneficial «canonical purge», remaining there for a few days, AFC, FG, 136493 (July 5, 1613). Member of the *familia* of the pope Paul V as *cameriere*, in 1616 Ridolfi was commissioned to carry the red hat to Cardinal Melchior Clesel and was advisor to the Emperor in Prague.

between the two, was used as an element to monitor the state of their ongoing personal relationship. Ingoli was sure, based on «scientia» – that is, an examination of the astral conjunctions, as indicated in his letter – that the cardinal could be persuaded of Ingoli's good faith and the correctness of the actions that he had questioned. Besides, «he saw much of my life in my homeland; he could hear many times from my fellow compatriots».[41] In the second half of 1614, the correspondence between the two dwindled to a trickle, and we have only two letters from Ingoli to Caetani, one dated 5 July and the other 12 November. In them he limits himself to recounting his activities in relation to the correction and approval of the synodal acts which Caetani had sent from Taranto. Here Ingoli's ever closer ties with Cardinal Lancellotti are evident, for he collaborated in the preparation of this document just as he had provided his expertise in checking the translation of Ptolemy for Caetani.[42]

4. Consultant to the Congregation of the Index

Turning away from this window onto the 'private life' of Francesco Ingoli, which was opened by his letters to Caetani, we can proceed to the reconstruction of his intellectual biography using information relating to his admission to the Congregation of the Index and to the condemnation and correction of Copernicus's text, which also involved Galileo. Light has been shed on these very significant events by Massimo Bucciantini, whose studies have clearly framed the role of Ingoli in the institutional context of the time.[43] If we briefly reconstruct what is known of the events, we first of all find that he has been appointed a Consultant to the Congregation of the Index, which we can consider to have been *in pectore* after Pallavicini's abovementioned words of praise, but this would not take place until Caetani himself became a member of the Congregation. The first session in which the cardinal participated, as has been stated, was that of 1 March 1616.[44] At that

---

41 AFC, FG 63609 (March 15, 1614): «si assicuri che pratticando con cardinali, prelati e signori sinceramente sempre procedo»; «havrebbe sempre disturbate tutte le operazioni pertinenti al culto di Dio»; «se li deve obbligo infinito perché aiuta e autorizza la persona mia in questa honorata piazza della città di Roma»; «perché ha veduta molta parte della mia vita e nella mia patria molte volte gl'haverà potuto sentire dai miei concittadini».

42 AFC, FG 63592 (July 5, 1614) e 130546 (September 12, 1614).

43 Bucciantini, *Contro Galileo* (cit. note 4), p. 25 and pp. 84-85 where he points out the mistakes that various authors have made concerning Ingoli's curial functions.

44 Mayaud, *La condamnation* (cit. note 2), p. 29.

time Ingoli was an ecclesiastic and had probably been admitted to the major orders, although we have no direct evidence of this. He had for some time been accepted into the Roman cultural milieu, and was seeking a position in the Curia. His credentials included a broad cultural background ranging from jurisprudence (in which he had a degree) to his scientific studies (testified to by his treatises on astronomy, which circulated in manuscript form), and his membership in important academies, as well as his constant study of astrology. He had also gained the friendship and esteem of various cardinals by virtue of his work in the Curia. Ingoli seemed to have the qualities necessary to act in the interests of the political-cultural agenda of the papacy during this difficult period, which saw the first titanic clash between religion and modern science. He did not embody, and perhaps was not even expected to represent – for the sake of appearances as it were – the figure of the scientist or philosopher, but he would have been an effective facilitator of the Church's governance, as will be shown and as was acknowledged by his peers during his twenty-seven years of tireless activity as the guiding force in the secretary's office of the Congregation *de Propaganda Fide*.

We would like to know more about the actual recruitment of Ingoli into the ranks of the Index, in particular on the role played by Caetani, who certainly constituted a link between the scientific culture of Rome and the Curia's patronage. Surely his competence as an astronomer was significant, as he demonstrated in the academy meeting that had been held a few months prior with Cardinal Magalotti, as well as in the dispute with Galileo mentioned at the beginning of this paper, which led to his writing the *Disputatio*. It seems plausible to consider this dispute as a sort of qualification test for the Index and, therefore, to entertain the possibility that Ingoli was recruited for his expertise on the matter.[45] We should here broaden our gaze to what has been recognized as an important stage in the Galilean affair, which just preceded the entrance of Caetani and Ingoli onto the scene: Cardinal Bellarmino's declaration on the 25th February 1616 that «the opinion of Galileo» was heretical.[46]

Given the close sequence of events, with Caetani joining the Congregation of the Index just a few days after Bellarmino (who was the pope's

---

45 The dispute was also an opportunity for a direct meeting between Ingoli and Galileo. Ingoli expressed himself in very respectful terms: «I've always been very happy to have had dealings with men like you are, cultured and with a restrained elegance in disputes» («cum viris doctissimis et in disputationibus modestissimis, qualis tu es, agere gratissimum semper mihi fuit», OG, V, p. 403). This respect was, in fact, lacking on the part of Galileo's response in the 1624 letter.

46 On this see Bucciantini, *Contro Galileo* (cit. note 4), *passim*.

intermediary in the Galileo affair) made his pronouncement and Ingoli joining very shortly afterwards, it may be hypothesized that Bellarmino was instrumental in the maneuver which entrusted Ingoli with the task of opposing Galileo through the refutation of «the opinion of Copernicus» and his followers. Still fairly young, and a known expert on astronomy who adhered to Magini's school and the Tychonic system, and moreover who was under the tutelage of a prelate such as Cardinal Caetani, who was well versed in these sciences, Ingoli must have appeared to be a trustworthy candidate.

As has often been observed, it was precisely in this period, around the year 1616, that the interest in the theories of Copernicus[47] passed from curiosity to the stage where the weight of the choice in such matters began to make itself felt, although there were still those who could put off taking a stand or remain open to curiosity, like Caetani himself, who did not hesitate to keep in touch with figures such as Tommaso Campanella, whom he visited in prison in Naples between 1614 and 1615. However, Caetani was of course a cardinal and in 1616 he passed Ingoli's anti-Copernican writings to Campanella, who informed Galileo that «a man from Ravenna [Ingoli] was contrary to his and Copernicus's» ideas and then wrote his *Apologia pro Galileo* in the same year.[48]

Instead, the versatile Agucchi was not favorable to the Copernican system, as he wrote to Galileo in 1613, when this «absurd» astronomical conception had become a daunting progressive philosophy that threatened to destroy «all philosophy, and much theology» while eliminating «astrology and its influences», as Girolamo da Sommaia, the *provveditore* (superintendent) of the Pisan Studium, said in 1616.[49]

Documents from this period suggest that at the theoretical level there was no clear, fixed division between astronomy and astrology. There seemed to be no categorical adherence to or rejection of either on the part of the individuals involved, but instead an oscillation between different positions. So, the debate took the form of opposing positions taken up and defended on the basis of scientific arguments rather than confrontations between individual scholars, at least until the choice became a question of orthodoxy imposed by the Church. Certainly, however, loyalty in personal

---

47 See the essays in *La diffusione del copernicanesimo in Italia* (cit. note 30).

48 Tommaso Campanella, *Lettere*, ed. by Vincenzo Spampanato, Bari, Laterza, 1927, p. 180 and *Apologia pro Galileo*, ed by Lerner, It. trans. by G. Ernst, Pisa, Edizioni Scuola Normale Superiore, 2006. Campanella's text was published in Frankfurt in 1622.

49 Bucciantini, *Teologia e nuova filosofia* (cit. note 29), pp. 423-25 with quotation at p. 424.

relationships persisted. In 1617, for example, Ingoli was commissioned by the Master of the Sacred Palace, Giacinto Petroni, to read Magini's text, *Confutatio diatribae Ios. Scaligeri de aequinotiarum praecessione* (Romae, apud Gulielmum Facciottum, 1617). Ingoli firmly approved of the text by Magini, which defended the traditional views on the subject.[50]

This all might point to Ingoli being the Congregation's spokesman on astronomical issues, but documents dating to the period when he was acting as consultant to the Index do not provide any support for this hypothesis. The picture that emerges is of his intense engagement in parallel activities, some of them relating to Copernicus and Kepler and the remainder to other responsibilities connected with the censorship.

Concerning the first point, a brief summary can be sketched based on the already mentioned studies by Mayaud, Lerner and Bucciantini. First of all, there was Ingoli's *Disputatio de situ et quiete terrae* written in 1616,[51] which was followed by the decrees issued by the Congregation of the Index on March 5 of that year and, a few days later, by Ingoli's appointment as a consultant to the Index with the task of amending Copernicus's text. Contemporaneously, news of the sentence together with the letter from Ingoli (June 1617), and later the condemnation decree in September 1617, reached Kepler, who was then in the process of publishing *Epitome astronomiae copernicanae*.[52] Kepler replied with his *Responsio ad Ingoli Disputationem* in May 1618.[53] Ingoli

---

50 Pizzamiglio, *L'astrologia in Italia* (cit. note 31), pp. 175-176. It must be taken into account that the approval reserved to Ingoli, who was a member of the Index, was in light of the decision taken simultaneously by the Index in the decree condemning Copernicus on 5th March 1616. On the basis of this decree the Congregation of the Index, and not the Master of the Sacred Palace, was entitled to express judgment on works related to Copernicanism. The Master of the Sacred Palace had drafted a decree on the condemnation of Copernicus, but which had been withdrawn; See Francesco Beretta – Lerner, *Un édit inédit. Autour du placard de mise à l'Index de Copernic par le maître du Sacré Palais Giacinto Petroni*, «Galilæana. Journal of Galilean Studies», 3, 2006, pp. 199-216.

51 BAV, Ottob. lat. 2700, ff. 55r-58r. Already published by Antonio Favaro, *La inedita «Disputatio de situ et quiete Terrae contra Copernici systema» indirizzata da Francesco Ingoli a Galileo Galilei con la inedita risposta di Giovanni Keplero*, in *Nuovi studi galileiani*, «Memorie del R. Istituto Veneto di scienze, lettere ed arti», 24, 1891, pp. 149-184, then in *Le opere di Galileo Galilei*, Firenze, G. Barbera, 1932, pp. 403-412 and translated now also in English with Latin text by Graney, *Setting aside all Authority*, Appendix A (cit. note 6).

52 Johannes Kepler, *Epitome astronomiae Copernicanae: Usitata forma quaestionum & responsionum conscripta inq. VII libros digesta, quorum tres priores sunt de doctrina sphaerica*, Lentijs ad Danubium [Linz], Johannes Plancus, 1618. About the dissemination of the news of Copernicus's condemnation see ACDF, Index, *Diarii*, I, 2, f. 98v (October 3, 1616) in which the secretary of the Index Maddaleni sums up letters from local inquisitors and Apostolic nuncio which were in response «to the reception of Copernicus's edict» («*circa receptum edictum Copernici*»).

53 Bucciantini, *Contro Galileo* (cit. note 4), pp. 103-105.

worked on Copernicus's text for the Index, suggesting corrections on the 2nd April 1618.[54] On July 3 the Index recorded the favorable response of two Jesuits of the Collegio Romano, Orazio Grassi and Christoph Grienberger, to Ingoli's proposed revisions and the Index seconded their judgment on the 1st July 1618. The amendments to Copernicus's work were therefore incorporated, being mentioned in the Index's *Diarii* on July 30, October 9, and May 10, 1619.[55] The secretary of the Index proposed that the revised version be published and this recommendation was approved on the 15th May 1620.[56]

Up until this point the interest of censors had centered on Copernicus, but with the publication of *Responsio* Kepler entered the scene. This work met with an immediate response from Ingoli in the form of a treatise dated 13th October 1618: the *Replicationes de situ et motu Terrae contra Copernicum ad Ioannis Kepleri impugnationes*,[57] which was sent to Ludovico Ridolfi in Prague. The Index also embarked on an examination of Kepler's *Epitome* (which provided the basis for his *Responsio*), and this time it was not a matter of corrections, but of banning the work by the imperial mathematician altogether. This decision was taken on February 28, 1619 on the recommendation of Ingoli,[58] who in a report summarized the points raised in his *Replicationes*. Moreover, in his proem Ingoli clearly explained to Ridolfi that *Replicationes* was written to provide the Index with an evaluation of the confessional matters raised by Kepler's work.[59] The Congregation emitted its definitive condemnation on May 10, 1619.[60]

Alongside this well-known chronology of the censorship of Copernicanism, one must keep in mind Ingoli's own astronomical activities. He conducted observations during the passage of the three comets of 1618, as

---

54 *De emendatione sex librorum Nicolai Copernici De Revolutionibus*, BAV, Barb. lat. 3151, ff. 58r-61r, published in Bucciantini, *Contro Galileo* (cit. note 4), pp. 207-209.

55 For this reference, see ACDF, Index, *Diarii*, I, 2, ff. 128r (April 2, 1618), 139r (July 3, 1618), 140v (July 30, 1618), 147r (October 9, 1618), 158r (May 10, 1619).

56 Beretta, *Le siège apostolique et l'affaire Galilée: relectures romaines d'une condamnation célèbre*, «Roma moderna e contemporanea», 7, 1999, 3, pp. 426 e 454, which reports Ingoli's original text conserved in the index archive. Bucciantini had published the *Monitum ad Nicolai Copernici lectorem, eiusque emendatio* issued by the Secretary of the Index, Francesco Maddaleni Capiferro in *Contro Galileo* (cit. note 4), pp. 211-212. On this document see also the textual analysis by Mayaud, *La condemnation* (cit. note 2), pp. 69-72 et 77-79.

57 BAV, Barb. Lat. 324, ff. 89r-113v, not found by Favaro and published by Bucciantini, *Contro Galileo* (cit. note 4), pp. 177-205.

58 ACDF, Index, *Diarii*, I, 2, f. 154r (February 28, 1619), *Protocolli* C 2, ff. 438r-v (new paging ex 436r-v).

59 For the text see in Bucciantini, *Contro Galileo* (cit. note 4), pp. 177-178.

60 *Ibid.*, pp. 138-139.

did many of his contemporaries, notably Agucchi and Campanella.[61] The meeting place for the discussion of their findings, which then shifted to the Copernican question in general, was the palace of Scipione Cobelluzzi, Cardinal of Santa Susanna in Viterbo, a humanist of vast learning.[62]

On these astronomical events Ingoli wrote a text, *De Cometa anni 1618 Tractatus*, which has come down to us in manuscript form with the author's corrections and the date February 21, 1619.[63] This work is divided into three broad sections.

The first provides a general introduction, looking into the possible causes of comets according to «the most famous opinions of the ancients and the moderns». Of the four types described – efficient, material, formal, final – he lays particular emphasis on the efficient cause as postulated by thinkers from Aristotle to the «perspectivists», from Ptolemy to Filippo Apiano (Philipp Bienewitz) and finally Tycho Brahe. All of these theories, in Ingoli's opinion, have weaknesses, including that of Tycho, to whom he nevertheless pays tribute with an elegy, for he «renders the cause of the comets metaphysically and supernaturally efficient, [even if] I could not see the reasons for his definition, since the books were never published which he says he had brought to light».[64] According to Ingoli, who also quotes Kepler's *Epitome*, the efficient cause is the sun and he refutes objections relative to the position and periodicity of the comets. On the material causes he refers to the theories of Aristotle, the «perspectivists» and to those of «almost all mathematicians of our time and especially Tycho».[65] Less extended is the discussion of the formal and final causes of comets. The first section continues with other considerations regarding comets: their crests; their tails or «seu barbae» (beards), regarding which Ingoli incorporates his own and Tycho's observations of the 1604 *nova*; the round shape of their bodies, the aspect and curvature of their tails, and the colors. Most importantly, he reviews the observations made by various astronomers on the motion of comets and the possible causes of such motion.

---

61 Ernst – Laura Salvetti Firpo, *Tommaso Campanella e la cometa del 1618. Due lettere e un opuscolo epistolare inedito*, «Bruniana e Campanelliana», II, 1-2, 1996, pp. 57-88.

62 Bucciantini, *Contro Galileo* (cit. note 4), pp. 152-153.

63 Rome, Biblioteca Vallicelliana, R 56 (A), doc. 61, ff. 776r-811r; quoted in Bucciantini, *Contro Galileo* (cit. note 4), p. 144. Also this text has never been specifically analyzed.

64 «Celebriores priscorum et recentiorum opiniones»; «cometarum causam efficientem metaphysicam et supranaturalem facit, eius sententiae rationes videre non potui, quia nondum libri, quos se compilasse dicit in lucem prodiere»; *ibid.*, ff. 778v-779r.

65 «mathematicorum fere omnium nostri temporis, et praesertim Tychonis»; *ibid.*, f. 782r.

This extensive introduction, which takes up almost half of the text, is followed by observations relative to the comet of 1618. Ingoli states that in this section he is primarily concerned with describing this particular comet and refers the reader to the previous section for more general explanations. He provides a warning «on the ambiguity and difficulty of the matter», adding that even in this part the things he states are «either sometimes true, or rather often probable».[66] Ingoli then goes on to describe various aspects of the comet as observed by himself and others around Europe,[67] noting the times of its appearance, its magnitude (calculated by observing its decline until 10 January 1619), shape and color, motion, and position with respect to the earth. On the latter point, rejecting the hypotheses of the Aristotelians, Ingoli is strongly aligned with the «mathematicians» and above all, relying on direct observation, he affirms that the comet is positioned in the heavens rather than in the atmosphere, specifically in the sphere of Venus; although the tail tended towards a lower sphere, he states that it was nevertheless «beyond the eighth sphere»: *infra 8.a sphera*.

A central point in the treatise is the tenth paragraph of the second section, which follows the author's considerations on its motion.[68] In this paragraph, which is entitled *Quod ex motu huius Cometae Copernicanus telluris motu redarguatur et quies probetur*, Ingoli tries to answer the question raised by Cardinal Cobelluzzi, who wanted to know from the consortium of scholars who met at his palace if it would be possible to derive from these phenomena evidence that could be used to disprove the Copernican theory of earthly motion («Copernici opinionem de annuo motu telluris») and support the condemnation handed down by the Congregation of the Index. Ingoli argued that the comet offers at least two proofs against the heliocentric hypothesis. Firstly, since the Copernican system postulates variations in the daily motion of all planets below the eighth sphere due to the variation in the parallaxes caused by the Earth's motion, this should hold all the more true for a comet, which is even closer to the Earth. The fact that this change in the comet's motion was not observed represented

---

66 «De materiae ambiguitate ac difficultate»; «ut vera aliquando, ut autem probabilia multoties». *Ibid.*, f. 793r.

67 Ingoli knows Agucchi's observations made in Rome and others from other places in Europe: from Antwerp, via Cardinal Ubaldini, from Innsbruck through Cardinal Cobelluzzi. He goes on to cite observations made in Venice and in Perugia as well as those made in Germany, which were actually in Prague, made by Kepler, referred to by Ludovico Ridolfi, the Emperor's counselor, who was well-known to Ingoli as we have already mentioned. The observations are shown in detail along with a drawing of the comet, *ibid.*, ff. 798r-802v.

68 *Ibid.*, ff. 803v-804r.

indirect proof of the fallacy of the Copernican assertion. Secondly, the same can be deduced from the impossibility of observing another specific variation in its daily motion. If one allows Copernicus's hypothesis of the movement of the earth, a latitudinal motion of the comet should also be observed, which would make sense of the data given by the displacement of the comet towards the boreal pole and of the eye of the observer placed on the ground, moving in the opposite direction. Since no such variation was recorded, one could conclude that Copernicus's hypothesis was not true.

The third and concluding section concerns astrology and the meaning and consequences of comets, with references to Ptolemy, Della Porta, Cardano and others. On this point as well Ingoli makes explicit comparisons with the *nova* of 1604 and with his astrological predictions based on the event, which he conveyed in his previous works.

Like Ingoli's other astronomical texts, this treatise on the 1618 comet remained in manuscript form, though apparently it had reached the stage of final correction. We do not know at present whether and in what manner it may have circulated, nor how much of a role the cultural environment in Rome played. In fact, in this same period the question of comets was explored by Orazio Grassi, a Jesuit priest and professor at the Collegio Romano, who published *De tribus cometis anni 1618 disputatio astronomica publice habita in Collegio Romano Societatis Iesu*, the work that triggered a debate with Galileo and the publication of the *Saggiatore*. It should not be forgotten that throughout this period Ingoli held an official position within the Congregation of the Index and it was precisely in the early months of 1619 that he was engaged in drafting the condemnation of Kepler and the relative decree. Based on the coincidence in timing with the publication of Grassi's treatise it would seem plausible that Ingoli ceded precedence to the Jesuits, and in particular Grassi, who with Christoph Grienberger had approved the corrections of Copernicus's text proposed by Ingoli in 1616. Now the Roman Collegio was to form the seat of the disputes, with the Jesuits promoting forth the Tychonian system in opposition to Galileo's heliocentrism, while the cultured but 'amateur' Ingoli had to step into the background.[69]

As we have already mentioned, parallel to his astronomical activities Ingoli was also involved with the Index as a consultant in the examina-

69 See LERNER, *L'entrée de Tycho Brahe chez les jésuites*, in LUCE GIARD (ed.), *Les jésuites à la Renaissance. Système éducatif et production du savoir*, Paris, PUF, 1995, pp. 145-185, especially p. 178 and FEDERICA FAVINO, *Contro Tycho: per una lettura contestuale del Discorso delle comete*, «Galilæana. Journal of Galilean Studies», 1, 2004, pp. 233-252.

tion of other works, which we have reconstructed from the *Diarii* of the Congregation. In October 1616 Ingoli requested the prohibition of a legal text, *Consiliorum seu Responsorum iuris* (Frankfurt, 1612) by Hieronimus Schurpf,[70] and in May 1617 he examined the *Practica criminalis* and the *Decisiones aureae*, both by Luis de Peguera (Barcelona, 1603 and 1605-11).[71] In July 1617 he reviewed the *Speculum aureum opinionum comunium contra comunes* by Jerónimo de Cevallos (Oberursel, Frankfurt, 1602; Venice, 1604; Strasbourg 1614) and in October of that year *La Vita e I Miracoli di Suor Juana de La Cruz* by Antonio Daça (Lerida, 1613; Florence [in Italian], 1616), both of which he concluded had to be corrected.[72]

In 1618 Ingoli scrutinized the work of political theory *Epimetron [...] aphorismorum politicorum* (Frankfurt, 1615) by Paul Buijs and found that it had to be purged of some errors.[73] The following year, 1619, he supported the prohibition of *Pallas Rhaetica, armata et togata* (Basel 1617) by the Protestant jurist and historian, Sprecher de Berneck (or Bernegg), administrator of Valtellina for the Grisons (the Swiss canton that ruled this valley in northern Italy),[74] and a volume about controversies *Acies disputationum politicarum* (Jena, 1619) by the Hessian Reinhard Koenig.[75]

In 1620 Ingoli proposed that *Collegium politico-juridicum* by the Protestant author Johann Rensler (Rostock, 1614) be banned, criticizing its many errors.[76] In particular, Ingoli objected to Rensler's claim that violence was necessary in the formation and governing of kingdoms, a theory which Ingoli stated was contrary to the doctrine of the church. In the same year he recommended the prohibition of *Templum iustitiae* (Tubingen, 1616) by the Protestant jurist Christoph Besold, a friend of Kepler and a defender of his ideas. During the same period Ingoli undertook to correct the *Tractatus de potestate episcoporum* by Etienne d'Alvin (Paris, 1607 and 1614) for subsequent review by Cardinal Bellarmino. Also in 1620, the Congregation

---

70 ACDF, Index, Diarii, I, 2, f. 96r (October 3, 1616).

71 ACDF, Index, *Diarii*, I, 2, ff. 110r-v (May 30, 1617), Ingoli's report is in *Protocolli* A 2, ff. 641r-v - 648r-v (new paging ex 643r-v - 650r-v) and 645r-648v (new paging ex 643r-646v).

72 ACDF, Index, *Diarii*, I, 2, ff. 111v-112r (July 10, 1617) and f. 118v (October 9, 1617) e 119v (November 28, 1617).

73 ACDF, Index, *Diarii*, I, 2, f. 126r (February 10, 1618) and also f. 134r (May 9, 1618). The text of the Ingoli report is in *Protocolli* A 2, ff. 663r-665v.

74 ACDF, Index, *Diarii*, I, 2, f. 170r (October 18, 1619). The text of the Ingoli report is in *Protocolli* A 2, ff. 685r-v.

75 ACDF, Index, *Diarii*, I, 2, f. 177r (November 18, 1619). The text of Ingoli's report is in *Protocolli* B 2, ff. 363r-v, 374r-v.

76 ACDF, Index, *Diarii*, I, 2, f. 185r (January 31, 1620). The text of Ingoli's report is in *Protocolli* C 2m f. 449r (new paging, ex 432r-v, 447r-v).

of the Index asked Ingoli to revised *Ius pontificium nouum* (Cologne, 1620) by the canon of Antwerp, Frans van de Zype, which had been submitted to them for review.[77]

From May to October 1621, under the new pope Gregory XV (born Alessandro Ludovisi), the Index entrusted to Ingoli the examination of more texts. Firstly, the volume *In theoriam, et praxim, beneficiorum ecclesiasticorum* (Tournon, 1616) by Antoine Faure des Bleins. Next an *Appellatio* submitted by the Nuncio in Flanders. Its authors were the British Benedictines Thomas Preston and Thomas Greene. Preston in particular was viewed in a poor light because he was one of the chief defenders of the oath of loyalty imposed on English Catholics by King James I and hence found himself engaged in a direct dispute with the Jesuits and particularly with Bellarmino.[78] Finally, there was the legal treatise by Sixtinus Regner, *Tractatus de regalibus* (Mulhouse, 1602).[79]

Ingoli managed to carry out this immense amount of work censoring texts on juridical, theological and political subjects for the Congregation of the Index in tandem with his scientific interest, which focused almost entirely on the Copernican question. That he was entrusted with the vetting of texts in such a wide range of fields clearly demonstrates that Ingoli possessed expertise which allowed him as a qualifier of the Index to judge works of jurisprudence and politics as well as astronomy.

5. A new life: secretary of the Congregation "de Propaganda Fide"

The year 1621 brought about great changes for all of those associated with the Bolognese family of the Ludovisi. In fact, it was this year that Ingoli's patron Cardinal Alexander came to the papal throne under the name

---

77 ACDF, Index, *Diarii*, I, 2, f. 199r (July 3, 1620); f. 202v (September 25, 1620); f. 204v (November 23, 1620).

78 ACDF, Index, *Diarii*, I, 3, f. 54v (May 22, 1621) and 59r (August 28, 1621). On the *Appellatio* see f. 54v (May 22, 1621) and *Protocolli* B 2, ff. 624-625 (new paging, ex 633r-v). On August 14, 1621 Ludovisi replied to Bentivoglio that the book would be examined and sent to the nuncio in France, *Correspondance du nonce Giovanni-Francesco Guidi di Bagno (1621-1627)*, ed. by Bernard de Meester, Brussels-Rome, Institut historique belge, 1938, p. 32; On the controversy, in which Preston wrote under the pseudonym of Roger Widdrington, see Sylvio Hermann De Franceschi, *Jésuites et bénédectins anglais: une conflictualité larvée au temps de l'absolutisme jacobéen. Concurrence missionnaire et affrontement doctrinal autour de l'affaire Thomas Preston (1611-1616)*, in *Religione, conflittualità e cultura. il clero regolare nell'Europa d'antico regime*, a monograph number of the journal «Cheiron. Materiali e strumenti di aggiornamento storiografico», ed. by Massimo C. Giannini, XII, 43-44, 2005, pp. 107-127.

79 ACDF, Index, *Diarii*, I, 3, f. 54v (May 22, 1621).

of Gregory XV, bringing Agucchi with him as his personal secretary. As for Ingoli, his commitments changed decisively. From the end of the year he no longer worked for the Index, as evidenced in the *Diarii* of the Congregation, where no further reports by him on books are to be found. We find that in October he sent a memorandum to the Congregation in support of the appointment of Paganino Gaudenzi as consultant. Gaudenzi was a native of Grisons, a theologian, and formerly a law professor at the Protestant university in Tübingen, who then converted to Catholicism and in these years was seeking employment in Rome.[80]

Was this a proposal for a substitute? In effect, Francesco Ingoli began a new life, in which he demonstrated considerable competence in carrying out his new duties. For a period of some months in 1621 he served as a trusted assistant to the Pope's nephew, the young Ludovico Ludovisi, when the latter took up his duties as cardinal of Bologna in a continuation of the family tradition of keeping control of the archdiocese within the family. During that period Ingoli wrote a treatise on canon law in relation to parishes.[81] In addition, he returned to Rome and was asked by the Pope to draft a reform of the conclave's election method, Gregory XV having appointed him Papal chamberlain.[82] To this end, a special congregation was created and Ingoli was nominated secretary.

He also participated in the Accademia dei Virtuosi at Monte Cavallo founded by the Ludovisi family with the probable collaboration of Agucchi. We know that Ingoli attended the second session, held on the 7th July 1621, and played the role of 'advocate' before twelve cardinals, defending the right of a prince to seek revenge, beginning with the example of King David.[83]

---

80 ACDF, Index, Diarii, I, 3, f. 61v (October 2, 1621), «Francesco Ingoli, by way of the secretary, proposed a certain Paganino Gaudenzi from the Canton of Grisons (Switzerland) to be admitted among the consultants» («[For] secretarium ex p[art]e D[omi] ni Fra[ncis]ci Ingoli ut adnumeraret inter consultores quidam Paganinus Gaudentius Grisonus»). The decision was postponed to a plenary congregation but it never came to fruition. On Gaudenzi, see Giampiero Brunelli, *DBI*, 52 (1999). Ingoli remained in contact with him for a long time, both when he was sent back controversially as a missionary to his home town (Poschiavo), and when, having unsuccessfully attempted a career in Rome, entered as a lecturer of humanities at the University of Pisa, where he unsuccessfully sought to return to the Pope's service until his death, which occurred in 1649 a few months before that of Ingoli.

81 Francesco Ingoli, *De Parochis, & eorum officio*, Bononiae, Typis Victorij Benatij, 1622.

82 Id., *Compendio delle cose più principali contenute nel Cerimoniale di Gregorio XV de electione Romani Pontificis*, Rome, Reverenda Camera Apostolica, 1623; Klaus Jaitner, *Die Hauptinstruktionen Gregors XV. Für die Nuntien und Legaten an den Europäischen Fürstenhöfen, 1621-1623*, Tübingen, Max Niemeyer, 1997, p. 88. On the reform of the conclave Ingoli worked with Prospero Fagnani, secretary of the Congregation of the Council. On the Conclave, see Maria Antonietta Visceglia, *Morte e elezione del Papa. Norme, riti, conflitti. L'età moderna*, Rome, Viella, 2013.

83 Ingoli carried out his task with the Dominican Girolamo Pollini (or maybe Alessandro

On January 6, 1622 the Congregation de Propaganda Fide was created and Ingoli served as secretary from the very first meeting. His appointment was ratified by a papal brief on the 1st September of the same year. From that moment onwards, the demands of his new position absorbed all of his time and apparently even Galileo was unable to draw Ingoli back into the maelstrom of astronomical issues. A letter that the Galileo wrote to him in 1624 in response to the *Disputatio* was never delivered, though Ingoli heard of its existence through Cardinal Cesi, who was trying to nudge Galileo into adopting a more prudent position.[84]

The extra-scientific dimension of the astronomical question in Italy did not escape either Galileo or Ingoli, and can perhaps be seen most clearly in its contrast with the scientific culture of Protestant Europe, at the time of the early stages of the Thirty Years War. The newly-born Congregation de Propaganda Fide, run by Ingoli as its secretary, was involved in this counter-reformist upsurge.[85] Consequently, Ingoli had to act in strict accordance with the pontiff's wishes and political agenda, and adapt his broad skills and experience to his new responsibilities in what would turn out to be the most important activity in his career. Of course as Secretary of the Congregation it was useful for him to have expert knowledge of the law so that he could oversee the judicial activities of the Congregation, and different languages, not only Greek, but also Arabic, enabling him to supervise the publication of works such as the Bible in Arabic and editions of liturgical books in Greek.[86] Another branch of expertise that was of fundamental

---

Pollini, a Florentine writer) and Girolamo Aleandro, a young member of the Accademia degli Umoristi, the lesson was held by Monsignor Girolamo de Pretis, Abbreviator for the Curia, see Jaitner, *Hauptinstruktionen* (cit. note 83), p. 158 and following which quotes BAV, Urb lat. 1090, f. 12v and Mario Rosa, *The 'World's theatre': the court of Rome and the politics in the first half of the seventeenth century*, in Gian Vittorio Signorotto – Visceglia (ed.), *Court and Politics in Papal Rome, 1492-1700* (1998), Engl. trans., Cambridge, Cambridge University Press, 2002, pp. 78-98: 83.

84 Admonitions to prudence also came to Galileo from his friend Mario Guiducci of Florence who, in a letter dated on June 21, 1624, did not hide his will to share with him the desire to respond to his opponents among whom they certainly included Ingoli, who now had transferred to Propaganda Fide: «I like the thought of V.S. [Galileo] to get rid of the likes of people who have courtesy and piety ascribed on their trophies. But we need to see the account without the mercy of anyone» (quotation by Bucciantini, *Galileo e Keplero*, cit. note 11, pp. 322-323). Guiducci, on which see Favino, *DBI*, vol. 61 (2004), was in Rome and, among other things, instructed by Galileo to distribute his response to Ingoli. On this phase of the events see also Egidio Festa, *Galileo. La lotta per la scienza*, Rome-Bari, Laterza, 2007, pp. 254-258.

85 On the relationship between the foundation of Propaganda and the Catholic reconquest see Eutimio Sastre Santos, *La fundación de Propaganda Fide (1622) en el contexto de la guerra de los Treinte Años (1618-1648)*, «Commentarium pro religiosis et missionariis», 83, 2002, pp. 231-261.

86 Pizzorusso, *I satelliti di Propaganda Fide: il Collegio Urbano e la Tipografia Poliglotta. Note di*

importance to the Congregation was his knowledge of geography, which proved useful in the assignment of missionary jurisdictions, as can be seen from the *Divisio Provinciarum* prepared in 1622 by his friend and mentor Agucchi. This envisaged the division of the world into different areas, each under the responsibility of an Apostolic Nuncio (or patriarchal vicar) and a Cardinal member of the Congregation de Propaganda Fide.[87]

Missionaries with scientific interests were encouraged to remain contact with Ingoli. Among them were Cristoforo Borri of Milan – mathematician, astronomer and Jesuit missionary – who was sent by his order to Cocincina (Vietnam) and then to Macau. While Ingoli was observing the 1618 comet from Rome (and Campanella was doing the same from his prison in Naples),[88] Borri described his observations of the event in Indochina (like Ingoli, he was an advocate of Tycho's system of cosmology). He was also interested in navigation and developed a method to determine longitude at sea. Borri fell out with the Society of Jesus, resigned and returned to Rome in 1631 where he published his *Relatione della nuova Missione delli PP. della Compagnia di Giesù al regno della Cocincina* (through Francesco Corbelletti, 1631). He then contacted Ingoli, who was always eager to receive information about the Jesuit missions, which were by tradition not subject to the Congregation's jurisdiction.[89]

One should not conclude that Ingoli wished to take advantage of his position and force his missionaries to carry out scientific observations. When, just a few years later, the French savant Nicolas Claude Fabri de Peiresc, who was then engaged in studies of longitude, asked Ingoli to assign missionaries on various continents with the scientific task of measuring the longitude in their locations based on eclipses, the secretary of

---

ricerca su due istituzioni culturali romane nel XVII secolo, «Mélanges de l'Ecole Française de Rome. Italie et Méditerranée», 116, 2004, 2, pp. 471-498.

87 APF, Acta, vol. 3, ff. 3r-6v, published in *Memoria Rerum* (cit. note 7), vol. III/2, pp. 659-661.

88 In the 1630s Ingoli was in correspondence with Campanella, then in Paris, after having met him in Rome. In the surviving letters the missionary argument prevails. CARLO LONGO, *Tommaso Campanella e la Congregazione de Propaganda Fide*, «Archivum Historicum Praedicatorum», 68, 1998, pp. 347-367.

89 The most recent study on Borri with bibliography, GUIDO MONGINI, *I rischi dei Nuovi Mondi. Tra 'Nuova Scienza' e missioni in Oriente: ambiguità e conflitti della vocazione del gesuita Cristoforo Borri*, in MICHELA CATTO – SIGNOROTTO (eds.), *Milano, l'Ambrosiana e la conoscenza dei Nuovi Mondi (secoli XVI-XVIII)*, Milan-Rome, Biblioteca Ambrosiana-Bulzoni, pp. 521-540; on Borri as a scientist, LUÍS MIGUEL CAROLINO, *Teoria e observação de cometas em Cristoforo Borri: o cometa de 1618 na Cochinchina*, «Revista brasileira de história de matemática», Especial, 1, 2007, pp. 179-198 and ID., *The Making of a Tychonic Cosmology: Cristoforo Borri and the Development of Tycho Brahe's Astronomical System*, «Journal of the History of Astronomy», 39, 2008, pp. 313-344.

the Congregation refused, saying that the priests were not sufficiently trained in mathematics nor in the use instruments of observation and concluded:

> I limit myself to saying to the missionaries that they should follow their itineraries, noting the places and distances in miles or in leagues, so at least I would get to know the above places and in some way be able to improve the maps which I certainly have found to possess defects and deceptive indications in the European Provinces, and especially those which are under the Turks.

Sensitive to the problem of how to gather knowledge in distant lands, he did encourage his missionaries to learn the local language, and left open the possibility of their participating in a scientific inquiry:

> I do believe however that if in the schools of languages at the seminars of the Regular Orders we had the students draw the sphere, and taught them the use of the compass following the rule given by Apiano [Peter Bienewitz] in his treatise *Cosmographia*, they could arrive at something close to the truth and this diligent application would be the quickest and most practical way [to go about establishing the longitude of the various places] and [furthermore] if among the students there were someone of particular intelligence, we could very well satisfy the request of M. Peiresc, from whose letters I see is a person of some account and who, if he were in Rome, could collaborate [with us] in this important objective [of teaching students the use of the sphere and the compass] at the school of San Pietro in Montorio, above all with the support of the Most Eminent Signor Cardinale [Francesco] Barberini Padrone. This is my sentiment which could be brought to the attention of the said M. Peiresc.[90]

In 1631 Ingoli wrote a long text, *Relazione delle quattro parti del mondo*,[91] on ecclesiastical and missionary geography that was intended to report on the first ten years of the Congregation and celebrate the spread of its influence across the world. This text was obviously connected with Ingoli's work, but Ingoli never did anything in a purely bureaucratic manner, so he

---

90 APF, Congregazioni Particolari, vol. 2, ff. 373r-378v. On Peiresc see Marc Fumaroli (ed.), *Peiresc et l'Italie*, Paris, Alain Baudry et Cie., 2009.

91 Published in 1999 edited by Tosi, Rome, Urbaniana University Press. The work is written in epistle style. The chapters are made up of letters addressed to the capuchin Valeriano Magni from Milan, resident in Prague, who in the years around 1640 carried out experiments on the vacuum, Michael J. Gorman, *Jesuit explorations of the Torricellian space: carp bladders and sulfur fumes*, «Mélanges de l'École française de Rome. Italie et Méditerranée», 106, 1994, 1, pp. 7-32 and Bucciantini, *Valeriano Magni et la discussion sur le vide en Italie*, in Festa – Vincent Jullien – Torrini (eds.), *Géometrie, atomisme et vide dans l'école de Galilée*, Fontenay-aux-Roses, ENS, 1999, pp. 129-152.

did not hesitate to express controversial judgments, which was probably the reason why publication of the volume was blocked and only a manuscript version survives. In addition, in a continuation of his past collaboration with Agucchi, who had since moved from the Curia to take up the office of nuncio in Venice, Ingoli sent the text to him for review, a section at a time, soliciting his comments and advice.

In the chapter on America, he recalls the providential theme of *Plenitudo temporis*, «the end and consummation of the world» which will come upon the completion of the conversion of humanity. To this end, he used images drawn from Ezekiel (1, 4-10) of the Tetramorphus, «the four animals [...] that drove the chariot in which sat the Divine Majesty»,[92] connecting each of the animals to one of the four continents, the four corners of the world and the four Evangelists, as described in the Apocalypse (4, 7) and as depicted by Cimabue in the upper basilica of Assisi in a fresco which hints at the Joachimite prophecy. The four winged creatures, the lion, the man, the bull (or ox) and the eagle, traditionally linked to the four evangelists and even to the four continents, now completed the quadripartite shape of the world since, as Ingoli states, America had finally been discovered. The association was established between the lion – Saint Mark and Europe (the North);[93] between man – Saint Matthew and Asia (the East); between the bull – Saint Luke and Africa (the South); and between the Eagle – Saint John and the Americas (the West). Moreover, the specific characteristics of each continent were interpreted allegorically through the winged animals. It is conceivable that Ingoli, when writing these pages, also directed his attention toward astrology, linking the four corners of the world with the points of the solstices and equinoxes in accordance with ancient tradition, and connecting their origins to the zodiac constellations of Taurus, Aquarius, Scorpion and Leo, the positions of which had changed since biblical times due to the phenomenon of the precession of the equinoxes – a phenomenon that Ingoli was quite familiar with, having examined Magini's book as cited above. Thus, while in his text Ingoli could not resist incorporating some of his astrological knowledge – the representation of continents as Tetramorph animals, as well as «various figures and images, and similarities, and parables and allegories» – in his position as secretary of the Congregation, the main purpose of *Relazione delle quattro parti del mondo* was to describe and celebrate the evangelical mission of his institution:

---

92 INGOLI, *Relazione delle quattro parti del mondo* (cit. note 7), pp. 231-241: 232.

93 In addition, the cardinal points are, as by tradition, indicated by the four winds which the angels restrain.

And all this can still be invoked to make the soul of the Missionary much more accentuated by his zeal in the propagation of the faith, seeing on the one hand the divine disposition, which is represented in the images of the four animals depicted; while on the other hand, in that part of the earth where the Gospel will be preached, the Evangelist will need to be seen as a protector, and need to use the means that, according to the properties of the animals, will help in the conversion of that [part of earth].[94]

In 1631, when writing this text, so rich in geographical knowledge as well as ethnographic information and descriptions of missionary activities, Ingoli put to good use his expertise and his scholarship as secretary of the Congregation de Propaganda Fide. In doing so, he used his learning as an instrument for the Curia, which employed him and in the safeguarding of whose interests he was deeply concerned. If one considers that Ingoli's text concludes with *Quinta lettera delle cose fatte in Roma per la propagatione della fede*, it is possible to complete the allegory of the Tetramorph by equating the chariot in which God sits, surrounded by the four Evangelists, with the Congregation de Propaganda Fide guided by the pope who governed the Church's universal apostolic effort.

6. Concluding Remarks

At the end of our attempt to pull together the still meagre data concerning Francesco Ingoli, how can we place the finishing touches on the biography of this man of the Curia who for almost twenty years observed the *novae* and comets and held discourse with the great European astronomers?

Certainly, in his youth Ingoli dedicated himself to gaining as broad and profound an education as possible during his years of study in Padua, Bologna and Rome. As far as his first and most intense passion – astronomy – was concerned, although he could not delve as deeply into the

94 Ingoli, *Relazione delle quattro parti del mondo* (cit. note 7), p. 233. Astrology was also subjected to the papal condemnation with Urban VIII's Bull *Inscrutabilis* on April 1, 1631 «contra astrologos iudiciarios» following the trial that had struck the Vallombrosian abbot, Orazio Morandi who had made pronouncements on death of the Pope, Brendan Dooley, *The Morandi Affair and Urban VIII's Rome*, «Roma moderna e contemporanea», 11, 2003, pp. 145-166. In the letter with Caetani of 1613, a certain Morandi is cited, who Ingoli recommends to the Cardinal as his substitute in the case of his transfer to the service of the Borghese. We are still unable to identify the character, even if 1613 is the actual year of the arrival of Orazio Morandi in Rome to serve in the Curia with the support of the Medici, AFC, FG 81937 (March 12, 1613) and 40316 (March 23, 1613); see also Ernst, *Scienza, astrologia e politica* (cit. note 31), and the article on Morandi by Stefano Tabacchi, *DBI*, vol. 76 (2012).

"profession" as other innovative scientists like Galileo, his involvement was sufficient to permit him to write five treatises between 1604 and 1619. These works, while not highly original, were of great breadth, marshaling arguments based on the astronomical theories of philosophers and men of science from Aristotle (whom he considered to be outdated) to Brahe (whose system instead he supported strongly) to refute the theories of Copernicus and Kepler and defend the authority of the Curia. As a scholar of astronomy and astrology, he reflected the interest in and knowledge of science exhibited by the men who made up the intellectual circles of Rome in this period and the osmosis that took place between cultural and curial activities.[95] Ingoli personified the level of astronomical knowledge that was widespread in this enlightened environment. He himself had been particularly interested in gathering information based on his own observations of *novae* and comets and in sharing his knowledge through a network of scientific contacts which branched out from Rome and Italy, beginning with his earliest «friends» who had encouraged him to write his first treatises to his active participation in various academies. For this reason he was well qualified to serve as an expert consultant for the Church's censorship office, which guaranteed him the much desired entry into the Curia. The meeting with Caetani in Ravenna, where Ingoli was teaching, allowed him to make the all-important career move to Rome, where he was able to exploit his knowledge at exactly the moment when the Copernican question was exploding. This scientific debate involved both theological and confessional issues. Between the years 1616 and 1620 the pursuit of astronomy was no longer a purely intellectual activity; it had become controversial and a target for censorship. A militant approach to scientific interests was required by the figures of power in the Church, the Church to which Ingoli owed his allegiance. Telling in this regard was the request by Cardinal Cobelluzzi that Ingoli draw arguments from his 1618 treatise on comets that could be used to refute Copernicus's heretical theories, which Ingoli obligingly did.

For this reason he placed himself at the service of the Curia, demonstrating that he could provide a vital contribution to the functioning of a pontifical machine whose power extended to the realm of culture and even controlled it, although discussion was still allowed within certain limits and the rules were occasionally relaxed. The demands of orthodoxy hid the political maneuvering and conflicts that were taking place behind the scenes in the Curia with the participation of additional figures who were

---

[95] For some examples see Romano (ed.), *Rome et la science moderne entre Renaissance et Lumières*, Rome, Ecole Française de Rome, 2008.

culturally equipped for the role. Ingoli was the embodiment of the Curia official who possessed eminent qualifications in terms of his solid cultural and scientific background and strong intellectual profile, but who could be counted on for absolute loyalty to the Church (in fact he abandoned his pursuit of astronomy and detached himself from the scientific controversy over Copernicanism), an example of the osmosis between culture and Curia. In his role as secretary of the Congregation, he placed his vast knowledge – legal, geographic, linguistic and even political – at the service of the missionary institution in which he played a very active role. Often drawn into the maelstrom of controversy and criticism, he emerges as a strong personality, one of great prestige among those engaged in working for the Curia and the papal government of the universal Church.[96]

[96] On the death of Urban VIII, he wrote a brief text with which he counseled the future pope about the government of the Church, who could reduce the administration of the Roman Pontificate with to most qualified in each argument to rule over it in his way, ASR, Fondo Santacroce, 61 B 6, edited by Josef Grisar, *Francesco Ingoli über die Aufgaben des kommenden Papstes nach dem Tode Urbans VIII. (1644)*, «Archivum historiae pontificiae», 5, 1967, pp. 289-324.

Édouard Mehl

KEPLER'S SECOND COPERNICAN CAMPAIGN
The Search for an Annual Stellar Parallax after the Roman Decree (1616)

For Miguel Ángel Granada

The brief text of the 1616 papal decree banning Copernicus' *De revolutionibus*[1] is significant, from the point of view also of what it *does not* say. Actually, it does not say a single word about the magnitude of the world according to the new astronomical hypotheses. It was issued more than fifteen years after Giordano Bruno's trial and execution,[2] but just a few years after the publication of *Sidereus Nuncius* in which, as the consequence of Bruno's trial and execution for heresy, Galileo carefully avoided making any cosmological statement on the infinity of the world. In the following decade such an oversight, whether deliberate or not, would provide further material for discussion and the continuation of a scientific enquiry, as Kepler, Galileo and Descartes noted.[3] But for the time being astronomers

---

1 See the new documents edited in Nicholas Copernicus, *Des Révolutions des orbes célestes / De Revolutionibus orbium coelestium*, ed. by Michel-Pierre Lerner, Alain Philippe Segonds, Jean-Pierre Verdet *et al.*, 3 vols., I, Paris, Les Belles Lettres, 2015, pp. 509-520; III, pp. 619-696; see also the "dossier iconographique", vol. III [22]. On the question of the immensity of the world, see vol. III, p. 640, the *emendatio* of Copernicus proposed by Francesco Ingoli, who does not modify the statement of *De Rev.* I, 8 (= DROC vol. II, p 29, ll. 5-13): «[...] mundus, cujus finis ignoratur, scirique nequit», whereas Ingoli removed the rest of the sentence.

2 As reminded by Miguel Ángel Granada in his introduction to the dialogue *De l'infinito, universo e mondi* (Giordano Bruno, *Opere Complete*, IV, Paris, Les Belles Lettres, 2006, p. lxxvi), the Nolan philosopher had been accused since 1592 to destroy the Christian Faith. Defending himself, he recognized that the doctrine *de infinito* (either infinite space or infinite plurality of worlds) was *indirectly* against the Christian Faith (see Luigi Firpo, *Il Processo di Giordano Bruno*, Roma, Salerno Editrice, 1993, p. 168: «Di sorte che è doppia sorte de infìnitudine de grandezza dell'universo et de moltitudine de mondi, onde indirettamente s'intende essere repugnata la verità secondo la fede»).

3 Descartes (*Œuvres*, ed. by Adam and Tannery [henceforth AT], new edition by Bernard Rochot and Pierre Costabel, 11 vols. Paris, Vrin, 1964-1974) pays attention to the position of

and scientists «are enclosed in certain limits, which a Catholic cannot transgress», as Tommaso Mingoni wrote to Kepler.[4] This is to be understood as the strongest paradox regarding the patent contradiction between the 'Catholic' – i.e. universal – truth of the Christian and Roman faith, and the impossibility of spreading it all over the world (*limites... extra quos catholicus non potest... expatriari*). A paradox that Kepler did not try to avoid, far from it, as he wrote in response to the Roman condemnation, suggesting that, on the contrary, the discovery of celestial immensity ensues from the *libertas philosophandi* that is only granted by the true Christian faith:

> Whether it should not have been permitted to Christians that they would be astronomers... or, once given that freedom to search for the truth of celestial things, Theology should not be allowed to use the right to enclose the search for the truth within such tight limits.[5]

Such a precise and carefully pondered critique can be found in Kepler's *Response* to the letter which Francesco Ingoli wrote in 1616 (perhaps at the behest of the Congregation of the Roman Inquisition) and sent to Galileo – *Disputatio de situ et quiete terrae contra copernici systema*.[6] This essay presented concise arguments against heliocentrism based on astronomy, physics, and theology, focusing on one or two points in each of these fields,

---

the Catholic Church on the question: see Descartes to Mersenne, 18 December 1629, AT I, 86, and later the epistolary exchange between Christina of Sweden and Descartes. See especially Chanut to Descartes, 11 May 1647, AT X, 620-621: «[Christine] ne refusa son consentement à aucune de vos opinions, cette ligne exceptée, où vous supposez le monde infiniment étendu».

[4] Johannes Kepler, *Gesammelte Werke*, ed. by Max Caspar *et al.*, München, C.H. Beck, 1937-2009 (henceforth *KGW*), KGW, XVII, p. 235 (Mingoni to Kepler, 27 September 1617), quoted by Massimo Bucciantini, *Contro Galileo. Alle origini dell'*Affaire, Florence, Olschki, 1995, p. 105.

[5] KGW, 20, 1, p. 180, ll. 33-38: «[...] Aut non fuit concedendum Christianis ut essent astronomi [...] aut libertate semel datâ veritatis rerum caelestium indagandae nullum ius sibi servabit Theologia coercendi veritatis indagationem intra angustas metas...». «Universality» is clearly attributed to the Copernican heliocentrism: see, *ibid.*, p. 172, ll. 11-12: «hoc pulchrior enitescit haec universalis hypothesis de situ Solis in mundo».

[6] Altough the *Disputatio* was adressed to Galileo, he would reply only eight years later with the *Letter to Ingoli* (1624); inbetween, there had been an important exchange with Kepler's own *Responsio* and Ingoli's *Replicationes... ad Kepleri Impugnationes*. Ingoli's *Disputatio* is now edited together with the galilean *Lettera a Ingoli* by Bucciantini – Michele Camerota: *Galileo Galilei. Scienza e Religione. Scritti copernicani*, Rome, Donzelli, 2009, pp. 165-188. Kepler's *Responsio ad Ingoli de Systemate*: KGW, 20.1, 168-180; Francesco Ingoli, *Replicationes... ad Kepleri Impugnationes* (henceforth: *Replic.*), dated from the 13th October 1618. Bucciantini, *Contro Galileo* (cit. n. 4), pp. 127, 177-205. On the *Letter to Francesco Ingoli*, see also Id., *Galilée et Kepler. Philosophie, Cosmologie et Théologie à l'époque de la Contre-Réforme*, Paris, Les Belles Lettres, 2008, pp. 403 sq. Bucciantini shows that the *Lettera* is a manifest – somewhat Brunian – against, finalism, finitism and anthropocentrism.

such as the absence of measurable annual parallax, the physical doctrine *de motu gravium* (and the 'laws of nature' more generally, which Tycho shared with Aristotle according to Ingoli[7]), and the problem of the *locus beatorum* in the Copernican system. Concerning the first point, Ingoli appeared quite confident in his *argumentum de parallaxi*, which he presented as proving the centrality and immobility of the Earth even though there was nothing really new about this, and assuming *grosso modo* the arguments against Copernicus laid out by Tycho twenty years earlier in his correspondence with Rothmann (1596) and in *Astronomiae Instauratae Progymnasmata*.[8] As a matter of fact, Ingoli applied to the parallax of the Sun the same kind of argument as Brahe did with *Mars acronychus* in his own 'Copernican campaign'.[9] But as we will see, Ingoli was a long way from grasping the technical difficulty of the problem and moreover, his knowledge and understanding of Tycho's observations was far inferior to that of Kepler. The opening remark by Kepler in *Responsio* regarding the adroitness of the argument presented, which Ingoli confidently described as demonstrative, was obviously ironical: Ingoli's argument was certainly clever for those not well grounded in the science of astronomy[10] – in other words, every member of the Congregation.

Tycho's objections to the Copernican model, which were repeated by Ingoli, dealt with – among other things – the theological argument of the

---

[7] INGOLI, *Replic.* (cit. note 6), p. 199: «Sed replicari potest argumentum Tychonis concludere secundum regulas naturae. Nam natura comparata est, ut uni corpori simplici unicus sit tantum motus naturalis, si ergo alii motus attribuuntur Terrae ultra unum violenti sunt, et ideo excludendi a corpore Terrae naturali, quia violenti motus destruunt naturam omnium corporum».

[8] The present essay can be read in the straight line of three others: 1°) OWEN GINGERICH – JAMES R. VOELKEL, *Tycho Brahe's Copernican Campaign*, «Journal for the history of astronomy», 29, 1998, pp. 1-34; 2°) PATRICK J. BONER, *Kepler's Copernican Campaign and the New Star of 1604*, in ID. (ed.), *Change and Continuity in Early Modern Cosmology*, Dordrecht, Springer, 2011, pp. 93-114; 3°) MIGUEL A. GRANADA, *Tycho Brahe's anti-Copernican campaign: his criticism of Maestlin and Thomas Digges in the "Astronomiae Instauratae Progymnasmata"*, in DARIO TESSICINI – BONER (eds.), *Celestial Novelties on the Eve of the Scientific Revolution* 1530-1630, Florence, Olschki, 2013, pp. 185-207.

[9] Ingoli's *argumentum de parallaxi* means that if the Sun were in the center of the Universe, it should show a greater parallax than the Moon because of the distance to the fixed stars; see INGOLI, *Disputatio* (cit. note 6), p. 167: «Nam si Sol esset in centro universi, maiorem admiterret parallaxim quam Luna: sed consequens est falsum: ergo, et antecedens». The syllogistic form of the *elenchos* is quite important: Ingoli aims to be demonstrative. The formal similarity of the argument with Tycho's (wrong) measure of the parallax of Mars acronychus is obvious (on this, see GINGERICH – VOELKEL, *Tycho Brahe*, cit. note 8). As Kepler made clear in the *Astronomia Nova* (ch. 11), Brahe had been deceived by his assistants, who did not at all *observe* a real parallax: they only tried to determine what the parallax of Mars *should be* according to Copernican hypotheses.

[10] KGW, 20, 1, 168, ll. 27-31.

symmetria mundi, i.e. the general assumption that in a world reflecting the perfection of its Creator, the parts of the world are proportionate to each other, whereas the immense void space between Saturn and the fixed stars necessarily implied by the Copernican hypotheses entirely destroyed this so-called «symmetry».[11] Symmetry here means nothing but the proportion of the celestial distances and, in that case, the impossibility of determining the proportion of the diameter of the *sphaerae fixarum stellarum* to the *orbis magnus* (diameter of the revolution of the Earth around the Sun). Tycho's line of argument against Copernicus was clear and unequivocal: one cannot claim the *harmonia et symmetria mundi* and at the same time the immensity of the world, or at least its 'indefinite magnitude' (Copernicus). Tycho would contend that this problem in modern astronomy, which could be referred to as the Aristarchean problem,[12] was equivalent to the emergence of irrationals in ancient Greek mathematics.[13] However, after reading

---

[11] INGOLI, *Disp.* (cit. note 6), p. 172: «[...] quae distantia adeo magna... asymmetrum esse universum ostendit [...]». Kepler responses (KGW, 20, 1, p. 170, l. 43): «Primo dicit tantam distantiam fixarum respectu telluris esse ἀσυμμετρον non geometrice loquitur de geometrica re». Kepler said exactly the same against Ursus (see KEPLER, *Contra Ursum*, ed. by Segonds and Nicholas Jardine, 2 vols., II/1, Paris, Les Belles Lettres, 2008, p. 114 and n. ad loc., p. 154, quoting the famous letter to Paul III: «mathemata mathematicis scribuntur»).

[12] As far as we know, the logical consequence of the immensity of the world in the heliocentric hypothesis first appeared with Aristarchus, as quoted by Archimedes: see for instance JAMES EVANS, *Histoire et pratique de l'astronomie ancienne* (Oxford, 1998; Fr. trans. by A. Segonds, Paris, Les Belles Lettres, 2016, p. 72). On Copernicus and Aristarchus, see DROC, vol. II, p. 476, and vol. III, pp. 596-598. Despite EDWARD ROSEN (*Aristarchus of Samus and Copernicus*, in ID., *Copernicus and his Successors*, London and Rio Grande, The Hambledon press, 1995, pp. 1-9), who claimed that Copernicus did not know Aristarchus, we take for granted that Copernicus, however, knew the treatise *The Sand Reckoner* (*De Arenae numero*). See RUDOLF VON ERHARDT – ERIKA VON ERHARDT-SIEBOLD, *Archimedes' "Sand-Reckoner": Aristarchos and Copernicus*, «Isis», 33/5, 1942, pp. 578-602, showing persuasively that the Copernican axiom of immensity, as stated in the *Commentariolus* (c. 1511) is formally identical with the second "equation" of the *Sand Reckoner* (p. 599): «Minorem esse comparationem distantiarum Solis et terrae ad altitudinem firmamenti quam semidimetientis terrae ad distantiam Solis adeo ut sit ad summitatem firmamenti insensibilis» (Eng. trans. by Rosen, *Three Copernican Treatises*, New York, Octagon Books, 1971^{3}, p. 58: «The ratio of the earth's distance from the sun to the height of the firmament is so much smaller than the ratio of the earth's radius to its distance from the sun that the distance from the earth to the sun is imperceptible in comparison with the height of the firmament»). Practically, that means that the annual stellar parallax must be smaller than the solar parallax. The original Greek text of Archimedes was first edited by Thomas Venatorius (Basel, Herwagius, 1544, pp. 155-163), after a copy made by Regiomontanus on a Greek manuscript possessed by Cardinal Bessarion. Venatorius was a common friend of Achilles Pirmin Gasser and Georg Joachim Rheticus (see KARL HEINZ BURMEISTER, *Magister Rheticus und seine Schulgesellen. Das Ringen um Kenntnis und Durchsetzung des heliozentrischen Weltsystems des Kopernikus um 1540-1550*, UVK, Konstanz und München, 2015, pp. 603-606).

[13] TYCHO BRAHE, *Opera omnia*, ed. by J.L.E. Dreyer, Copenhagen, Gyldenhal, 1913-1929, VI, p. 222, ll. 26-31: «Faceßat ista ageometrica, et asymmetra, inordinataque philosophandi ratio, a sa-

Mysterium Cosmographicum, Tycho was – as we find in a letter that he wrote to Kepler in December 1599 – less skeptical about Copernicus's 'imaginations' as such, than about the attempt to demonstrate them *a priori*, as Kepler does in what Tycho called, in a condescending way, his «ingeniosum scriptum».[14] In the same letter Tycho explicitly noted his own failed attempt to observe an annual parallax of the Earth and for the first time mentioned an experimental protocol that could lead to a verification *a posteriori* of the Copernican 'imaginations'.[15] Surprisingly, as we will see later, Kepler had achieved this on his own the previous year, so that the noble Dane did not teach anything to this rustic Swabian that he did not already know.

Digges, Mästlin, Kepler, Galileo, Descartes, and in a word every Copernican after Tycho Brahe, had to try and solve this problem,[16] a challenge that would lead to some decisive philosophical conclusions, which may be summed up in three main points:

i/ Sensible perception is a necessary but not sufficient criterion of reality; the lack of *sensible* parallax does not mean that an annual parallax does not exist.

ii/ Observational instruments are not tools designed to *replace* the device of our senses, but to enhance them.

iii/ As a direct consequence of the debate on the immensity of space beyond Saturn, Kepler refutes the conclusion that the world's immensity should be seen as a *miserabile refugium* (Ingoli) and radically transforms the meaning of the term itself: immensity loses its negative connotation of a magnitude that exceeds any perception and cannot be measured, but now means the magnitude itself, in so far as it makes the right measurement *possible*.

Paying particular attention here to Kepler and to the chronology of the dispute, we will show how the German astronomer substantially devel-

---

pientia, providentiaque Divina alienissima». See ÉDOUARD MEHL, *"Novum struam mundum. Kepler's rebuilding of the symmetria mundi"*, in BONER – GRANADA – TESSICINI (eds.), *Unifying Heaven and Earth: Essays in the History of Early Modern Cosmology*, Barcelona, PU, 2016, pp. 197-216.

14 Tycho Brahe to Kepler, 9 December 1599, KGW, 14, letter n° 145, p. 94, ll. 186-202.

15 *Ibid.*, p. 95, ll. 228-232: «E stella polari vel quauis alia subtiliter eruere, an orbis annuus terrae à Copernico excogitatus, si is esset octauae sphaerae, stellis aliquantulam insinuet Parallaxin, iam ante, saepius et diligenter experiri tentaui, uerum nihil eiuscemodi, diligenti facta animadversione hactenus comperi».

16 On Galileo, see JUAN CASANOVA, *The Problem of the Annual Parallax in Galileo's Time*, in GEORGE V. COYNE – MICHAEL HELLER – JOSEF M. ZYCINSKI (eds.), *The Galileo affair: a meeting of faith and science*, Città del Vaticano, Specola Vaticana, 1985, pp. 67-74. On Descartes, see MEHL, *Note complémentaire sur une observation astronomique et la recherche d'une parallaxe stellaire (septembre 1642)*, «Bulletin Cartésien XLV», *Archives de Philosophie*, 79, 2016, pp. 189-193.

oped his works and writings during the year 1618, and how they outstandingly converged into the discovery of the third law, which can be regarded in a way as Kepler's response to the problem, insofar as the law of periodic times could be applied to demonstrate the validity of the heliocentric hypothesis. Therefore we should not be surprised that Book Five of *Harmonice Mundi* (1619), which recounts his recent discovery of periodic times (March-May 1618), starts on its front page with a formal concession to Tycho's hypotheses[17] but ends with an *Epilogus de Sole, conjecturalis*, where Kepler deals – obviously against all the neo-scholastics (including Ingoli) – with the *sedes beatorum* located, according to his conjecture, in the very center of the Sun. In fact, this *Epilogus* seems to be the consequence of the censorship of Kepler's *Epitome* by the Congregation of the Index (February 1619), in which the author was accused by Ingoli of repeating the errors of Origen concerning the life of the stars and the animation of the Sun.[18]

---

[17] KGW, 6, 287 = Eng. trans. *The Harmony of the world*, ed. by E.J. Aiton, A.M. Duncan, J.V. Field, Philadelphia, The American Philosophical Society, 1997, p. 38: «On the most perfect harmony of the heavenly motions... According to the precepts of the most thoroughly corrected astronomical teaching of the present day and the hypotheses of Copernicus, but also those of Tycho Brahe, one or the other of which are today publicly accepted as true (*ut verissimae, publice recipiuntur*), superseding those of Ptolemy». This formal concession is reminded in Kepler's *Admonitio ad Bibliopolas* (DROC, vol. III, Appendix, text 7, p. 652). Whether Kepler thinks about the frontispiece of the *Harmonice V* or the *Epitome V*, we follow Alain Segonds (Kepler, *Le Secret du Monde*, ed. by A. Segonds, Paris, Les Belles Lettres, 1984, p. 268), rather than DROC, p. 690 n. 16. Let me recall that the *Epitome* had been announced since 1614, with the following title: *Joannis Kepleri, Math. Caes. Epitome Astronomiae sine fictis Hypothesibus, brevibus et dilucidis quaestionibus et responsionibus ex commentariis de Motibus stellae Martis petitis, vetustissimam Aristarchi Samii et recentissimam Copernici de sideribus doctrinam ob oculos ponens* (see KGW, 7, 574). Of course, such a triumphalist title could not be maintained after the decree of 1616.

[18] See Pierre-Noël Mayaud S.J., *La condamnation des livres coperniciens et sa révocation à la lumière des documents inédits des Congrégations de l'Index et de l'Inquisition*, Rome, Pontificia Universitas Gregoriana, 1997, p. 59. Both Cardinal Bellarmin and Barberini took part in the session in which Ingoli pronounced the censorship of the *Epitome* (Book I). Ingoli had to read the *Epitome*, which Kepler more than once referred to in his *Responsio*. See the text edited by Mayaud (*ibid.*, pp. 65 sq.): Ingoli first recalls the perfect morality of the Imperial Mathematician (*summa quoad mores... integritate*), but, of course, also the fact that he was deceived by the Lutheran heresy (*miserabiliter a Ministris Lutheranis deceptus in haeresi Lutheri detinetur*). In his *Letter to Ingoli* (1624), Galileo does not explicitly name Kepler, but there is no doubt that the whole section dealing with the relation of "heretics" with heliocentrism is referring to this session (see *Lettera*, cit. note 6, p. 193). Ingoli claims that Kepler did the same error as Origen, and suggests that the Congregation should take position against the error of the animation of the celestial bodies, in order to prevent all attempts to persist in this «swarming» error and to go further in that way. Ingoli's original statement about Kepler's origenism does not come from the reading of the *Epitome* but from the *Responsio ad Ingoli* itself: «[...] Patres, qui ipsi non in superficiè fixarum, sed in Sole, quem etiam Ptolemeus longissimè à fixis interius (sc. *inferius*?) ponit, praecipuam agnoscunt maiestatis divinae officinam in naturalibus: Cum secundum hanc centri et superficiei oppositionem Sol rectius empyreo iungeretur [...] essetque omnibus modis Sol sedes beatorum, et sit in centro mundi» (KGW, 20.1, 176, 12-21).

Such a 'conjecture', presumably written hastily just before Kepler delivered his manuscript to the editor, should be regarded *politically*, so to speak, as the triumph and apotheosis of heliocentrism against the Church's ban.

In this regard, the immediate result of the Roman decree of March 1616 was that Kepler launched a *second* Copernican campaign in 1618, which he obviously thought would be decisive. Indeed, with *Epitome astronomiae Copernicanae*, *Harmonice mundi*, the treatise *De cometis*,[19] and this unpublished *Responsio* to Ingoli – what more could be accomplished by a single man? But, incredible as it might seem, Kepler did more: he produced the second edition of *Mysterium Cosmographicum* (1621), and may even have been planning, with his former professor Mästlin – who had been, as shown by Granada, a target of Tycho's anti-Copernican campaign thirty years earlier – a new edition of *De Revolutionibus*. Unfortunately, this edition would never be published, for reasons that still remain unclear. Nevertheless, we have good reason to presume that Kepler intended to lead this ultimate effort with the help of Mästlin, who himself wrote a harsh criticism of the Roman decree, giving an important account of it in the 1621 edition of *Narratio Prima* that did not appear in the earlier 1596 edition.[20] His account, as we will see, can be read in a way that is quite in keeping with Ingoli's statement on the annual parallax.

This context explains why five years later the *Letter to Ingoli* (1624) contains, as Koyré rightly pointed out, the most important and precise statement by Galileo concerning the inability of human understanding to determine whether the world is finite or not,[21] and at the same time the most clear and affirmative claim concerning the possibility of discovering an annual parallax that would prove the Earth's movement, which was said to be «*notabile e notabilissimo*»[22] despite the lack of factual evidence. Of

---

19 See the conclusion of the *De Cometis* (I), KGW, 8, p. 220, ll. 17-20: «Denique quot sunt in Coelo Cometae, tot sunt argumenta (praeter ea quae à Planetarum motibus deducuntur), Terram moveri motu annuo circa solem. Vale Ptolemae, ad Aristarchum revertor duce Copernico».

20 Nevertheless, Max Caspar had put it in the first volume of the KGW (*Mysterium Cosmographicum*, 1596), instead of the eighth (*editio altera*, 1621). See *infra*, n. 51.

21 Alexandre Koyré, *From the closed world to the infinite*, Eng. trans., Baltimore, Johns Hopkins University Press, 1957, ch. IV, pp. 54-56. Galileo's *Letter to Ingoli* is quoted by Jean Seidengart, *Dieu, l'univers et la sphère infinie. Penser l'infinité cosmique à l'aube de la science classique*, Paris, Albin Michel, 2006, pp. 335-337, without any explicit reference to Koyré. Bucciantini, *Contro Galileo* (cit. note 4), pp. 161-163; Philippe Hamou, *La mutation du visible. Essai sur la portée épistémologique des instruments d'optique au XVIIe siècle*, 2 vols., I, Villeneuve d'Ascq, Presses Universitaires du Septentrion, 1999-2001, pp. 43 sq.; Id., *La nature est inexorable. Pour une reconsidération de la contribution de Galilée au problème de la connaissance*, «Galilæana», 5, 2008, pp. 149-177.

22 Galileo Galilei, *Letter to Francesco Ingoli* (cit. note 6), p. 235. We can quote here the

course this claim regarding the prominent *possibility* of finding a parallax denotes the strength of the Copernican convictions of Galileo. But underlying this claim a new conception of the growth and progress of human knowledge over time is implied. While Ingoli denies the existence of any parallax, Galileo considers that it simply has not been found *yet*, and is still to be found – "possible" meaning "in the future". At the same time Galileo obliquely but pointedly criticized the theoretical grounds of Keplerian heliocentrism as laid out in *Harmonice mundi*, which he saw as springing from a form of (in Kantian terms) transcendental realism.

The explanation for this divergence between Kepler and Galileo goes back to the time of their first exchange in 1597, after Galileo had read *Mysterium Cosmographicum* and the shared project to find the proof (or at least a *demonstratio quia*) of Copernican heliocentrism.

1. Kepler's strategic alliance after 1596

Before enquiring into the immediate effect of the Church's ban on Kepler's thought and work, we have to look back at the beginning of his career and his theological deduction of heliocentrism, as it appears in *Mysterium Cosmographicum* (1596) and in the Preface to *Astronomia Nova*.[23]

Let us briefly come back to the context of Galileo and Kepler's first communication. The first letter of Galileo to Kepler (4th August 1597) is very short: Galileo mentioned his formal adhesion to Copernicus, and declared that he would publish his refutation of the common hypotheses if he could find some help and support, not only from Kepler but from several men like him (*si plures, qualis tu es, extarent*).[24] In his response, Kepler made a very concrete proposal, offering to collaborate on measuring the annual parallax and to publish the results, whatever they were.[25] As Kepler

words quoted by Kepler in ch. 11 of the *Astronomia Nova* (KGW, 3, 121, ll. 13-14): «Cum autem Braheus passim viris doctis affirmaverit, ex hujus anni observatis inventam esse parallaxim Martis *notabiliter* majorem Solari...» (our italics).

23 Kepler, *Mysterium Cosmographicum* = *The Secret of the Universe*, Eng. trans. by A.M. Duncan. Introduction and commentary by E.J. Aiton, with a preface by I. Bernard Cohen, New York, Abaris Books, 1981; see also *Le Secret du Monde* (cit. note 17). Let me recall here that the extensive theological considerations in the *Introductio* to the *Astronomia Nova* (1609; KGW, 3, pp. 18-35) were initially written for the *Mysterium*, and removed at the request of the Tübingen authorities.

24 On this letter, see Bucciantini, *Galilée et Kepler* (cit. note 6), pp. 59 sq.

25 Kepler to Galileo Galilei, n° 76, KGW, 13, 146, ll. 58-70: «Nunc abs te placet aliquid observationum postulare: scilicet mihi, qui instrumentis careo, confugiendum est ad alios.

had shared his intentions regarding this project with Mästlin as well,[26] one can conjecture that he envisioned the three of them forming a scientific community, or at least directing one, as asked by Galileo. Of course Kepler did not at all give up the idea that, with the inscription of the five regular geometric bodies, *Mysterium Cosmographicum* provided the best possible confirmation of the Copernican hypotheses, demonstrating *a priori* not only the fact (*quia*), but above all the reason why and the *sufficiens ratio* for the heliocentric arrangement.[27] Nevertheless, a demonstration *a posteriori*, such as the discovery of an annual parallax of the Earth, would have a more persuasive effect on the many unbelievers around them, whether skeptics or dogmatists. And if Kepler asked for Galileo's assistance, the only reason was that Galileo seemed to have an instrument – a quadrant – that was accurate enough to reveal the parallax. In fact Kepler did not wait for Galileo's answer, which never came, instead fashioning a simple instrument himself, as he scrupulously reported in his letter to Herwart dated 16th December 1598.[28]

At this time was Kepler already aware of the contents of Tycho's unpublished manuscript *Astronomiae Instauratae Progymnasmata*, which would appear a few years later? It is very unlikely, though a striking coincidence is to be found here. In the third part, where Tycho criticizes both Thomas Digges and Michael Mästlin,[29] he insists on the importance of having ac-

---

Habes quadrantem in quo possis notare singula scrupula prima et quadrantes primorum? Observa igitur, circa 19 Decembris futurum, altitudinem eductionis caudae in Ursa maximam et minimam eadem nocte. Sic circa 26 Decembris observa similiter utramque stellae polaris altitudinem. Primam stellam observa etiam circa 19 Martii anni 98, altitudine nocturna, hora 12; alteram, circa 28 Septembris, etiam hora 12. Nam si, quod opto, differentia quaedam inter binas observationes intercedet unius atque alterius scrupuli, magis si decem aut quindecim, rei per totam astronomiam latissime diffusae argumentum erit; sin autem nihil plane differentiae deprehendemus, palmam tamen demonstrati nobilissimi problematis, hactenus a nemine affectatam, communiter reportabimus».

26 The project had been sent to Mästlin, in a letter of October 1597 (n° 75, KGW, 13, 142, ll. 88-101), a few weeks before the one to Galileo (n° 85, 6 January 1598, KGW, 13, p. 165, ll. 136-212): «Elegi aliam quamvis huic negocio minus idoneam, solam tamen post polarem, quae mihi non nihil satisfaciat, Eductionem Caudae in ursa Majore».

27 See Daniel A. Di Liscia, *Kepler's* A Priori *Copernicanism in his* Mysterium Cosmographicum, in Granada – Mehl (eds.), *Nouveau Ciel, Nouvelle Terre. La révolution copernicienne dans l'Allemagne de la Réforme (1530-1630)*, Paris, Les Belles Lettres, 2009, pp. 283-317.

28 Kepler to Herwart von Hohenburg, 16 December 1598, KGW, 13, p. 266, ll. 90 sq.

29 See Granada, *Tycho Brahe's anti-Copernican campaign* (cit. note 8), p. 198. Digges thought that the decreasing apparent diameter of the Nova was due to the increasing distance from the Earth, moving apart on its *orbis magnus*. If that explanation had been correct, the star should have been growing in the six following months of 1573, which, of course, did not happen (the book of Thomas Digges appeared in London at the beginning of 1573: *Alae*

curate and reliable instruments (Digges claimed that he had some new and very efficient ones, promising some wonders among other *res inauditae*, but as Tycho ironically wrote, «it would have been more appropriate to *do* rather than to *speak* about it» [30]). Immediately following this passage Tycho mentions another point that was at the time (circa 1586-1588) under discussion between himself and Christoph Rothmann, the first Copernican astronomer in Germany. Rothmann had noticed a slight seasonal variation in the altitude of the Pole that could be imputed, according to him, to the annual revolution of the Earth. Tycho refutes this, and reiterates that the error is not *in natura rerum*, but in the limits of human observation.[31] Tycho, who prided himself on having the very best observational instruments available, declared that despite repeated attempts he never succeeded in observing such an annual parallax.

We know that Kepler and Tycho both directly spoke of this problem, in which Tycho found the major *obstaculum*, if not the only one, against Copernicus:[32] the immense space of the universe was, according to the Dane, *para phusin*. Kepler contradicts Tycho, showing that his comparisons were wrong from a logical point of view: comparing the size of a man to the movements of the stars was a grave fallacy of *metabasis eis allo genos*.[33] What is absurd is

---

seu Scalae Mathematicae, quibus visibilium remotissima Coelorum Theatra conscendi, et Planetarum omnium itinera novis et inauditis Methodis explorari). Therefore, what Digges thought to be the best opportunity (*occasio maxime opportuna*) to measure the annual movement had been on the contrary an excellent opportunity to disprove the Copernicans and make them stand in contradiction with the phenomena.

30 TBOO 3 (*Astronomiae Instauratae Progymnasmata*, III), 197, ll. 1-10: «Promittit postmodum Instrumentum quoddam nouum, quo Parallaxes, aliaque eiuscemodi inaudita facillime perscrutari liceat. At qualenam futurum sit hoc Instrumentum, cuius solius adminiculo, Parallaxes absque demonstratione & supputatione ulla demetiendae veniant, equidem non video: & vix eiusmodi quippiam, ulla inuentionis aut artificij subtilitate dari posse, re omni satis circumspecte deliberata, affirmare non dubito. Sin id praestare Diggeseus, vel quispiam alius confidit, agedum proferant eiuscemodi organum in medium, & spectentur agendo. Hic enim non saltem dicere, sed potius facere operae pretium est».

31 *Ibid.*, p. 197, ll. 28-34: «Nos sane sive hyeme, sive aestate, Poli Altitudinem et Fixarum quarumvis distantiam, per nostra Instrumenta... scrutati sumus, nullam prorsus in his diversitatem, crebra et diligenti adhibita exploratione, adinvenimus». NB «nullam *prorsus* diversitatem» (see *infra*, n. 59).

32 Kepler to Herwart von Hohenburg, May 1603 (n° 256, KGW, 14, p. 395, ll. 2-7): «[...] immensitas unius Sphaerae fixarum. Solum hoc Tycho mihi solebat opponere. Negabat exempla similia in natura, ubi tam ingens et frustraneum vacuum uti digitus instar montis ad reliquum corpus hominis. Ego contra proportionem ita requiram: sicut diameter motoris (Solis) ad diametrum mobilium (Saturni), ita haec ad diametrum loci in quo motus, scilicet Sphaerae fixarum».

33 *Ibid.*, ll. 81-83: «Non peccat qui hominem comparat cum Elephante, sed si quis hominem comparet cum motibus stellarum, genera transcendit et ineptit».

not the immense space but the comparison itself. One reason why we may assume that Kepler had still not heard of this in 1597 is that he does not mention it at all, although on the other hand he comes back to the point in *Epitome IV* (1620), once he had gained access to Tycho's *Thesaurus observationum*. Kepler reminds the reader that Tycho attempted to discover an annual parallax during the years 1586-1588 through accurate observations of the stars around the Pole (as, for instance, the three stars forming the tail of the Great Bear).[34] However, the conclusion drawn by Kepler was crisp: Brahe's *instrumentarium* was not sufficient, that is, not at all suited to the purpose so that, despite Brahe's authoritative reputation, the question remained undecided and open.

Let us return now to the letter to Herwart, written after Kepler tried to measure the seasonal variation in the altitude of the Pole. The failure of this experiment did not lead him to skepticism; instead it strengthened his philosophical and theological convictions. We find him paraphrasing Cicero: «What could seem big to those who...?»[35] However, the main context of the argumentation refers to the theology of the *immensitas dei*, which was directly and deeply influenced by Mästlin and Rothmann. Since the 1580s, both of them had frequently rested their argument on the *infinita potentia dei*, showing that nothing is impossible to Him and that, on the contrary, the Copernican distances and the magnitude of the world do not confuse the human imagination but rather reveal the *immensitas dei*. Mästlin claims that human imagination cannot be the measure of what is possible, whether logically or physically. Asking therefore, «ad quas leges [Ingoli] examinet opera manuum Dei, ut ea *improportionata* dicat», Kepler repeats the same locus *de immensitate potentiae divinae* that Rothmann used against Tycho («Whatever the size and the magnitude of the universe you assume, they won't have any proportion to the Creator's infinity»[36]), and that can be found in Mästlin's writings: one cannot deny that immensity of the world, unless by measuring the *opera dei* according to the weakness of our senses.[37]

---

34 KGW, 7 = TBOO 11 (*Thesaurus observationum astronomicarum*), 108; 117. Galileo made the same attempt with the star Alpha Lyrae, as quoted in the *Dialogo*.

35 Editors of the KGW did not identify the source, i.e. *Tusculanae Disputationes*, IV, 37. Kepler does not quote *ad literam*, and forgets some important words: «Quid enim videatur ei magnum *in rebus humanis* (omissed by Kepler), cui aeternitas omnis totiusque mundi nota sit magnitudo?» (What could seem big *in human affairs* to whom knows the eternity and the magnitude of the whole world?).

36 Christoph Rothmann to Tycho Brahe, 18 April 1590, TBOO 6, p. 216, ll. 30-34 (quoted in DROC, vol. I, pp. 452-462: 458): «Et quantumcumque etiam Mundi Vastitatem & magnitudinem concedas, nullam tamen proportionem ad infinitudinem habebit».

37 Michael Mästlin (*Narratio Prima*, ed. 1621), quoted in KGW, 1, 438: «Quid, quaeso, aliud est, quam Omnipotentiam Creatoris impotentiae arguere, et Omnisapientiae legès praescribere?» (on the context, see *infra*, p. 207).

2. Kepler's astronomy and the rebuilt astrology

In terms of theology Kepler belonged to the school of Philipp Melanchthon; his teacher was Jacob Heerbrand, who had studied under the Lutheran reformer.[38] Like his mentor Michael Mästlin, Kepler's work in the field of astronomy cannot be separated from his theological motivations and aims.[39] Fundamentally, according to Mästlin and Kepler astronomy had the prominent as well as the ancillary function of a natural theology, as the 'father of German astronomy' Johann Stöffler defined it a few years after the excommunication of Luther.[40] Astronomy not only reveals the rationality of the laws of nature – which do not lead to a real knowledge of God, since pagan astronomers knew these laws but still ignored the one true God – but also divine Providence and, first and foremost, the Creation of the world. And contrary to the 'philosophers', who only reach the metaphysical concept of a first cause by natural means, true astronomy, grounded in biblical chronology, should be able to determine an absolute beginning of the World.[41] This is the main reason why according to the Melanchthonian school astronomy and chronology are intertwined,[42] and from a structural or architectonic point of view this reciprocal implication cannot be separated from the link between astronomy and astrology.[43]

Actually, knowledge of the absolute beginning of the world was, until Kepler, more a task for astrology; it is therefore not very surprising that Copernicus does not say a single word about the problem and that only his

---

38 See Charlotte Methuen, *Kepler's Tübingen. Stimulus to a Theological Mathematics*, Aldershot, Ashgate, 1998.

39 On the theologian defense of Copernicus and the so-called «new» hypotheses, see Kepler, *Contra Ursum* (cit. note 11), vol. I, p. 230, n. 2. (see also II/1, pp. 53-55).

40 *Ioannis Stoefleri iustingensis mathematici eruditissimi, facileque omnium principis, in Procli Diadochi, authoris gravissimi sphaeram mundi, omnibus numeris longe absolutissimus commentarius*, Tübingen, Morhart, 1534, fol. 2 B: «Et ideo astronomia haud indigne naturalis theologia nominatur. Quia sicut superior theologia ad Dei cognitionem per supernaturalem fidem inducit: sic ista, tanquam inferior ancilla, eidem subserviens, ad divinae cognitionis introductionem per naturalem rationem manuducit. Nam per effectus in cognitionem causae descendimus».

41 See also Kepler, *Epitome IV*, KGW, 7, p. 252: «... nec fieri posse, vt non à certo temporis initio creatus sit Mundus».

42 On this topic, see Mehl – Nicolas Roudet (eds.), *Le Temps des astronomes. L'astronomie et le décompte du temps de Pierre d'Ailly à Newton*, Paris, Les Belles Lettres, 2017.

43 A perfect illustration of this conception of astronomy and its task is to be found in Helisaeus Röslin, as several studies by Miguel Ángel Granada have shown. See for instance Granada, *Helisaeus Roeslin contre Raymarus Ursus après la publication du "De astronomicis Hypothesibus": les lettres à Herwart von Hohenburg de 1597*, in Caroline Noirot – Nuccio Ordine (eds.), *Omnia in uno. Hommage à Alain-Philippe Segonds*, Paris, Les Belles Lettres, 2012, pp. 425-451.

pupil Rheticus tries to reconstruct the history of the world from its very creation on the grounds of the Copernican hypothesis (the cycle of the anomaly of the precession).[44] If we take a closer look at the philosophical and theological purpose of *Mysterium Cosmographicum* – without passing too hastily to the Platonic theory of the five regular bodies, a hypothesis developed *ad hoc* for that purpose – we only need to read the opening pages. The contemplation of the magnitude of the universe, which cannot be compared to anything else (insofar as «Nihil est hoc vniuerso mundo maius neque amplius»), is the only way «to honor God, to venerate him, to admire him [*Deum vero cultu celebrare, venerari, admirari*]».[45] In the mind of the theologian, this statement cannot only refer to the space separating Saturn and the fixed stars. Nonetheless it concerns the astrology of the "great conjunctions" and the necessity claimed by astrologists of reaching a limit or a definite term in time. In other words, Kepler understands the Copernican axiom of the world's incomprehensibility («mundus, cujus finis ignoratur, scirique nequit»[46]) in terms of the temporal aperture of physical reality, regarding which human understanding and science have no grasp. The *Mysterium Cosmographicum* should also be read in the light of its conclusion (ch. XXIII), regarding the impossibility of determining a «finis» of celestial motions: they would never come to a point that could be regarded as a *finis motus*. Here Kepler quotes the final words of *De Rev.* I, 10: «Tanta nimirum est divina haec Opt. Max. fabrica» – words that Ingoli and the 1620 decree censored and asked to be deleted,[47] although, as we have seen, the sentence «cujus finis ignoratur...» would remain intact.[48]

---

44 Peter Barker and Bernard R. Goldstein do not pay much attention to this problem in *Theological foundations of Kepler's astronomy*, «Osiris», Second Series, 16, 2001, pp. 88-113, although it is fundamental. As one of the first readers of the *Mysterium Cosmographicum*, Heliaseus Roeslin understood the importance of the final chapter (MC, ch. XXIII) about chronometry and Creation, but obviously he had more interest in the positive assertion according to which the world would have been created in the autumnal equinox, than he cared about the conjectural and hypothetical status of the reasoning. See Helisaeus Röslin to Herwart von Hohenburg, 20/30 July 1597, Letter n° 72, KGW, XIII, p. 129, ll. 45 sq.

45 Kepler, *Mysterium Cosmographicum*, *Epistola Dedicatoria*, KGW, 1, p. 5, l. 28. *The Secret of the universe* (cit. note 23), pp. 38-39; see also *Le Secret du Monde* (cit. note 17), p. 12.

46 See *supra*, n. 1.

47 *Mysterium Cosmographicum* (cit. note 45), ch. XXIII, p. 79, ll. 27-35.

48 DROC, vol. III, p. 151, n. 62; p. 686, n. 11. The editors note that Ingoli adds the word "Deus" in the sentence to be deleted («Tanta nimirum est divina haec *Dei* Optimi Maximi fabrica»). In this sense, the quotation of this sentence in the *editio altera* of the *Mysterium Cosmographicum* (1621: KGW, 8, p. 125, l. 32) is clearly to be understood as a defiance of the Roman interdiction one year before. We have to keep in mind Descartes' *Principia Philosophiae*: «Opera Dei nimis ampla cogitari non posse» (AT VIII-1, 80).

This conclusion to *Mysterium Cosmographicum* is not of secondary importance. It leads *a posteriori* to the underlining of the fact that the very starting point for Kepler in 1595 – the schema that would be *used* to figure the distances between Jupiter and Saturn – originally did not involve any sense of spatial reality, nor any relationship to the astronomical measure of celestial distances. The inner circle created by the intersections of the lines figuring the motion of the great conjunctions on the Zodiac only appears if we want to schematize the infinity of time, as Nicole Oresme did in order to calculate the incommensurability of the celestial motions themselves; here we find a sort of geometric evidence that the intersections of the lines forming the inner circle will never cross at the same point, so that the perfection of that circle is the figuration of an infinite *time*.[49]

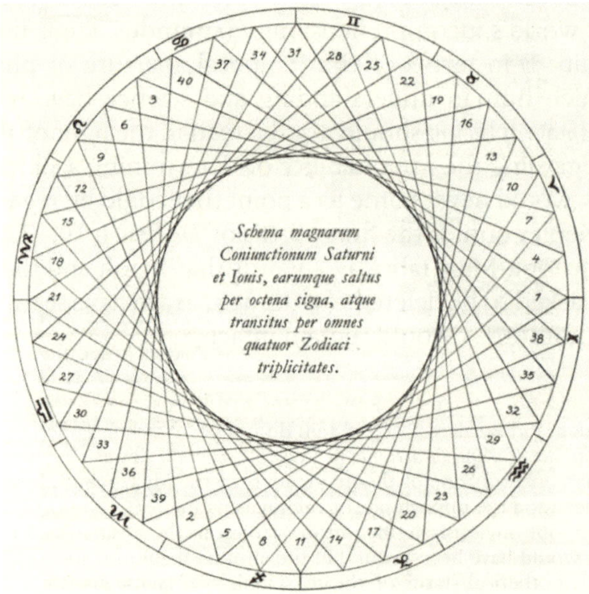

Kepler's schema representing the great conjunctions of Saturn and Jupiter (1596).

[49] On this matter, see Edward Grant, *Nicole Oresme and the kinematics of circular motion. Tractatus de commensurabilitate vel incommensurabilitate motuum celi*, Madison, Milwaukee, London, The University of Wisconsin Press, 1971. Grant reminds Duns Scot, *ibid.*, p. 122, n. 98: «prima opinio, quod omnia scilicet redeant eadem numero secundum circulum non est tenenda, quia est haeretica».

This does not at all mean that *Mysterium Cosmographicum* has an esoteric background or purpose. On the contrary, the aim of Kepler was to refute the vain attempts of 'heretic' astrologers to proclaim astronomical predictions of the end of the world – including in its 'orthodox' Melanchthonian version (Rheticus).[50] With the theory of regular bodies, Kepler found an original way to kill two birds with one stone: demonstrating the Copernican hypotheses and refuting the sheer lunacies of the astrologists. These two aims had to meet, leading Kepler to an idealistic turn – or, at least, a phenomenalist one – in the field of astrology. Denying the aspects any reality, Kepler claims that aspects do not exist *formaliter*, i.e. effectively, in the celestial bodies but in the *perception* of them. Hence the importance of his naturalized psychology, extended to every living being.[51] Thanks to these new *fundamenta*, Kepler could reject Ingoli's objection as inept, and conclude that the Copernican magnitude of the world would rule out any possibility of an *influxus astrorum*, because distances have nothing to do with the *operatio* and *effectus astrorum* (which is not real, but rather due to illumination). However, he did not even try, for he surely knew how severely Ingoli dealt with the supposition of an *anima Terrae*.[52]

3. The book nobody read: the uncensored edition of *De Revolutionibus*

As Gingerich and Voelkel had shown, Kepler and Mästlin's discussion of Tycho's argument about the diurnal parallax of Mars continued until

---

50 See Paola Zambelli (ed.), *"Astrologi hallucinati": Stars and the End of the World in Luther's Time*, Berlin and New York, Walter de Gruyter, 1986. For the whole topic see also Robert S. Westman, *The Copernican Question. Prognostication, Skepticism, and Celestial order*, Berkeley and Los Angeles, University of California Press, 2011. One can still read the suggestive essay by Aby Warburg, *La divination païenne et antique dans les écrits et les images à l'époque de Luther*, 1920, in Warburg, *Essais Florentins*, Fr. trans. by S. Müller, Paris, Hazan, 2015, pp. 245-294.

51 See Boner, *Kepler's Imprecise Astrology*, in Mehl, *Kepler: La physique céleste. Autour de l'Astronomia Nova (1609)*, Paris, Les Belles Lettres, 2011, pp. 115-132, and Id., *Kepler's cosmological Synthesis. Astrology, Mechanism and the Soul*, Leiden, Boston, Brill, 2013, chaps. 1 and 2.

52 See Ingoli, *Replic.* (cit. note 6), p. 201: «Quod vero Terra motricem habeat animam, absurdissimum esse...». Ingoli presents himself as a partisan of Christian astrology that Copernicans could not support, in spite of all their attempts (since the distance of the stars would destroy the possibility of their action on the sublunary world). See *ibid.*, p. 184: «Nam certum est ex literis sacris Deuter. 4 Dominum Deum Solem, Lunam et astra fecisse ut essent in ministerium cunctis gentibus quae sub coelo sunt»; the proportion of the magnitudes and the distances is necessary to guarantee the possibility of that *ministerium*: «...inter agentia et passa proportionem esse debere, et in magnitudine et in distantia...» (*ibid.*). On Ingoli and his project to write a defense of astrology, and, so to speak, to christianize the Ptolemaic *Tetrabiblos*, see Bucciantini, *Contro Galileo* (cit. note 4), pp. 142-144, and Tommaso Campanella, *Apologia pro Galileo*, ed. by Lerner, Paris, Les Belles Lettres, 2001, p. xxxv.

1621 and the publication of the second edition of *Mysterium Cosmographicum* and *Narratio Prima*, where Mästlin told the end of the story. But this was not only the story of the diurnal parallax of Mars. Let us rather say that this story was part of a larger drama, in which the problem of the annual parallax was equally important, as we can see by comparing the two additions from Mästlin's hand in this new edition of the *Narratio Prima*.[53] Within the same period, Kepler published the second part of *Epitome astronomiae Copernicanae* (IV-VII), where he reveals that Tycho unsuccessfully tried to discover an annual parallax in 1586-1587.

> Observavit ille altitudinem maximam stellae polaris [...] anno 1586. in media nocte post aequinoctium autumnale, fuitque gr. 58. m. 51. Eandem observavit etiam circa solstitium hiemale 26. Decembris vesperi hora circiter sexta, invenitque rursum 58. 51. Itaque differentia non fuit ulla [...] Ergò diameter ista orbis in quo tellus circumfertur, per instrumenta Braheana non est sensibilis.[54]

Here Kepler has changed his mind, or at least his strategy; he is no longer trying to refute Tycho, but to enroll Tychonians by showing that Copernican distances, whatever they may be, do *not* conflict with Tycho's hypotheses. This means that the immensity of the universe can be accommodated to Tychonian hypotheses, so long as Tycho's observations do not disprove or prove anything, except that his instruments are not accurate enough to prove anything! At the same time, Kepler transforms that non-provable immensity into a principle of astronomy. In the beginning of the fourth part, he claims something new – and probably unbearably vexatious for Tychonians: i/ as the fourth hypothesis or principle of astronomy: the proportion of the great orb to the diameter of the fixed stars is «undetectable, where from the fixed stars appears immense»; ii/ astronomers can omit one or the other principle of the science; iii/ Tycho fully *omits* this (fourth) hypothesis or principle.[55]

The reality, of course, is very different, as we can see through Ingoli's vehement protest and claim that the insensibility of the Earth's deferent with respect to the eighth sphere cannot in any way be considered as a principle, because of its incomprehensibility and because of the lack of the

---

53 KGW, 1, pp. 439-440 (on Mars acronychus and Tycho's Copernican campaign), and *ibid.*, pp. 436-439 (against Tycho's own judgment on the distance and the magnitudes of the stars). French translation of these two important texts in *Georg Joachim Rhetici Narratio Prima*, Henri Hugonnard-Roche – J.-P. Verdet *et al.* (eds.), *Studia Copernicana*, XX, Wroclaw, Ossolineum, 1982, pp. 165-168 (n. 120), 186-187 (n. 227).

54 Kepler, *Epitome IV* (cit. note 41), p. 286, ll. 26-37.

55 *Ibid.*, p. 257, ll. 8-9; 25-26.

praecognita that could allow this insensibility to be understood or perceived as real.[56]

These clever tactics, however, are hardly compatible with Mästlin's desire to succeed in demonstrating the falsity of geo-heliocentrism, as well as his dangerous tendency to make the cosmological conflict a confessional one. This tendency had been obvious since the beginning of the 1580s and the debate on calendar reform in which Mästlin was involved, although he did not succeed in drawing in[57] Kepler. It was also obvious in the second Preface that Mästlin wrote for the 1621 edition of *Mysterium Cosmographicum*, which remained unpublished.[58] Kepler and Mästlin had also planned a new edition of *De Revolutionibus*. It was supposed to be published in Basel, but the reasons why this never happened remain unknown. As the main part of the correspondence between Mästlin and Kepler in these years is lost, we are compelled to conjecture. According to us, there is a high level of probability that Mästlin wrote this text as *one* Preface to the *De Revolutionibus*. Moreover, the very long addition published in the *Narratio* (1621, KGW, 1, 436-439) can be read as a discussion with Tychonians, if not with Ingoli himself, on the Copernican significance of the undetectable parallax. Examining every Copernican *locus* in *De Revolutionibus*, Mästlin gives a new demonstration that, according to Copernicus *himself*, the parallax is not non-existent in itself, but only to our *senses*, which are not, in any case, a criterion for reality.[59] From this point of view the statement in *Lettera a Ingoli* regarding the weakness and inadequacy of human cognition (insofar as it is measured by the senses) can be regarded as strictly Copernican, and

---

56 Ingoli, *Replicationes* (cit. note 6), pp. 193-194: «Deinde hoc principium insensibilitatis deferentis Terram respectu octavi orbis ad defensionem motus Telluris ab absurdis assumptam incomprehensibile videtur esse [...] nam intellectus humanus ex praecognitis ad incognita obiecta procedit [...] sed in nostro casu non adsunt ista praecognita pro investigatione huiusmodi insensibilitatis, igitur intellectus humanus non potest // (104r) hanc insensibilitatem comprehendere».

57 On Mästlin and the calendar Reform, see Jürgen Hamel, *Landgraf Wilhelm IV. und Michael Mästlin in der Polemik um die Gregorianische Kalenderreform*, in Mehl – Roudet (eds.), *Le Temps des astronomes* (cit. note 42), pp. 231-268.

58 DROC III, pp. 654-663. Notes pp. 691-696. The editors reject the hypothesis of Frisch that this text had been written for a new edition of the *De Revolutionibus* (on this project, see DROC I, pp. 612-617).

59 Mästlin, *Narratio prima* (cit. note 37): «Therefore there is no doubt that Copernicus does not exclude *absolutely* any parallax (of the fixed stars) but only regarding to the judgment of our eyes» (KGW, 1, p. 438: «Ergo Copernicum ad oculorum iudicium, non ad universalem omnis parallaxeos exclusionem respexisse, extra dubium est...»). This conclusion and the whole demonstration is a refutation of a single word quoted in the *Progymnasmata* (TBOO 2, p. 430, l. 23): according to Tycho, if Copernicus was right, the magnitude of the great orb would *totally* disappear («*prorsus* evanesceret»).

comparable to the positions of Mästlin, although Galileo obviously omits any theological statement, like those of the 'non-Catholic' Copernicans.

As we know, this project of re-editing Copernicus did not succeed. The editor in Basel who was originally supposed to publish the work may have been responsible for this. In any case we would suggest that in the end Kepler and Mästlin could not fully agree on a strategy with regard to the Roman decree and the interdiction. Kepler became, so to speak, *politically* more efficient than he had ever been before. *Epitome* radically changed the problem. Immensity was up to then a problem or paradox that Copernicans had *volens nolens* to deal with. But after the publication of *Epitome* it became a principle of astronomy and, as such, *notum per se*, whereas the 'old' Mästlin still fought using theological arguments to prove that immensity or whatever may appear *ignotum* and impossible to *us* was, nonetheless, not impossible for *God*. However, he did not recognize the evidence of this. This would be, to conclude, a huge paradox: although Mästlin was absolutely opposed to the Roman decree, he still thought within the same theological and philosophical framework as Ingoli did, while Kepler was a length ahead of both of them.

4. Conclusion

Discovering an annual stellar parallax of the Earth would have provided a *demonstratio quia* of the Copernican hypotheses. Tycho had been of this since at least 1586 and attempted to do so within the frame of his own Copernican campaign. It was still a challenge for the young Kepler, who shared a similar project with Galileo from the very beginning of their epistolary communication. It would have been the fruit of their cooperation. If Galileo did not answer Kepler's letter, it may be due to the fact that their experiment had failed and the measurement of a parallax was still to be achieved. Because of this silence, Kepler does not mention Galileo in ch. XV of *De Stella Nova* (1606) where, as shown by Boner, he deals with the parallax and pushes the case for Copernicus. The discoveries reported in *Sidereus Nuncius* in 1610 forced astronomers to reconsider their assumptions, and we may presume that the observations of Mizar made by Benedetto Castelli in 1613 had initially been planed in order to reveal the annual parallax – but in vain.[60]

---

60 See Leos Ondra, *A New View of Mizar*, «Sky and Telescope», July 2004, pp. 72-75 (http://leo.astro-nomy.cz/mizar/article.htm). Christopher M. Graney, «Galileo's Double Star: The

But the manner in which Galileo and Kepler reacted against Ingoli reveals a striking opposition between them. It is certain that this opposition can be understood through the obvious discrepancy between their cosmological views concerning the existence of the *orbis stellarum fixarum* and the centrality of the Sun. But more deeply and fundamentally, it stemmed from the purely philosophical problem of the (in)commensurability of the celestial motions, and consequently the different ways in which they understood number and *time*. Indeed, Galileo's claim in his *Lettera* that the problem of cosmological infinity lay beyond the reach of human understanding clearly had an Oresmian background, whence Galileo's disagreement with Kepler's harmonies. But this purely philosophical approach requires further investigation.

Experiment That "Proved" the Earth Did Not Move» (online: arxiv.org/pdf/physics/0606255), and Harald Siebert, *The Early Search for Stellar Parallax: Galileo, Castelli, and Ramponi*, «Journal for the History of Astronomy», 36, 2005, pp. 251-271.

Rienk Vermij

COPERNICANISM AND THE BIBLE IN THE DUTCH REPUBLIC AROUND 1616: A NON-DEBATE

The fact that the Amsterdam publisher Willem Blaeu published a third edition of Copernicus' *De revolutionibus* in 1617, less than a year after the condemnation of the heliocentric theory by the Roman See, appears to have been purely coincidental. The Dutch Republic was officially a Protestant country. Catholics still formed a sizeable part of the population, but did not have a say in public affairs. Of course, even Protestants could not deny that the pope was an important player in European politics. The Dutch followed the Venetian interdict crisis closely. But when it came to points of doctrine, scholars and authorities did not take the slightest interest in anything the pope decreed. The editor of this new edition, the professor of astronomy Nicolaus Mulerius, did add a note to Copernicus' dedication to the pope, but it concerned historical details only: he explained that Pope Paul III was the great-grandfather of Philip II of Spain's famous general Alexander Farnese, a well-known figure in the history of the Netherlands.[1] The irony that *De revolutionibus* had been dedicated to a pope apparently escaped him. As far as we can tell, the decree of 1616 went completely unnoticed in the Dutch Republic.

That is not to say that the question on which the decree focused, whether the motion of the earth is in agreement with Holy Scripture, was unknown or deemed uninteresting. Scholarly works in which this issue was discussed, like those by Kepler or Tycho, were readily available. Still, the question initially does not appear to have drawn much attention. Most scholars were not very outspoken about the system of the world any-

---

The author thanks Dr. Kathleen Crowther for linguistic improvements.

1 Nicolaus Copernicus, *Astronomia instaurata, libris sex comprehensa, qui de revolutionibus orbium coelestium inscribuntur*, ed. by Nicolaus Mulerius, Amsterdam, Blaeu, 1617. See on this edition Djoeke van Netten, *Koopman in Kennis. De uitgever Willem Jansz Blaeu in de geleerde wereld (1571-1638)*, Zutphen, Walburg Pers, 2014.

way. They kept aloof, like Willebrord Snellius, or they simply ignored the new ideas, like Rudolf Snellius or Adriaan Metius. They therefore had little reason to discuss the biblical arguments. But even those scholars who had strong opinions on the motion or rest of the earth referred to the biblical arguments casually at most. Before the second quarter of the seventeenth century, the debate on Copernicanism and the Bible was hardly taken up by the Dutch. This makes an interesting case for international comparison. Before passing in review the few known contributions to the debate, I will briefly discuss some authors who in spite of their outspoken views on the constitution of the universe did not refer to the Bible at all.

1. The first Copernicans

The first Dutch person who openly defended the motion of the earth was the mathematician Simon Stevin. He gave an elaborate defence of the diurnal and annual motion of the earth as really existing in nature in his *Hemellloop* ('Course of the heavens'), a part of *Wiskonstige gedachtenissen* ('Mathematical memoirs'), published in 1608. Although he discussed at length the arguments for and against the heliocentric system, he was completely silent on the arguments from Holy Scripture. He must have been aware of the debate as he made extensive use of William Gilbert's book on the magnet. As Westman pointed out, this book had a preface by Edward Wright which argued that the motion of the earth was compatible with scripture. Stevin appears to have been simply uninterested in the issue. Actually, there is not a single reference in any of his works to the Bible.[2]

Equally silent was the philosopher Isaac Beeckman. From his private notebook, the *Journael*, it appears that by 1616, he felt that a heliocentric constitution of the universe was the most probable. In later notes, he consistently presupposed a heliocentric world. Beeckman (unlike Stevin) was a pious Christian and an active member of the Reformed Church, so the problem of how the Copernican system agreed with the Bible must have occurred to him. His *Journael*, however, is devoted to scientific speculations and theology is consistently left out. He did not explain how his scientific ideas could be harmonized with the Bible. Moreover, the *Journael* had a

---

[2] Simon Stevin, *Wisconstige gedachtenissen*, Leiden, Jan Bouwensz, 1605-1608. See Rienk Vermij, *The Calvinist Copernicans. The reception of the new astronomy in the Dutch Republic*, Amsterdam, Edita KNAW, 2002, pp. 58-68; Robert S. Westman, *The Copernican Question. Prognostication, Skepticism, and Celestial order*, Berkeley and Los Angeles, University of California Press, 2011, pp. 426-429.

purely private character. Beeckman never published his ideas on the motion of the earth, or any other subject, and therefore did not need to justify them publicly.[3]

2. Anti-Copernicans

Proponents of the traditional world-system also were not very vociferous about the religious arguments. However, some of them did refer to the Bible. An early example is the Leiden professor Paullus Merula. In his *Cosmographia generalis*, published in 1605, he stated that the earth was the center of the universe: «But the authority of Holy Writ and of the philosophers, supported by the most powerful reasons I indicated, is not capable to defend the truth in this matter in such a way, that meanwhile there have not been people in the past, nor are lacking nowadays, who with great daring and doing violence to the nature of things, come to attack it».[4] Merula argued for the central position and the immobility of the earth using physical reasons and philosophical authorities, but also a long list of biblical sentences.

The most vehement denunciation of the Copernican system was ventured by Ubbo Emmius, the professor of history at the university of Groningen. However, it should be noted that his reaction was not in any published work, but in some private letters to his learned friends. His outburst was occasioned by the publication of Stevin's book on the heavens, mentioned above. In the most extensive of these letters, to the theologian Sibrandus Lubbertus, he fulminated against the «absurd» opinions that Stevin defended in his book: «that the earth moves and emits light, and would be an eighth planet. That it displays the same phenomena as the moon: that there is full and new earth like there is full and new moon, and the other phases, and also that the moon is inhabited, and similar things like that. Good God, how far will this impiety go? What better could one do to bring down all of our doctrine and religion? For these things demolish the very foundations. Were they true, as our author I learn argues vigorously, then Moses would be a liar and the whole of Scripture untrue...».[5]

---

[3] On Beeckman, see Klaas van Berkel, *Isaac Beeckman on matter and motion. Mechanical philosophy in the making*, Baltimore, John Hopkins University Press, 2013. See p. 98 for his Copernicanism and pp. 42-45 and 140 for his religion. See also Vermij, *Calvinist Copernicans* (cit. note 2), pp. 113-117.

[4] Paullus Merula, *Cosmographia generalis libri tres*, Leiden, Plantijn, 1605, p. 211.

[5] Emmius to Lubbertus, 1608 Sept.ember 29, in Hajo Brugmans (ed.), *Briefwechsel des*

Lubbertus, a staunch defender of orthodoxy, appears to have tried (in a letter that has not been preserved) to calm his friend down by putting forward the familiar argument that Copernicus' theories were useful as hypotheses to explain the phenomena, for in a later letter Emmius freely admitted this. However, he felt that arguing these theories to be true was a different thing. «For without doubt this is accusing the Holy Ghost, as it speaks through Scripture, of lying».[6]

It deserves notice that the motion of the earth takes only a small place in Emmius' attack. His main problem was with the earth-like nature of the moon. Stevin had indeed made a point of the basic similarity of the moon and the earth as celestial bodies, but had not suggested that the moon was inhabited. Emmius' mistake is somewhat understandable from the fact that, as he conceded, he himself had not seen Stevin's work. He relied on the report of someone else. The fact that he brought up the inhabitants of the moon indicates that this idea was regarded as a logical consequence of Stevin's ideas, and a very bad consequence at that. Most likely, Emmius got his information from his friend Nicolaus Mulerius, the professor of astronomy and medicine at Groningen, with whom we will deal next and who deserves a somewhat more elaborate discussion.

Mulerius was a prominent scholar who published several works on astronomy and chronology. Unlike his colleagues at other universities, he felt compelled to take a clear position on the system of the world. In 1611, he published his Frisian tables. This is a work on mathematical astronomy and as such in principle not concerned with the constitution of the universe. Yet, in its *isagoge* Mulerius did include a passage on that issue. Having stated that the earth is the center of the universe, he admitted that the ancient Pythagoreans had defended a heliocentric system, and that their theory had been revived by some eminent mathematicians of his own time. He did not want to discuss this question at this point, «testifying only this, that so great with us is the authority of Holy Writ, and to such a degree is our mind moved by reverence for it, that against its express statement we do not dare to descend to the opinion of the Pythagoreans. So, following the principles of Ptolemy and the best philosophers, which are in agreement with the sacred words, let us discard this hypothesis».[7]

Ubbo Emmius, II, *1608 bis 1625*, The Hague, Nijhoff, 1923, pp. 51-52. Other persons to whom Emmius wrote on the matter were Witten (beginning of October) and Dohna (1609 January 7 and 30): see *ibid.*, pp. 54, 60-62.

[6] Emmius to Lubbertus, 1608 October 20, in *ibid.*, p. 55.

[7] NICOLAUS MULERIUS, *Tabulae Frisicae lunae-solares quadruplices*, Alkmaar, Jacob Meester, 1611, p. 318.

Mulerius' religious scruples are evident, but he does not explain what «express statement» in the Bible he actually had in mind.[8] His position became somewhat clearer when, in 1616, he published his textbook on astronomy. Interestingly, in the preface to this work, he appeared willing to accept the diurnal rotation of the earth as a hypothesis that «does not seem to contain anything repugnant to the authority of the Holy Scriptures».[9] However, Mulerius did have very serious problems with the earth's annual motion, and thus with the heliocentric universe. In the very beginning of the preface, he gave an explanation of the debate between heliocentrists and geocentrists. As he said, having examined the arguments of both parties over and over again for twenty-five years, he could not find a valid reason that would compel him to reject an opinion (the geocentric one) that had been approved by so many great men. He fully conceded that the common astronomical hypotheses contained many absurdities, but in his view the system of Copernicus was marred by other absurdities. In particular, Mulerius objected to the immense size that Copernicus required for the cosmos, «and that in that way there would be more suns in the universe. That this merits to be called absurd and against Christian piety, has to be taken into account by those who are given to defend the opinion of Copernicus».[10]

The enormous size of the universe in the Copernican system appears to have seriously bothered Mulerius, for the theme turns up again in later work. There are several instances in his annotations to his 1617 edition of Copernicus' work. None of these states explicitly that this idea is contrary to religion, but it seems probable that Mulerius' objections were religiously inspired. So, in his annotation to book I chapter VI, «On the immensity of the heavens compared to the size of the earth», Mulerius admitted that the idea that the earth is a point compared to the heavens was nothing new and was accepted by the ancients. «But when he [Copernicus] adds that that immense space between the earth and the sun is made in a similar way, he says something new that will not well find belief among the common mathematicians».[11]

In an annotation to book II chapter I, Mulerius again rejected the idea that there could be more suns in the universe: «From this it would seem to

---

8 On pp. 321-322, he is quoting Psalm 19: 5 and 1, and Psalm 8: 4, but in the context of wonder about the immense size of the universe and the rapidity of its movements.

9 Nicolaus Mulerius, *Institutionum astronomicarum libri duo*, Groningen, Johannes Sassius, 1616, preface.

10 *Ibid.*

11 Mulerius in Copernicus, *Astronomia instaurata* (cit. note 1), p. 11.

follow that there are more suns in the world, with each of them pouring their light into the part of the universe that is close to them. This however diverts from Copernicus' idea, as he said above "everything is illuminated by the sun". But I do not see nor grasp how someone could maintain this opinion with this paradox».[12] Still another note is on book III chapter XV. After having again pointed out that Copernicus' hypotheses cannot be sustained without assuming that the distance between earth and sun is nothing compared to the distance to the fixed stars, he added:[13]

> The most ingenious Copernicus was well aware that this opinion would seem an absurdity, but he excused himself saying: "I think it is much easier to concede this than to distract the understanding with an almost infinite multiplicity of spheres, as those who have kept the Earth in the middle of the universe have been compelled to do". From which can be seen the weakness of the human mind, which is incapable of explaining the admirable work of God, or the machine of the world, without recourse to the absurd.

The imperfection of human knowledge and our incapacity to understand God's handiwork was Mulerius' ultimate argument in his rejection of the heliocentric arguments. «Why are we staying with hypotheses, that are mere human figments, by which we try in vain to solve analytically the system of the world?» he wrote in 1616.[14] In his 1617 edition of Copernicus, he also added a note to Osiander's preface. Referring to Ptolemy book 13 chapter 2, Mulerius agreed that one should not imagine that the heavens really are as the mathematicians imagine. Human things (the models we frame) are not comparable to Divine things (the heavens themselves).[15]

All in all, Mulerius' criticism was well in line with Ubbo Emmius' earlier outburst. Although both did have problems with the (annual) motion of the earth, the main bone of contention was the cosmology. Emmius rejected the idea that the moon is like the earth, with the consequence that the moon could be inhabited. Mulerius rejected a plurality of suns. The basic problem seems to be their fear that acceptance of a heliocentric universe lead to the belief in a plurality of worlds. The real butt of their anger seems to be Bruno, not Copernicus. Still, although Mulerius' reasons

---

12 *Ibid.*, p. 64. The quote (not very accurate) is from *De revolutionibus* book I chapter 10.

13 *Ibid.*, 203. The quote is again from *De revolutionibus*, book I chapter 10. I follow the translation by A.M. Duncan: Nicolaus Copernicus, *On the revolutions of the heavenly spheres*, New York, Barnes & Noble, 1976, p. 49.

14 Mulerius, *Institutionum* (cit. note 9), preface.

15 Mulerius in Copernicus, *Astronomia instaurata* (cit. note 1), *Ad lectorem*.

to denounce such a world view as absurd appear to have been at least partly religious, in his refutations he only rarely brings the Bible into play.

3. Lansbergen

Things are somewhat different with Philips Lansbergen, along with Stevin the most important Dutch advocate of the Copernican system at this time. Lansbergen was a Calvinist minister until he was deposed after a conflict with the authorities. Whereas Stevin and Beeckman explained the natural world by physical or mechanical principles, and thereby at least implicitly created a distance between God and nature, Lansbergen's universe was heavily tinged by religious and neo-Platonist ideas. Still, even he initially spent little time discussing the biblical arguments about the cosmos.[16]

The first indication of Lansbergen's Copernicanism is in his *Cyclometria nova* of 1616. The title page of this work contained a heliocentric diagram, with an inscription, in Greek: «God always cyclometrizes», an obvious allusion not only to Plato's famous dictum «God always geometrizes», but also to the Copernican axiom that all celestial motions happen in regular circles or combinations of these.[17] In 1619, he then published *Progymnasmatum astronomiae restitutae*, the first volume of a projected multi-volume work wherein he intended to restore the whole of astronomy. The book is rather technical; most of it consists of tables. However, the last three chapters dealt with the motion of the earth. Chapter 22 argued that the daily rotation that we see in the heavens is actually in the earth; chapter 23 argued that the annual motion of the sun was equally in the earth; and chapter 24 dealt with these two motions together.[18] In these chapters, Lansbergen listed his arguments in favour of the motion of the earth and refuted counter arguments, most of the latter taken from Ptolemy and Tycho. Here, he also referred to Tycho's arguments taken from the Bible. This discussion took up only one single paragraph, at the very end of chapter 22. I will quote it in full:[19]

---

16 On Lansbergen, see Vermij, *Calvinist Copernicans* (cit. note 2), pp. 73-97.

17 Philips Lansbergen, *Cyclometriae novae libri duo*, Middelburg, Richard Schilders, 1616, titlepage.

18 Id., *Progymnasmatum astronomiae restitutae liber I. De motu solis*, Middelburg, Richard Schilders, 1619, pp. 102-106, 107-111, and 112-115 respectively (I used the re-edition of 1628, but this is identical to that of 1619).

19 *Ibid.*, p. 106.

Nor does, as Tycho and others feel, the authority of Holy Scripture form a problem, which attributes the motion to the sun. For Scripture also attributes ascent, descent, and other changes of place to God, whereas it is obvious that He is without those. Just as it is argued, from what Scripture teaches about God being omnipresent and that heaven and the earth are full of him, that those sayings are figurative; so too we have to acknowledge that what is read about the motion of the sun, is said in a human way so that it can be understood by humans. For nature and reason teach us that the earth is moving. If someone is not satisfied with this answer, he has also to state against geometrical truth that [in a circle] the proportion of the circumference to the diameter is 3:1, what Scripture affirms at other places, to wit, 1 King 7: 23 and 2 Chronicles 4: 2. Whereas one should rather feel that the Holy Ghost has spoken not according to strict accuracy, but according to the popular and rude opinion that at that time in the country of Israel was held for true. But I do not want to make a category mistake by mixing the sacred with the profane. So I will say no more.

The biblical reference here is to King Solomon's molten sea, «round, ten cubits from brim to brim [...]. A line of thirty cubits would encircle it completely.» In the dedication to his *Cyclometria* of 1616, Lansbergen had already highlighted this text as a testimony to the rude and imperfect knowledge of geometry in biblical times, especially of the proportion of the diameter to the circumference of a circle.[20]

It is interesting that Lansbergen, when refuting biblical objections, only paid attention to arguments against the diurnal rotation and did not speak of any biblical arguments against the annual motion. As we saw above, Mulerius was willing to accept the diurnal rotation, but on biblical grounds rejected the annual motion. Lansbergen's arguments apparently were a reaction to what he read in the literature by Tycho and others, rather than to his colleagues in the Netherlands. The reasons why Lansbergen and Mulerius felt that the rotation of the earth was acceptable were quite different, however. According to Mulerius, a rotating earth could rightly be called immobile, «as it does not change its place, and as its motion cannot be perceived by the senses».[21] In other words, he felt that it was in agreement with a literal reading of the text, whereas Lansbergen argued that the text should not be taken literally.

Lansbergen, writing in 1619, must have been aware of Mulerius' doubts about the annual motion, so one may well wonder why he only referred to biblical objections in his chapter on the daily rotation. As a matter of

---

20 Lansbergen, *Cyclometriae* (cit. note 17), dedication.

21 Mulerius, *Institutionum* (cit. note 9), preface.

fact, however, although in the chapter on the annual motion he did not refer to biblical sentences, he did close with an appeal to religion. Having concluded that the heliocentric system was genuine, he stated: «For what some contend, that the Copernican system is not true, because it supposes that the orbit of the earth is like a point if compared to the sphere of the fixed stars, is very frivolous. [...] This is our opinion: which does not at all derogate from the Divine Majesty and Its infinite power, but rather illustrates and exalts it, and teaches that, since this work of the Almighty God is so immense, we must stand in awe for his infinite power, and must humbly and piously worship and adore Him».[22] This looks very much like a direct reply to Mulerius' strictures, albeit in a way that could offend nobody.

Ten years after *Progymnasmatum*, Lansbergen wrote another book on the constitution of the universe, *Considerations on the diurnal and annual motion of the earth, as well as on the true image of the visible heaven*. Whereas *Progymnasmatum* was a scholarly Latin work directed at fellow astronomers, the new book was written in the vernacular and addressed a lay audience with the aim of convincing them of the heliocentric theories. This time, Lansbergen felt compelled or obligated to defend himself at greater length against biblical objections.

That he expected contradiction is already indicated by the title page, which quotes as an exhortation from the gospel according to John 7:24: «Do not judge by appearances, but judge with right judgment». In the preface, he referred to Aristarchus of Samos, whose opinions, he claimed, were banned by the Greeks for religious reasons, as they did not agree with the appearances. The same happened now in his own time: «For as soon as one starts to talk about the motion of the earth, and the immobility of the sun, one immediately gets this reply, to wit, that it is outrageous to say that the sun would remain at rest and that the earth would be moved, as Holy Writ clearly says that the earth stands fixed and that the sun runs as a hero in the heavens».[23] Lansbergen's words seem to indicate that he had met with resistance from people in his surroundings and was not just reacting to the works of Tycho or other learned authors. In the book itself, Lansbergen discussed the biblical argument at the same place as in 1619: at the end of the part on the diurnal rotation. Again, with respect to the annual mo-

---

22 Lansbergen, *Progymnasmatum* (cit. note 18), p. 112.

23 Id., *Bedenckingen op den dagelyckschen, ende iaerllijckschen loop van den Aerdt-kloot. Mitsgaders op de ware af-beeldinge des sienelijcken hemels; daerin de wonderbare wercken Godts worden ontdeckt*, Middelburg, Zacharias Roman, 1650, preface. The text of the 1650 edition shows no real differences with the text of 1629.

tion he did not discuss any biblical arguments. Of course, many lay people probably did not really distinguish between the two motions, let alone that they were aware of the consequences for the size of the universe.

After having refuted the reasons Ptolemy had advanced against the diurnal motion of the earth, Lansbergen stated that in principle, now he could move on to discussing the annual motion. But since «in the present day» people advanced still another reason against the diurnal motion, taken from Holy Scripture, «and that by these almost everybody is moved to believe that not the earth but the heaven turns around every day, and that the earth is at rest», he felt obliged to discuss this argument as well. Lansbergen then pointed out that disciplines like geometry or astronomy «are not built on the foundation of Holy Scripture, but have special foundations proper to these disciplines».[24] After all, Moses took his wisdom from the Egyptians and Daniel from the Chaldeans. Next, Lansbergen adduced various examples to show that mathematical or astronomical knowledge cannot be derived from the Bible. Apart from the molten sea of Solomon, he discussed the cases of Lactantius, who rejected the spherical shape of the earth, and of Augustine, who did not believe in the antipodes, both on biblical grounds. Finally, he discussed the various texts that were most commonly used to prove that the earth is at rest and argued that in these cases, the Bible speaks according to the appearances, not the truth of the matter.[25]

Lansbergen was clearly more concerned with the problem of biblical accommodation in 1629 than in 1619. Whether this means that the issue had actually grown more urgent, or that he had become more aware of it, or that he just gave it more attention because he was addressing a different audience, is hard to say. In any case, one is hard pressed to find the resistance Lansbergen referred to in any written sources of the period. There may have been concern, but Copernicanism was not the subject of any sustained attack.

4. The character of the debate

So, although the religious implications of the system of the world were generally (though not by all) considered important and people had differ-

---

[24] *Ibid.*, p. 18.

[25] *Ibid.*, pp. 17-22. The texts he mentions (apart from those on the molten sea) are Eccl. 1: 5, Psalm 19: 6-7, Matth. 5: 45, Joshua 10: 13, 2 Kings 20: 11, Psalm 24: 2, Psalm 104: 5. More broadly discussed are Psalm 19: 5-6 and Psalm 24: 2.

ent ideas about how to interpret the relevant biblical sentences, in the first three decades of the seventeenth century this did not lead to any heated controversy. To the contrary, the debate was rather muted. It is admittedly a somewhat dubious historical practice to ask why something did *not* happen, so the question of why there was not more debate seems anachronistic. Still, knowing that the issue gave rise to serious tensions elsewhere or at a later date, it may be illuminating to inquire why the situation was so different in the Netherlands in the early seventeenth century.

One reason for the lack of real controversy may have been that even where people disagreed, there was often agreement on more fundamental principles. Both Mulerius and Lansbergen saw the universe as an expression of God's power and felt that the Bible was a legitimate source of natural knowledge. True, Lansbergen claimed that astronomy and mathematics had their own principles, but the idea of the universe as sketched in *Considerations* draws on the Bible as well.

However, the consensus on the divine nature of the universe was not unanimous. Simon Stevin ignored it completely. Lansbergen wrote deprecatingly about Stevin's model of the universe that it was disordered and lacking in harmony and therefore could not be true.[26] It is certainly significant that, as far as we know, Stevin was the only one who for religious reasons drew a vehement attack from one of his colleagues. But it is also significant that this attack remained private.

The principal reason that the controversy remained smoldering under the surface rather than flaring up, was that it never became part of a wider theological debate. Not that such wider debates were lacking. In 1616, the Arminian controversy was raging in the Dutch Republic and brought the country to the brink of civil war – in 1618, it would be decided by military force. However, the motion of the earth played no role in this conflict whatsoever. Theologians left the subject alone.[27] The constitution of the universe remained the concern of mathematicians and other secular scholars. The members of the 'republic of letters' were used to ignoring or downplaying confessional and religious differences in their scholarly exchanges. The constitution of the universe was a controversial issue, but it

---

26 Lansbergen, *Bedenckingen* (cit. note 23), pp. 55-56.

27 At a later date, Lansbergen's Copernicanism by some was seen as connected to Arminianism (or Remonstrantism), as pointed out by Tabitta van Nouhuys, *Copernicanism, Jansenism, and Remonstrantism in the Seventeenth-century Netherlands*, ed. by John Brooke and Ian Maclean, *Heterodoxy in Early Modern Science and Religion*, Oxford, Oxford University Press, 2005, pp. 145-168: 158-160. Lansbergen's own views with regard to the Remonstrants remain unclear, *ibid.*, pp. 152-155.

was a legitimate subject of scholarly debate. Biblical arguments did play a role, but scholars, even if they disagreed, respected each other's positions. As Mulerius wrote, the scholarly debate between geocentrists and heliocentrists was «sharp and learned, but without hatred».[28]

This situation changed only in the second half of the century, when factional strife within the Dutch Reformed Church and tension between Church and state flared up again. This time, for some theologians the literal interpretation of the Bible became a touch-stone of orthodoxy. The motion (or rest) of the earth thereby became an important rallying point and an element in the self-definition of the confessional parties as they formed in the 1650s. Henceforth, Copernicanism was the subject of political controversy instead of purely scholarly discussion. The question of the motion of the earth was a sensitive one, but like a mass of tinder, it would flare up only when someone actually lit a spark.

---

28 Mulerius, *Institutionum* (cit. note 9), preface.

Steven Vanden Broecke

AN ASTROLOGER IN THE WORLD-SYSTEMS DEBATE. JEAN-BAPTISTE MORIN ON ASTROLOGY AND COPERNICANISM (1631-1634)

1. Introduction

For a long time, historians assumed that astrology and Copernicanism were theoretically and historically incompatible, and that the early modern rise of the latter automatically implied a downfall of astrology's credibility. Indeed, it was often held that heliocentric cosmologies tended to destabilize astrology "as its theories were based on a Ptolemaic finite universal order".[1] There is sufficient evidence to question this assumption.[2] The English Copernican astrologer Christopher Heydon (1561-1623), for instance, simply observed that:

> whether they be as Tycho would have it but one continued orb, or many, or whether as Copernicus saith, the Sun be the center of the world, and the earth be in the Sun's place between the sphere of Mars and Venus, the astrologer careth not.[3]

---

[1] Liana Saif, *Astrology: Homocentric Science in a Heliocentric Universe*, in Nicholas Campion – Dorian Gieseler Greenbaum (eds.), *Astrology in Time and Place: Cross-Cultural Questions in the History of Astrology*, Newcastle upon Tyne, Cambridge Scholars Publishing, 2015, pp. 159-172: 160. Similar claims can be found in: Bernard Capp, *Astrology and the Popular Press. English Almanacs, 1500-1800*, London & Boston, Faber and Faber, 1979, p. 278; Kocku von Stuckrad, *Geschichte der Astrologie. Von den Anfängen bis zur Gegenwart*, München, C.H. Beck, 2003, pp. 265-266.

[2] Historiographic disavowals of the negative impact of heliocentrism on astrology can be found in: Gérard Simon, *Kepler astronome astrologue*, Paris, Gallimard, 1979, pp. 93-96; Jim Tester, *A History of Western Astrology*, Woodbridge, The Boydell Press, 1987, pp. 214, 225; Nicholas Campion, *A History of Western Astrology*, vol. 2, *The Medieval and Modern Worlds*, London, Bloomsbury, 2009, p. 168.

[3] Christopher Heydon, *Defence of Judiciall Astrologie* (1603), quoted in: Saif, *Astrology* (cit. note 1), p. 171.

Likewise, Johannes Kepler's prized 'aspect astrology' continued to operate with a geocentric rather than a heliocentric reference-frame, despite his cosmological Copernicanism.[4] Bernard Capp, on the other hand, has emphasized «the pioneering role of almanacs in accepting and popularizing the new astronomy».[5]

It was only with the publication of Robert Westman's *The Copernican Question* (2011), however, that a substantially different approach to the relation between astrology and Copernicanism was proposed. According to Westman, astrology was a stimulus, not an obstacle, for the exploration of new conceptions of planetary order until the second quarter of the 17th century. It was only with the coming of a new generation of modernizing natural philosophers like Descartes, Gassendi, or Mersenne (who happened to be opponents of astrology as well) that the world systems debate became motivated by «questions of agreement with [...] physical principles and issues of biblical compatibility» rather than astrological concerns.[6]

Like all great books, Westman's study calls attention to many phenomena which would otherwise remain invisible or insignificant. One of these phenomena is the precise way in which astrologers themselves engaged the Copernican question after the 1620s. Westman's narrative seems to imply that there was no more room for astrological concerns in the newly configured world systems debate. This would also lead us to expect that astrologers were indifferent about the Roman condemnations of Copernican doctrine, and that their voice was not represented in the Copernican controversies of the late 1620s and early 1630s.

This paper focuses on an apparent exception: Jean-Baptiste Morin (1586-1656), a Parisian physician, astrologer, and professor of mathematics who published two anti-Copernican treatises in 1631 and 1634.[7] Several of Morin's anti-Copernican arguments were derived from astrological consid-

---

4 Patrick J. Boner, *Kepler's Cosmological Synthesis. Astrology, Mechanism and the Soul*, Leiden & Boston, E.J. Brill, 2013, p. 50.

5 Capp, *Astrology* (cit. note 1), p. 191.

6 Robert Westman, *The Copernican Question. Prognostication, Skepticism, and Celestial Order*, Berkeley, University of California Press, 2011, pp. 495-496.

7 For an excellent bio-bibliographic introduction to Morin, see Monette Martinet, *Jean-Baptiste Morin (1583-1656)*, in Pierre Costabel – Martinet, *Quelques savants et amateurs de science au XVIIe siècle: Sept notices biobibliographiques caractéristiques*, Paris, Société Française d'Histoire des Sciences et des Techniques & Éditions Belin, 1986, pp. 69-87. Also see Robert Alan Hatch, *Between Astrology & Copernicanism: Morin-Gassendi-Bouillau*, «Early Science and Medicine», 22, 2017, pp. 487-516.

erations, and the case can be made that Morin's opposition to a heliocentric ordering of the planets was strongly informed by his astrological convictions. By studying the precise nature of Morin's perceived conflict between astrology and Copernicanism, this paper hopes to contribute to our understanding of their parting of the ways in the period 1620-1640.

2. Jean-Baptiste Morin and the world systems debate (1628-1634)

As Robert Westman observed, one cannot speak of a 'world systems debate' before the 1580s, while its real consolidation can be dated to the period between the publication of Kepler's *Epitome* (1618) and Galileo's *Dialogo* (1632).[8] Alongside these two towering contributions, this period also saw the publication of several other texts arguing the merits of the Ptolemaic, Tychonic and Copernican systems. In 1629, the Calvinist minister Philips Lansbergen published *Bedenckinghen op den daghelijkschen ende iaerlijkschen loop vanden aerdt-cloot*. The first extensive vernacular defense of Copernican cosmology, Lansbergen's *Bedenckinghen* soon received wider exposure through a Latin translation by Maarten van den Hove (1630).[9]

In early 1631, a Catholic theologian from Louvain called Libert Froidmont rebutted Lansbergen by publishing an anti-Copernican *Ant-Aristarchus sive orbis-terrae immobilis liber*. Later that year, Jean-Baptiste Morin, physician, astrologer, and professor of mathematics at the Collège Royal in Paris, also prepared an anti-Copernican *Famosi et antiqui problematis de telluris motu, vel quiete, hactenus optata solutio*, which came off the presses in December 1631.[10] Morin's *Solutio* did have a connection to the Low Countries debate. On the one hand, it attacked some of Lansbergen's pro-Copernican arguments in the Latin *Commentationes*; on the other hand, Morin had received a copy of Froidmont's *Ant-Aristarchus* through Gassendi, which allowed him to devote a few pages to its arguments in chapter 11 of *Solutio*.[11]

---

8 Westman, *Copernican Question* (cit. note 6), p. 492a.

9 Rienk Vermij, *The Calvinist Copernicans. The Reception of the New Astronomy in the Dutch Republic, 1575-1750*, Amsterdam, KNAW, 2002, p. 83.

10 Jean-Baptiste Morin, *Famosi et antiqui problematis de telluris motu, vel quiete, hactenus optata solutio*, Paris, Jean-Baptiste Morin, 1631. *Solutio*'s dedicatory letter to Cardinal Richelieu was dated 8 December 1631. In a letter to Joseph Gaultier (9 July 1631), Gassendi wrote that Morin had stopped by on 6 July while leaving for the country-side, where he planned to finish the *Solutio*. See Marie Tannery-Prisset – Cornelis de Waard, Jr. – René Pintard (eds.), *Correspondance du P. Marin Mersenne, religieux minime*, 17 vols., Paris, Gabriel Beauchesne & fils, 1932-1988, vol. III, pp. 173: 1-174: 16.

11 Morin, *Solutio* (cit. note 10), pp. 14 sq., 98-99.

Nevertheless, it should be pointed out that Morin's public anti-Copernicanism antedated Lansbergen's *Commentationes*.[12]

Meanwhile, back in the Low Countries, Philips Lansbergen's son Jacob authored an *Apologia pro commentationibus Philippi Lansbergi* (1633) against both Froidmont and Morin. According to Isaac Beeckman, this book was finished by 30 May 1633.[13] Lansbergen's *Apologia* elicited a counter-reply from both Froidmont and Morin. Froidmont's *Vesta, sive Ant-Aristarchi vindex* came out in early 1634, while Morin's *Responsio pro telluris quiete* was probably published in the summer of 1634.[14] After this, none of the three parties continued their exchanges.

Nevertheless, it is clear that the debate drew wider interest. In a letter to Diodati of 15 January 1633, Galileo expressed his regret that Morin's *Solutio* and Froidmont's *Ant-Aristarchus* only reached him six weeks after the publication of the *Dialogo* in February 1632.[15] Through Gassendi, Morin had indeed sent an incomplete copy of his *Solutio* to Galileo, while it was probably also Gassendi through whom Galileo obtained a copy of Lansbergen's *Commentationes* and Froidmont's *Ant-Aristarchus* in March 1632.[16] Morin personally sent Galileo a copy of his *Responsio* on 15 November 1634.[17] Jacob van Lansbergen's *Apologia* and Froidmont's *Vesta*, on the other hand, reached Galileo around the beginning of 1635.[18]

The overt intention of Morin's *Solutio* (1631) was to bring closure to the world systems debate. His intervention set out from a perceived gap between what was true and what was credible. For his own part, Morin knew perfectly well what was true: the earth was perfectly at rest in the middle of the universe, and Scriptural references overwhelmingly supported this.

---

12 Id., *Ad australes et boreales astrologos pro astrologia restituenda epistolae*, Paris, Jean Moreau, 1628, p. 19.

13 On 30 May 1633, Beeckman wrote to Mersenne that Lansbergen jr. had now responded to Froidmont and Morin, Beeckman also mentioned that it was not yet published. See Tannery – De Waard, *Correspondance* (cit. note 10), vol. III, p. 405: 50-52.

14 The privilege of Philip IV, for a period of nine years, for Froidmont's *Vesta* was issued in Brussels on 15 November 1633. Morin signed his dedication of the *Responsio* to Cardinal Richelieu on 24 June 1634, and sent the treatise to Galileo on 15 November 1634. See OG, XVI, p. 158: 2-4.

15 Galileo to Diodati, 15 January 1633, OG, XV, p. 23: 13-15.

16 Morin to Galileo, 15 November 1634, OG, XVI, p. 159: 13-19; Galileo to Diodati, 9 April 1632, OG, XIV, p. 340: 30-31; Pierre Gassendi, *Opera omnia*, 6 vols., Lyon, Laurent Anisson and Jean-Baptiste Devenet, 1658, vol. VI, pp. 45-46. For this last letter, see Gassendi to Galileo, 1 March 1632, OG, XIV, p. 334: 14-17; Tannery – De Waard, *Correspondance* (cit. note 10), vol. III, p. 263.

17 Morin to Galileo, 15 November 1634, OG, XVI, p. 158: 3-4.

18 Pier Battista Borghi to Galileo, 30 December 1634, OG, XVI, p. 185.

However, without common ground about *how* to determine truth, being true was not quite the same as being credible. As Morin observed in a prefatory letter to his readers, part of the trouble with the world systems debate was the impossibility of settling on shared principles.[19] On both sides of the debate, he specified, arguments were deployed from astronomy, Scripture, and natural philosophy.[20] Concerning Scripture, Morin stated that it «confirms the rest of the earth with far more certainty and clarity than its motion», while also accepting the fact that this argument would not sway Copernicans as long as the relevant exegetical principles could be questioned.[21] As for astronomy, Morin remained convinced that observation of the visible heavens would never provide a secure basis by which to discern whether the earth or the heavens moved.[22]

Only physical arguments would work. Already in his initial address to the reader, Morin disclosed that he would introduce astrological criteria into this category:

> Because astrology deals with the natures, powers and modes of action of universal causes – the stars along with the first physical cause, the first mobile –, these fix the earth at rest in the center of the heavens, as will become very clear from this book. Indeed, this book will be a forerunner of *Astrologia Gallica*, which we will publish in the next few years, if the necessary funds for reckoners and scribes are not wanting.[23]

Morin's intervention in the world systems debate, then, simultaneously sought to recall astrology to its rightful disciplinary place as «the head of all physics, to whom it belongs, above all other sciences, to define the place of the earth, that receptacle or passive subject of all celestial influences».[24]

One must be careful in characterizing Morin's experience of the world systems debate. In his mind, this was not a case of two distinct bodies of

---

19 MORIN, *Solutio* (cit. note 10), p. [vii].

20 *Ibid.*, p. 5.

21 *Ibid.*, p. 11. Michel-Pierre Lerner has pointed out that Morin's engagement of Rome's censorship of Copernicanism was limited to the *Responsio*, where Morin evoked the recent condemnation of Galileo while refusing to include the relevant texts (Morin would do this, however, in his *Tycho Braheus in Philolaum, pro telluris quiete* of 1642). See MICHEL-PIERRE LERNER, *La réception de la condamnation de Galilée en France au XVIIe siècle*, in JOSÉ MONTESINOS – CARLOS SOLIS (eds.), *Largo Campo di Filosofare. Eurosymposium Galileo 2001*, La Orotava, Fundación Canaria Orotava de Historia de la Ciencia, 2001, pp. 513-547: 520.

22 MORIN, *Solutio* (cit. note 10), pp. 14, 21-23.

23 *Ibid.*, p. [viii].

24 *Ibid.*, p. 80.

knowledge vying for intellectual dominance through the use of arguments. Instead, Morin interpreted Copernicanism on the traditional model of religious heresy.[25] It was not so much a positive doctrinal phenomenon, but a negative deviation from the truth. This is also why Morin interpreted his discursive activity as an act of restoring Copernicans to intellectual health. A stark example of this attitude can be found in Morin's relation to Galileo. On 15 November 1634, Morin decided to send Galileo a copy of his *Responsio*, in which he qualified the *Dialogo* as a «veiled defense of terrestrial motion» while evoking Galileo's condemnation as a warning to other *savants*.[26] The imprudence of this move is striking, but is mitigated by considering Morin's habit of thinking of truth in pastoral rather than political terms. In his subsequent letter to Galileo of 4 April 1635, Morin fully adopted this pastoral posture by inviting Galileo to consider the example of St.-Augustine, who «wished to triumph over himself, not to be won over by someone else».[27] Members of Galileo's circle clearly understood the epistemological angle Morin was coming from, but were disturbed by it all the same. On 2 July 1635, Campanella wrote to Peiresc that Morin «exhorted [Galileo] to convert to the truth by means of the arguments of his book. I will say no more».[28]

Morin's interpretation of Copernican convictions as a mere sign of the absence of truth also explains another feature of their portrayal in *Solutio*. Consistently presented as an intellectual seduction leading men into a labyrinth of errors and lies, it was easy for Morin to make the final step of associating Copernicanism with diabolic agency.[29] Likewise, Morin did not simply approach Scripture as a unique epistemic resource which could only speak truth (as Fromondus would have it).[30] Instead, Morin took the even stronger interpretation of Scripture as a divinely granted protective shield against human errors which the Holy Spirit had foreseen from eternity.[31]

---

25 *Ibid.*, pp. 4, 54.

26 MORIN, *Responsio pro Telluris quiete. Ad Jacobi Lansbergii doctoris medici Apologiam pro Telluris motu*, Paris, Jean-Baptiste Morin & Jean Libert, 1634, pp. 54, 56. For a more complete discussion, see LERNER, *La réception* (cit. note 21), pp. 527-528.

27 Morin to Galileo, 4 April 1635, OG, XVI, pp. 251-253; esp. *ibid.*, p. 252: 39-40.

28 Campanella to Peiresc, 2 July 1635, OG, XVI, p. 288.

29 Seduction: MORIN, *Solutio* (cit. note 10), pp. [iv], 8, 37. Errors and lies: *ibid.*, pp. 13, 23. Copernicanism as diabolism: *ibid.*, p. [viii].

30 See STEVEN VANDEN BROECKE, *Copernicanism as a religious challenge after 1616: self-discipline and the imagination in Libertus Fromondus's anti-Copernican writings (1631-1634)*, «Lias. Journal of Early Modern Intellectual Culture and its Sources», 42, 2015, pp. 67-88: 78-80.

31 MORIN, *Solutio* (cit. note 10), p. 138.

3. Morin's astrology as a tool of spiritual self-monitoring

As early as 1619, in *Nova sublunaris anatomia*, Jean-Baptiste Morin embarked on a lifelong pursuit to restore astrology's accuracy and legitimacy. From 1623 onwards, Morin explored specific topics, such as astrological house division or the universality of celestial effects, in various publications on astrological reform. By 1631, Morin's published works began to announce an astrological *summa* in which his results would be systematized. This massive *Astrologia Gallica* (1661) was only published posthumously, but seems to have been largely completed by 1648.[32] Its content amply confirms Morin's claim that it was the outcome of decades of astrological research, and provides a singularly complete means of understanding the broader contours of Morin's project.

Even by pre-modern standards, *Astrologia Gallica* stood out by the extent to which it located the meaningfulness of astrology inside a Christian framework of sin and redemption. For Morin, the powers of the inferior soul were equated with the Christian flesh. Celestial influence was the agent which gave individual form to the shared effects of man's original sin.[33] The stars were identified as one of the four enemies which man had to battle unto death:

> God doesn't want us to be lazy in this world, but to constantly fight in His honour against the most forceful enemies of the world, the flesh, and the Devil. In His wisdom, he allowed man's infirmity to be tempted by these, so that we would merit the crown of eternal beatitude through our victory against them. [...] the constitution of the heavens at every man's birth is the fourth enemy besides the world, the flesh, and the Devil.[34]

Unsurprisingly, the 'Apologetic preface' of *Astrologia Gallica* described astrology as the study of natural causes which bore a double function in the economy of divine grace. On the one hand, astrological influence led men into sin. On the other hand, they served as a tool of divine retribution,

---

[32] According to Morin's prefatory letter to the reader, the book was finished by 1648. See Morin, *Astrologia Gallica Principiis et Rationibus Propriis Stabilita*, The Hague, Adrianus Vlacq, 1661, p. i (my numbering). The *Praefatio apologetica* to *Astrologia Gallica*, however, still managed to attack Gassendi's *Syntagmata* (1649) and to discuss the 'Black Monday' eclipse of 1652. Beginning with the *Solutio* (1631), announcements of *Astrologia Gallica* can be found throughout Morin's published works.

[33] Id., *Astrologia Gallica* (cit. note 32), p. 264b.

[34] *Ibid.*, p. 264b.

leading those who embraced sinful influences to be «punished by evil influences so as to come to their senses».[35] Moreover, knowledge of astrology also assisted men in more accurately discerning God's supernatural action against the background of regular natural causes.[36]

Against this background, it is easy to understand how Morin privileged astrology as an art of moral self-diagnosis and self-monitoring. Not unlike Thomas Aquinas before him (*Summa Theologiae* Ia. q. 115.a4), Morin connected the relevance of astrologers to the human tendency to follow native inclinations rather than the will. However, Morin was far more pessimistic about man's natural ability to overcome such inclinations. On the one hand, he painted a bleak picture of the interminable victory of celestial inclinations over human laws and *exempla* in the battle for human mores.[37] On the other hand, Morin also qualified the very possibility of fighting such inclinations as preternatural or supernatural.[38] Indeed, without divine grace, he figured that hardly one tenth of humanity would escape eternal damnation.[39]

To a greater extent than most astrologers, then, Morin integrated the physical realities of celestial influence in the cosmic drama of individual sin and redemption. On the one hand, celestial bodies were thus approached as God's «representatives in nature, through which He distributes and governs the fate of natural effects». On the other hand, these celestial bodies were also created to allow man «to inspect the powers and modes of action, both theirs and His».[40] However, Morin also built in a strict distinction between 'physical' and 'moral' actions, or between natural and supernatural economies.[41]

It seems likely that the importance of astrology to Morin's interpretation of Christian spirituality was also instrumental in shaping his unconditional trust in astrology's forensic, diagnostic, and predictive powers. Guillaume Tronson's contemporary biography of Morin thus ascribed his wondrous ability to escape dangerous situations to his attentiveness to 'divine providence'.[42] Morin even claimed that divine providence frequently shaped

35 Morin, *Praefatio apologetica*, in *ibid.*, p. xxvi.

36 Id., *Epistolae* (cit. note 12), p. 13.

37 Id., *Astrologia Gallica* (cit. note 32), pp. 267a-268a.

38 *Ibid.*, pp. 265b-266a.

39 *Ibid.*, p. 269a.

40 *Ibid.*, p. 531b.

41 *Ibid.*, pp. 143a, 269b.

42 Guillaume Tronson, *Vita*, in Morin, *Astrologia Gallica* (cit. note 32), p. iii.

external contexts in such a way that they allowed innate astral inclinations to emerge.[43] This was especially the case in cities and courts, which Morin presented as specific *milieux* brimming with potential triggers of desire.[44]

4. Morin and the New Piconians

Morin's emphasis on astrology as an essential tool of spiritual self-monitoring goes a long way towards explaining his unshakeable faith in the powers of the art. However, his was also an age of, in the felicitous phrase of Robert Westman, 'new Piconianism': a new wave of astrological criticism which took up the example of Giovanni Pico's original *Disputationes adversus astrologiam divinatricem* (1496) in the newly confessionalized context of late Renaissance and Baroque Europe.[45]

Morin's own position towards the Piconians was remarkably ambivalent. The 'Apologetic Preface' of *Astrologia Gallica*, for instance, candidly announced that Morin's book would be almost as critical of the astrologers Ptolemy and Cardano, as it would be of the critics Pico della Mirandola and Alessandro De Angelis.[46] In his treatment of the traditional male and female zodiacal signs, Morin stated that Pico had been quite right to ridicule this theory, which Ptolemy and Cardano had sought to defend.[47] Unlike Pico, however, Morin did not grant the vanity of true astrology. Carefully distinguishing the art from its historical incarnations, Morin claimed that critics like Pico had never attacked this true astrology, but a «most stupid vanity, suggested by the Devil for the ruin of man in this world, and his damnation in the next».[48]

Earlier anti-Piconians also made a distinction between good astrologers and abusers of the art. They too diagnosed astrological criticism as targeting the latter instead of the former. Francesco Giuntini had done this in his

---

43 Morin, *Astrologia Gallica* (cit. note 32), pp. 249a-250b, 266b, 270b.

44 *Ibid.*, p. 267b.

45 Westman, *Copernican Question* (cit. note 6), pp. 226-228. On the seventeenth-century tradition of astrological criticism, see Vanden Broecke, *From Cosmic Governance to Governmentality. Shaping Sublunary Order in Seventeenth-Century French Critiques of Astrology*, in Miguel Angel Granada – Patrick J. Boner – Dario Tessicini (eds.), *Man and cosmos from Copernicus to Kepler*, Barcelona, Universitat de Barcelona, 2016, pp. 315-340.

46 Morin, *Praefatio apologetica* (cit. note 35), p. v.

47 Id., *Astrologia Gallica* (cit. note 32), p. 325a. Other examples of Morin's criticism of Ptolemy can be found in *ibid.*, chs. XIII.1, XIV.7, XV.1, XV.5-6, XV.12, XVII.3, XX.2.

48 Id., *Praefatio apologetica* (cit. note 35), p. iii.

massive *Speculum astrologiae*.[49] In 1629, Campanella too deployed Giuntini's distinction between different ways of appropriating the art,[50] in a context which singled out Pico as a formidable opponent.[51] Nevertheless, Morin's version of this argument was different in two ways. First of all, it turned the phenomenon of astrological criticism into an objective measure of the art's past imperfect state, rather than of the imperfections of specific practitioners. Secondly, it diagnosed this history of astrological failure as the effect of a diabolical conspiracy against astrology and its practitioners, and hence as embedded in a Christian soteriological framework. In a very real sense, then, Morin did not see critics like Pico as astrology's main challenge. Their presence merely formed a symptom of the diabolical debasements which plagued the art for most of its history, and it was this history which required closure:

> Everybody mentions Ptolemy as the prince of astrology. Cardano's commentary says the same thing about *Tetrabiblos* as one does about the Apocalypse of St. John: as many mysteries are contained therein as there are words. [...] If I am not mistaken, it is enough to call Ptolemy the prince of astrology in its current state, which contains many absurdities and dreams, [...] not of astrology as such. This great man should not be spurned, for he did what he could in such a sublime and abstruse art, depraved by the Devil's arts.[52]

Morin's halfway position between the astrologers and their critics is, of course, very reminiscent of the attitude of Johannes Kepler. Nevertheless, there are crucial differences in the way that Kepler and Morin understood the precise nature of astrological reform. Kepler saw himself as engaged in a technical activity of separating out the good from the bad in a given art, with Pico as one of his main inspirations and natural philosophy as one of his main assaying tools. Once extracted from the dung heap of humanly invented beliefs and practices, this philosophically sound kernel would then receive a monopoly on the name of astrology.

Morin, on the other hand, was carrying out a historical activity of bringing *prisca astrologia* back to life: his point was to reform the art by going back to the pristine forms which history had deformed. Indeed, Morin considered certain parts of the art, particularly the doctrine of 12 celestial

---

49 Francesco Giuntini, *Speculum astrologiae*, Lyon, Filippo Tinghi, 1573, p. 2v.

50 Tomasso Campanella, *Astrologicorum libri VI*, Lyon, Jacques, André & Mathieu Prost, 1629, p. 3.

51 Id., *Astrologicorum Libri* (cit. note 50), p. 8.

52 Id., *Astrologia Gallica* (cit. note 32), p. 444; see also p. 208.

houses, to have been divinely inspired in Adam.[53] At the same time, he claimed that astrology's extraordinary potential had been obscured since its pristine existence at the time of Adam and Noah.[54] Only in his own *Astrologia Gallica* would Adamic astrology finally be restored. Accordingly, Ptolemy's *Tetrabiblos* now figured an antiquated stage in the history of astrology, not the site for an encounter with the art's original purity. Likewise, Morin distinguished modern astrologers like Offusius and Kepler from 'the ancients' of the art and their Renaissance epigones, among whom Morin particularly singled out Cardano.[55]

In *Ad australes et boreales astrologos pro astrologia restituenda epistolae* (1628), Morin had already listed the central *desiderata* towards a restored astrology.[56] Particularly important was a larger base of empirical astrological data, corrected by superior astronomical knowledge. Interestingly, Morin also included 'solving the world systems debate', about which he claimed: «[...] the truth of this matter is of such importance to astrology, that Tycho and Kepler considered it necessary to overturn astrology in order to stabilize their systems».[57] The end-points of Morin's project for astrological reform were:

> [...] the first causes of effects, which are also the first principles of this science. These are few in number, both in this science and others. When our universal science is stabilized through these principles, combining them will easily lead to innumerable conclusions.[58]

Apparently, Morin's emphasis on the necessity of astrological self-analysis (see section 3) also led him to approach astrological reform as a matter of securing a small number of theoretical principles of astrological judgment – especially in the realm of natal astrology.

5. Astrology vs. Copernicanism in Morin's *Solutio* and *Responsio*

Morin's *Solutio* (1631) consisted of six parts: [1] a *prodromus* in which Morin laid out the historical lineage and present following of the geo-kinetic

---

53 Id., *Praefatio apologetica* (cit. note 35), p. v.

54 Id., *Astrologia Gallica* (cit. note 32), p. 531b.

55 *Ibid.*, p. 349. Morin adopted Tycho Brahe's calls for astrological reform as a precursor of his own enterprise. See Id., *Praefatio apologetica* (cit. note 35), p. xvi.

56 Id., *Epistolae* (cit. note 12), pp. 28-30.

57 *Ibid.*, p. 29.

58 *Ibid.*, pp. 29-30.

hypothesis (chapters 1 and 2); [2] in chapter 3, Morin distributed existing arguments for and against a moving earth over three categories (Scriptural, astronomical, and physical); [3] next, *Solutio* argued for the a priori insufficiency of Scriptural (chapter 4) and astronomical (chapters 5 and 6) arguments, leaving only natural-philosophical arguments with the capacity to settle the issue (chapter 7); [4] having established the latter claim to his own satisfaction, Morin discussed the physical arguments for the eccentricity and motion of the earth (chapters 8 and 9), as well as [5] those for the centrality and stability of the earth (chapters 10 and 11); [6] *Solutio* then ended with an epilogue (chapter 12). For our present purposes, it is to part [5] of *Solutio* that we should turn first. As we already saw, four of the eight physical arguments for the earth's centrality in chapter 10 drew on astrological considerations.

Morin's first argument attributed a double action to the planets, one by themselves and another with the first heaven. The first mode of action, Morin stated, occurs in any direction (*orbiculariter, sive undequaque*). The second mode, however, operates rectilinearly in the direction of the center of the first heaven (*directe versus Caeli centrum*), and only with that part of the first heaven that is obscured by its visible diameter.[59] Morin's 'first heaven' referred, quite simply, to the outermost sphere of the universe, situated just beyond that of the fixed stars.

For Morin, it was precisely this terrestrial goal-directedness of planetary effects, in tandem with the first heaven, which Copernicans seemed to deny.[60] More specifically, Morin associated Copernicanism with two theses on celestial influence. On the one hand, Copernicans portrayed celestial influence as moving in any direction from celestial bodies.[61] On the other hand, they tended to approach rays of celestial influence as affecting every part of the earth simultaneously.[62] In both cases, Morin saw considerable challenges for the rigor and precision of astrological analysis.[63]

---

59 Id., *Astrologia Gallica* (cit. note 32), pp. 83-84.

60 Id., *Solutio* (cit. note 10), p. 84: «Non ergo Saturnus cum Coelo aget in Terram? Quod tamen asserere falsissimum est, quandoquidem experientijs constat Saturnum in Tellurem agere pro ratione partis Coeli, quam visibili sua diametro occupat».

61 Id., *Responsio* (cit. note 26), p. 28: «Ergo planeta suo situ corporeo tantum agit influentialiter radio ad centrum Coeli directo, cum ea parte Coeli quam abscindit sua visibili diametro».

62 See Jacob Lansbergen, *Apologia pro Commentationibus Philippi Lansbergii in Motum Terrae Diurnum & Annuum*, Middelburg, Zacharias Romanus, 1633, pp. 107-108; Morin, *Responsio* (cit. note 26), p. 28: «nec volui quamlibet Telluris partem affici rectitudine eiusdem radij, quod mihi imponit Lansbergius; longeque abfui ab eius sententia qui Terram in Ecliptica collocans procul a concursu radiorum influentialium, ait nihilominus eam eodem modo affici, ac si esset in centro ob semidiametrum magni orbis, respectu Coeli ut voluit insensibilem».

63 Morin, *Responsio* (cit. note 26), pp. 27-28.

The second astrological argument drew on astrology's empirical data of terrestrial effects when the Sun is in a specific sign of the zodiac, such as Leo or Scorpio. Morin argued that these experiences did not agree with the effects which should occur on earth under the Copernican interpretation of these phenomena.[64] To this, Jacob van Lansbergen replied with a version of the argument which we previously encountered in Heydon's *Defence*. More specifically, Lansbergen's *Apologia* asserted the under-determination of astrological experience by cosmic distances in the planetary system:

> The law of action's dependence on distance does not obtain here. Whether the earth is in one part of the ecliptic or another [on the Copernican hypothesis], it is always equidistant from the eighth heaven to the senses and in appearance. For it has often been repeated that the ecliptic is the apparent center of the world.[65]

Morin was not impressed. In *Responsio*, he observed that this was the Copernicans' only remaining defense against his astrological arguments, while reasserting the reliability of astrologer's experiential data as one indicator of cosmic distances.[66]

In his third astrological argument, Morin claimed that a Copernican arrangement destroyed the theory of astrological aspects, «whose force is experienced most forcefully on earth, even by Kepler, hater of astrology».[67] More specifically, Morin emphasized that the Copernican destruction of something like a first heaven, also destroyed a secure frame of reference for unambiguously defining aspects like 'conjunction', *pace* Jacob Lansbergen's alternative account.[68]

Morin's final argument in chapter 10 was by far the most adventurous, developing his astrological emphasis in ways that were «partly physical, partly theological».[69] Morin began by advancing the earth's position in the center of the world as a suitable explanation why there were so many and such perfect generations in the sublunary world. Now, what would follow if one were to situate the Sun in the center instead? First of all, this

---

64 Id., *Solutio* (cit. note 10), pp. 88-89.

65 Lansbergen, *Apologia* (cit. note 62), pp. 108-109: «Respondeo, legem actionis in majus vel minus distans hic locum non habere, quia sive terra in una Ecliptica parte sua sit, sive in altera, ad sensum et quoad apparentiam, aeque a caelo octavo semper distat, quia Ecliptica est mundi centrum apparens ut jam saepius inculcatum fuit».

66 Morin, *Responsio* (cit. note 26), p. 30: «cum tamen experientia doceat contrarium, differentia etiam maxime sensibili Astrologorum iudicio».

67 Id., *Solutio* (cit. note 10), p. 90.

68 See Lansbergen, *Apologia* (cit. note 62), p. 110; Morin, *Responsio* (cit. note 26), p. 34.

69 Id., *Solutio* (cit. note 10), p. 94.

would mean that it would be superior to the earth in terms of quantity and quality of generation. Secondly, there would be no reason not to assume that other planets had perfect generations too and would, indeed, be inhabited.

Assuming that God indeed created corporeal intelligent creatures on other planets (albeit it non-human), then these would have to be as naturally sinful as man or the angels, and would also be subject to diabolic seduction towards sin. This would necessitate either that Christ also died on earth for the extraterrestrials, or that he had been incarnated and sacrificed in their form in their worlds as well. This, Morin opined, was in firm contradiction of the tenets of faith. Accordingly, the earth could only be in the center of the universe, and the planets could not be inhabited: all of material creation was there for humans alone.[70]

Having determined the issue of the earth's centrality to his own satisfaction, Morin proceeded, in chapter 11 of *Solutio*, to settle the debate concerning the earth's rest or motion. Of the six arguments given in favor of the earth's rest, the last one («more illustrious than the others») was again taken from "the supreme physical science of astrology".[71] Morin built his argument on the 'Copernican' interpretation of the zodiac signs as superfluous imaginary divisions of an immobile sphere of the fixed stars, not as real divisions of a first heaven beyond the sphere of the fixed stars.[72] Once again, Morin invoked the collective experience encoded in astrological aphorisms to argue that planetary rule over specific zodiac signs did produce very real effects on earth.[73]

6. The first heaven and the restauration of Adamic astrology

How should these arguments be interpreted? At first sight, they seem to support the traditional thesis that Copernicanism undermined a traditional astrological cosmos in which superlunary influences were directed to a central, immobile earth. Morin repeatedly emphasized that God's primary purpose in creating the world was to guarantee the generation of individual species. Indeed, the material universe was designed as a giant incubator for terrestrial being, making it a matter of *decorum* that the earth should be situ-

---

70 *Ibid.*, pp. 95-96.

71 *Ibid.*, p. 130.

72 *Ibid.*, pp. 131-132.

73 *Ibid.*, pp. 132-133.

ated in the center.[74] At the beginning of chapter 10 of *Solutio*, he categorically stated that «astrological reasons contradict the eccentricity of the earth».[75]

But whose reasons were this, exactly? A closer look suggests that Morin's argumentation was not so much informed by a more or less timeless 'astrological world-view', but that it was consciously designed to rebut neo-Piconian astrological criticism. We can discern this deeper layer by focusing on one of the central notions which Morin deployed throughout his aforementioned astrological arguments: the first heaven (*primum caelum* or *primum mobile*). Although Morin's theory of the first mobile received its full development in *Astrologia Gallica* (which he began to announce in 1631), it had been publicly defended by him as the key to sound astrology since 1623.[76] Morin's 'first heaven' lay just beyond that of the fixed stars, and was carried around with a daily motion around the poles of the celestial equator.[77] Its importance as a cornerstone of sound astrology was situated on at least three different levels.

First of all, it secured the natural-philosophical foundations of the art. In this respect, Morin expected his first mobile to kill two birds with one stone. On the one hand, it addressed the natural-philosophical deficiencies which Morin diagnosed in earlier astrologer's theoretical accounts of the zodiac signs, including those of Ptolemy, Bellanti, and Cardano.[78] Indeed, Morin's earlier attempts to establish a scientific correspondence with southern astrologers in the vice-kingdom of Peru was designed to gather empirical evidence against the physical implications of existing astrological theories.[79] On the other hand, Morin's recourse to the first mobile seemed tailor-made to counter the critiques which Giovanni Pico, Johannes Kepler and Alessandro de Angelis launched against the zodiac signs.[80]

---

74 *Ibid.*, pp. 28, 38, 94.

75 *Ibid.*, p. 81.

76 MORIN, *Astrologicarum domorum cabala detecta*, Paris, Jean Moreau, 1623.

77 ID., *Astrologia Gallica* (cit. note 32), pp. 319b-320a and 322b.

78 *Ibid.*, pp. 321b-322a and 323b.

79 *Ibid.*, p. 321b: «Quod autem falsa sit haec de natura signorum sententia, inde probatur quia ex rationibus Ptolemaei et Cardani erit signum ♉ in Mundi parte australi frigidissimum, atque siccissimum». The universal validity of astrology is again defended, now especially against Cardano and Campanella, in book 15 of *Astrologia Gallica*, where Morin draws on the evidence of genitures from the southern hemisphere.

80 *Ibid.*, p. 322b: «Idque probare contendit de Angelis, tum authoritate Kepleri et Abraham Avenesra, tum ratione, quod (inquit) primum Coelum debeat esse corpus simplex ac homogeneum, proindeque in partes diversae naturae dividi non possit». For Morin's use of the first heaven against Kepler's 'Copernican' critique of the zodiac signs, see the illuminating discussion in ID., *Solutio* (cit. note 10), pp. 130-137.

Secondly, Morin's emphasis on the primacy of the first heaven expertly unified the superlunary realm as a separate agent which operated on the sublunary realm: «The first cause or Heaven only acts in these inferior things to the extent that she is determined by all subordinate causes towards a particular effect, as is most fitting to a first cause».[81]

True to his concern for the soundness of astrology's natural-philosophical foundations, Morin was careful to portray his first heaven as a perfectly homogeneous substance which simultaneously acted as a celestial chameleon: specific parts of this *caelum* were extrinsically activated towards specific kinds of celestial agency by the lower fixed stars or planets.[82] When Morin, in book 14 of the encyclopedic *Astrologia Gallica*, finally made the transition to astrology proper,[83] he began with «the first and supreme secret of judicial astrology», the system of the zodiac signs (and their relation to planetary virtues in books 15 and 16).[84] Book 17 completed this with «the second fundament of astrology», the astrological houses.[85] In both cases, the doctrine of a first heaven provided Morin with an essential basis for theorizing these foundations.

In Morin's hands, the *primum mobile* became the most universal and most powerful of all efficient causes, directly affecting anything generated in the world.[86] On the one hand, it was a permanent and lifelong incubator of being.[87] On the other hand, it also brought universal change: «any sublunary effect, in as far as it can be considered as a whole, has to be referred to the entire Heaven in its beginning, vigor, decline, and death», and never to any single part of the first mobile.[88] Indeed, the diurnal motion of the first mobile was nothing more than a steady 'persuading' of each of the zodiac signs to play their part in the fashioning of «the whole effect [...] and whatever happens to it celestially during the time of its existence». Clearly, Morin defended a strong unity and regularity of sublunary history, in which even a thing's being was a permanent effect of its relation to the *primum coelum*.[89]

---

[81] Id., *Astrologia Gallica* (cit. note 32), p. 324a.

[82] *Ibid.*, pp. 314b-315a.

[83] *Ibid.*, p. 314.

[84] *Ibid.*, p. 384.

[85] *Ibid.*

[86] *Ibid.*, pp. 314b and 384a.

[87] *Ibid.*, p. 315a.

[88] *Ibid.*, p. 384a.

[89] *Ibid.*, p. 384a-b.

Morin defended a remarkably historical and contingent interpretation of the modalities of this relation. It was because the Sun *happened* to be on the cusp of the sign of Aries at the creation of the world, and because God *happened* to move the Sun towards Taurus rather than Aquarius, that Aries was first determined to heat.[90] The distribution of different natures over the signs of the zodiac was thus premised on a historically contingent beginning. *Mutatis mutandis*, Morin added, the same pattern applied to individual births.[91]

Thirdly, and most importantly, Morin expected this interpretation of the first heaven to restore the traditional rigor and precision of astrological analysis. Especially in book 21 of *Astrologia Gallica*, Morin patiently uncovered a frequent methodological error in the practice of astrological judgment. In their handling of 'universal signifiers' (that is, planets associated with specific aspects of human life, like the Sun signifying both honors, kings, fathers, etc.), Morin claimed, astrologers often judged such matters from the signifier alone, regardless of the astrological house which it occupied or ruled. Morin's theory of the first mobile was a solution for this problem, in as far as it sought to remind working astrologers that «the stars, although they can illuminate and heat, cannot impart a specific influence without the concourse of the first heaven».[92] Of themselves, Morin repeated again and again, the planets were indifferent to the specific effects which they caused on earth (e.g., on one's health or professional life). Only further determination by twelve astrological houses, themselves determined by the first mobile, unlocked this level of causal specificity.[93]

It was this basic doctrine which constituted the quintessential key to the secrets of the Adamic *prisca astrologia*.[94] It was also this doctrine which Morin felt to be undermined by astronomical *novatores*. This conviction was particularly prevalent in the earlier *Ad astrologos epistolae* (1628). There, Morin claimed that for all of their differences, Tycho Brahe and Johannes Kepler were united in their opinion that ancient astrology could not coexist with their astronomy, «as if their astronomy was to be erected on the ruins of astrology».[95] Apparently, Morin changed his mind about Tycho by the time he wrote his *Solutio* (1631). Not only did Morin come out as a staunch

90 *Ibid.*, pp. 318b-319b, 320b, 324a.

91 *Ibid.*, pp. 320b, 324a.

92 *Ibid.*, p. 497b.

93 *Ibid.*, p. 499b.

94 *Ibid.*, pp. 496, 500a, 503b.

95 Id., *Epistolae* (cit. note 12), pp. 16, 29.

defender of the geo-heliocentric system; he now found additional proof for the truth of that system in the precise order of planetary associations with the zodiac signs, concluding that «the true system of the world was known by reasoning to the astrologers, before it was known by observation to the astronomers».[96]

Morin's views of Brahe's assistant, however, did not change: the *Solutio* still presented Kepler as a firm opponent of astrology.[97] For Morin, Kepler's astrological theorizing in *De stella nova* (1606) amounted to the suggestion that only planetary aspects, not the first heaven, had any real effect on earth.[98] Morin even suggested that Kepler, following Plotinus, defended astrology as an art of significations without a corresponding causal mechanism.[99]

7. Conclusion

The *novatores* who would soon monopolize the world-systems debate remained ambivalent about Morin's astrological reform. In his letter to Joseph Gaultier of 9 July 1631, Gassendi listed Jacques de Valois, Mersenne, and himself as Morin's regular interlocutors on the motion of the earth and the principles of astrology. On both topics, Gassendi seemed more disturbed about Morin's stubborn tendency to dismiss counter-arguments than he was about the content of his astronomical and astrological opinions, even adding that «to be honest, I believe him in this».[100] Descartes, on the other hand, simply dismissed Morin's astrological arguments for the immobility of the earth in a letter to Mersenne, adding that there were better arguments available for this opinion.[101] Finally, Galileo's comments, in a letter to Diodati of 15 January 1633, on Morin's project to «stabilize the certainty of astrology» consisted of equal measures of skepticism and fascination.[102] Diodati clearly transmitted Galileo's interest to Morin: on 4

[96] Id., *Solutio* (cit. note 10), p. 136.

[97] *Ibid.*, p. 81: «Quod autem Kepplerus libro de Trigono igneo, alibique passim Astrologiam è scientiarum numero reiecerit, cum alijs plerisque virtutum coelestium ignaris Philosophis».

[98] Id., *Astrologia Gallica* (cit. note 32), p. 501b.

[99] *Ibid.*, p. 505a-b.

[100] Tannery – De Waard, *Correspondance* (cit. note 10), vol. 3, pp. 173: 1-174: 16.

[101] René Descartes, *Œuvres*, ed. by Charles Adam and Paul Tannery, 11 vols., Paris, Léopold Cerf, 1964-1974, vol. I, p. 258: 4-10. Another reference to Morin's astrology is found in: *ibid.*, vol. 1, p. 289: 2-4.

[102] Galileo to Diodati, 16 January 1633, OG, XV, p. 24: 18-24. Galileo's personal notes on

April 1635, Morin made good on his request to inform Galileo about the *Astrologia Gallica*, sending him a *précis* which had recently been printed for private distribution among Morin's friends.[103]

Despite this ambivalence, Morin's astrological critique of Copernicanism could be seen as the exception which proves the rule that astrologers had become marginal to the world-systems debate by the late 1620s. Nevertheless, Morin's interventions underlined the ongoing complexity of the relation between astrology and Copernicanism in the late 1620s and early 1630s. For Morin's commitment to a rotating first heaven surrounding a central earth was not the manifestation a traditional astrological worldview, but a piece of astrological theorizing which was developed against the background of an ongoing dialogue with the Renaissance tradition of astrological criticism.

We have seen how Morin internalized this tradition to an unusual extent. Astrologers with a more modest interpretation of the need for astrological reform, such as Christopher Heydon, easily circumvented Copernicanism by claiming an under-determination of astrology's practical traditions by the true order of the heavens. They insisted on the ongoing homocentricity, for all practical purposes, of a heliocentric universe. This was not an option for Morin, precisely because he felt that astrology's practical traditions were in need of a (meta-)physical framework that could provide them with secure principles. Underlying Morin's emphasis on the first heaven, then, was a deeper concern for the epistemic order of astrological judgment, not so much cosmological inertia. In a published letter to Christian Longomontanus of 21 March 1627, Morin already regarded the Copernicans as men who subverted the "purely celestial" nature of equator and ecliptic, and thereby also the a priori derivation of the twelve astrological houses.[104] From the very beginning, it was Morin's commitment to a deeper form of astrological reform which shaped his experience of an incommensurability between Copernicanism and astrology.

Morin's *Solutio* suggest a rather superficial engagement with his astrological arguments. See Galileo Galilei, *Note per il Morino*, OG, VII, pp. 567-568.

103 Morin to Galileo, 4 April 1635, OG, XVI, pp. 251: 5-7 and 251: 12-15 sq.

104 Morin, *Epistolae* (cit. note 12), p. 19.

INDEX OF NAMES

FINITO DI STAMPARE
PER CONTO DI LEO S. OLSCHKI EDITORE
PRESSO ABC TIPOGRAFIA • CALENZANO (FI)
NEL MESE DI MARZO 2018

MUSEO GALILEO

BIBLIOTECA DI «GALILÆANA»

I. Filippo Camerota, *Linear Perspective in the Age of Galileo. Lodovico Cigoli's* Prospettiva pratica. 2010, xx-360 pp. con 246 figg. n.t. e 19 tavv. f.t. di cui 18 a colori.

II. *Il caso Galileo: una rilettura storica, filosofica, teologica*. Convegno internazionale di studi (Firenze, 26-30 maggio 2009). A cura di M. Bucciantini, M. Camerota e F. Giudice, con DVD allegato. 2011, xiv-520 con 21 figg. n.t.

III. *Celestial Novelties on the Eve of the Scientific Revolution, 1540-1630*. Edited by Dario Tessicini and P.J. Boner. 2013, xvi-284 pp. con 19 figg. n.t.

IV. Federica Favino, *La filosofia naturale di Giovanni Ciampoli*. 2015, xviii-364 pp. con 4 tavv. f.t.

V. Paola Zambelli, *Alexandre Koyré in incognito*. 2016, xxii-288 pp.

VI. Gregorio Baldin, *Hobbes e Galileo. Metodo, materia e scienza del moto*. 2017, xxiv-244 pp. con 9 figg. n.t.

VII. Annibale Mottana, *Galileo e la* bilancetta*: un momento fondamentale nella storia dell'idrostatica e del peso specifico*. 2017, xvi-208 pp. con 30 figg. n.t.

VIII. *Copernicus Banned. The Entangled Matter of the anti-Copernican Decree of 1616*. Edited by Natacha Fabbri and Federica Favino. 2018, xxvi-252 pp. con 5 figg. n.t.